TIME: FROM EARTH ROTATION TO ATOMIC PHYSICS

In the 21st century, we take the means to measure time for granted, without contemplating the sophisticated concepts on which our timescales are based. This volume presents the evolution of concepts of time and methods of timekeeping up to the present day. It outlines the progression of time based on sundials, water clocks, and the Earth's rotation, to time measurement using pendulum clocks, quartz crystal clocks, and atomic frequency standards. Timescales created as a result of these improvements in technology and the development of general and special relativity are explained. This second edition has been updated throughout to describe 20th- and 21st-century advances and discusses the redefinition of SI units and the future of Coordinated Universal Time (UTC). A new chapter on time and cosmology has been added. This broad-ranging reference benefits a diverse readership, including historians, scientists, engineers, and educators, and it is accessible to general readers.

DENNIS D. McCARTHY is former Director of Time at the US Naval Observatory and the leading authority in the United States for astronomical and timing data. He has led and been a member of various commissions and working groups within the International Astronomical Union and has authored and edited numerous publications dealing with fundamental astronomy, time, and Earth orientation.

P. KENNETH SEIDELMANN is a research professor of astronomy at the University of Virginia and was Director of Astrometry at the US Naval Observatory. He has led and been a member of a division, various commissions, and working groups of the International Astronomical Union. He has coauthored two other books – *Fundamentals of Astrometry* and *Celestial Mechanics and Astrodynamics* – and is a coeditor of the *Explanatory Supplement to the Astronomical Almanac*.

TIME: FROM EARTH ROTATION TO ATOMIC PHYSICS

SECOND EDITION

DENNIS D. McCARTHY
US Naval Observatory (Retired)

P. KENNETH SEIDELMANN
University of Virginia

CAMBRIDGE
UNIVERSITY PRESS

University Printing House, Cambridge CB2 8BS, United Kingdom

One Liberty Plaza, 20th Floor, New York, NY 10006, USA

477 Williamstown Road, Port Melbourne, VIC 3207, Australia

314–321, 3rd Floor, Plot 3, Splendor Forum, Jasola District Centre, New Delhi – 110025, India

79 Anson Road, #06–04/06, Singapore 079906

Cambridge University Press is part of the University of Cambridge.

It furthers the University's mission by disseminating knowledge in the pursuit of education, learning, and research at the highest international levels of excellence.

www.cambridge.org
Information on this title: www.cambridge.org/9781107197282
DOI: 10.1017/9781108178365

First Edition © 2009 WILEY-VCH Verlag GmbH & Co. KGaA, Weinheim
Second Edition © Dennis D. McCarthy and P. Kenneth Seidelmann 2018

First published 2009
Second Edition 2018

Printed in the United Kingdom by TJ International Ltd. Padstow Cornwall

A catalogue record for this publication is available from the British Library.

Library of Congress Cataloging-in-Publication Data
Names: McCarthy, Dennis D., author. | Seidelmann, P. Kenneth, author.
Title: Time : from Earth rotation to atomic physics / Dennis D. McCarthy (United States Naval Observatory (retired), P. Kenneth Seidelmann (University of Virginia).
Description: Second edition. | Cambridge ; New York, NY : Cambridge University Press, [2018] | Includes bibliographical references and index.
Identifiers: LCCN 2018030821 | ISBN 9781107197282
Subjects: LCSH: Time measurements. | Time. | Earth (Planet) – Rotation.
Classification: LCC QB213 .M385 2018 | DDC 529/.7–dc23
LC record available at https://lccn.loc.gov/2018030821

ISBN 978-1-107-19728-2 Hardback

In honor of our wives, Diane McCarthy and Bobbie Seidelmann, and our families.
Dedicated to the scientists who preceded us and taught, mentored,
and inspired us.

Contents

Preface

This second edition is an updated version of the first edition with various additions that reflect recent developments in timekeeping as specified in what follows.

Everyday use of time in one form or another is a common experience for everyone throughout their lives. The availability of a means to measure the passage of time with the required accuracy is taken for granted. However, the concepts on which timescales are based and the requirements for accuracy in many applications can be both sophisticated and complex. Time is not a simple subject.

During the 20th century the variability of the Earth's rotational speed was established. The basis for time that had served for so many centuries was no longer adequate to meet the more demanding needs for time. A search for the definition and introduction of a uniform second and timescale led to Ephemeris Time, based on the orbital motions of solar system bodies. At that time atomic clocks were being developed that offered a more convenient and accurate basis for time. Time measurement progressed from timescales based on astronomical phenomena to atomic physics. In addition, improvements in the accuracy of planetary positions required the introduction of dynamical timescales that recognized the role of general relativity in timekeeping. Over the same period of time the accuracies of timekeeping and time transfer improved significantly, and requirements for time have become even more demanding. The atomic Système International (SI) second quickly achieved recognition as the most accurate and fundamental unit of measure.

Although the Earth was no longer the basis for the most precise timekeeping, the demands of new technologies made it even more critical to observe, analyze, and predict the actual variations of its rotation. The motions of its rotational axis, both in space and in the Earth itself, also required a parallel effort of observations, analysis, and prediction. These activities pushed the improvement of celestial and terrestrial reference frames by orders of magnitude and encouraged new developments in the study of the dynamics of the Earth, including the core, mantle, atmosphere, oceans, etc., and the forces acting on it due to the Sun, Moon, and planets. These studies have gone on to spur the further development of even more accurate methods of observations.

This book is intended to tell the story of the progress in timekeeping over the past century. It begins with time solely based on the rotation of the Earth, and proceeds through the discovery of the variations in Earth rotation and motions of the Earth's pole. During that time clocks progressed through improvements in mechanical clocks to the development and improvements of atomic clocks. The availability of atomic time, the routine observations of the variable Earth rotation, and the development of the theory of relativity led to the introduction of Universal Time, International Atomic Time, Coordinated Universal Time, and a family of dynamical timescales. In the process there have been a number of scientific discoveries, significant improvements in accuracy, the development of new applications of accurate time, and the growth of the scientific field of Earth dynamics.

Additions in this edition include a chapter on Time and Cosmology, developments in optical frequency standards, possible redefinition of the second, and the future of UTC and leap seconds, as well as discussions of the difference between UT1 and mean solar time, the "move" of the Greenwich prime meridian, geomagnetic jerks and their effects on Earth orientation, possible timescales constructed from pulsar and white dwarf observations, and applications of time and frequency for intelligent highways and self-driving cars.

A list of acronyms and a glossary are included to ease the use of a number of specialized terms that have developed over the years in this field.

It is our pleasure to acknowledge and thank our colleagues: Professor Chris Impey, whose presentation at the Science of Time Symposium in June 2016 provided the inspiration and outline of the chapter on Time and Cosmology; Professor Mark Whittle corrected and improved our draft of that chapter; and Paul Hughes assisted in literature searches to update this book.

1

Time Before the 20th Century

1.1 In the Beginning

The earliest people on the planet surely recognized the cycles of the most basic astronomical motions. The Sun and Moon rose and set each day, and moved through the sky with predictable patterns, and the weather followed a cycle related to the movement of the Sun with respect to the stars. The units of days, months, and years naturally followed. With further observations, those watching the sky could visualize patterns in the stars and distinguish comets and "stars" that "wandered" among the others. Dramatic events such as solar and lunar eclipses were seen, and, in some cases, they were recorded and predicted, and their "meanings" interpreted. The observation and measurement of these cycles were important for daily life, religious practices, and agriculture, and they became the bases for timekeeping and calendars. Apparently, some of these observations even affected the orientations of early constructions of tombs or stone circles, such as Stonehenge. Naturally different cultures developed different customs for both keeping time and developing calendars (Aveni, 2002). Variations in accuracy of observations and application led to changes in understanding of the motions and ability to make predictions. These improvements in knowledge and accuracy even continue to drive changes in our definition and use of time in the present and into the future.

1.2 Characterizing Time

Much has been written about the topic of time, and so it is necessary to acknowledge the distinctions between the concepts of time, idealized timescales, definitions of timescales, requirements for different timescales, practical realizations of timescales, and the applications of timescales. In addition, there are time units that we can measure, and nonmeasurable time units that can only be calculated.

Traditionally, we have recognized the desirability of a kind of "uniform time" with basic units that always remain the same. Isaac Newton (Newton & Motte, 1934) distinguished between an idealized "absolute time" and the time provided by physical measurements. The use of the word "uniform," however, implies the existence of a standard of comparison to establish that it is truly uniform. The failure of observations to agree with prevailing instrumentation and models based on established philosophy or theory crafted to provide that standard has led to changes in the theories as well as the means of determining time.

The search for a practically realized uniform time drives the hunt for an ideal unit of time and the means to access it. As measurement accuracies improve, it has been, and will continue to be, necessary to develop better concepts, definitions, and practical realizations of time. Today general and special relativity along with theories regarding the origin of the universe impose further considerations regarding time in the coordinate systems of the future.

1.3 Calendars

The development of calendars varied, depending largely on religion, culture, politics, and economics. Religious practices and holidays along with agriculture cycles have been defined in terms of lunar sightings, solar motion, and the appearance of the stars in the sky. Hence, calendars have been based on lunar or solar motions, or a combination of the two. Unfortunately, years, months, and days are not integral multiples of each other, and this has led to complications in creating calendars. For example, the year as measured by the length of time for the Sun to return to the same place along its path in the sky (ecliptic) is currently equal to 365.2421897 days of 86,400 seconds, and the length of the month measured by the Moon's phases is 29.53059 days. The year cannot be composed of an integral number of months or days. Historically, the counting of years in different calendars has been based on the reigns of rulers, the lives of religious leaders, and the traditional beginnings of cultures.

Today, while a number of calendars remain in use for religious or national reasons, the Gregorian calendar, introduced by Pope Gregory XIII in 1582 and adopted by various countries over the next 340 years, is the calendar used internationally for civil purposes (*Explanatory Supplement to the Astronomical Ephemeris and the American Ephemeris and Nautical Almanac*, 1961). It will be of satisfactory accuracy for thousands of years to come. A number of books and references on calendars have been published, and software has been developed to convert between them (Seidelmann, 1992; Richards, 1999; Urban & Seidelmann, 2013). A set of chronological "eras" exist based on the various calendars, and these along with current years of various calendars are tabulated annually (*The Astronomical Phenomena*).

1.4 Astronomical Observations

Astronomical observations throughout human history have led to catalogs of star positions and theories of motions, all for the purpose of being able to predict future phenomena. In order to catalog the positions of stars, planets, the Moon, and the Sun, it was necessary to develop a reference frame. A natural choice devised in antiquity was a set of measures based on the equator, defined by the apparent diurnal motion of the stars, and the ecliptic, which is the apparent path of the Sun in the sky. The intersection of these two "circles" provides a natural origin useful for making angular measurements. That origin was called the *equinox*, since days and nights are approximately equal when the Sun is located in that direction. *Declination*, measured north (positive) or south (negative) of the equator in degrees, minutes, and seconds of arc up to 90 degrees, provides one angular coordinate of the direction to a celestial object. The other is the measure along the equator from the equinox, designated as *right ascension* and measured in hours, minutes, and seconds of time up to 24 hours. The sexagesimal system of measure, currently in use, probably is of Babylonian origin. It was widely used for scientific purposes in antiquity and is retained today in angular measurements. Hipparchus discovered the motion of the equinox in about 129 BC from comparisons of his star positions with positions determined about 150 years earlier (Neugebauer, 1975). This motion is called *precession* and is about 50″ per year. Further contributions to our fundamental astronomical knowledge were made by Ptolemy and various Chinese, Mayan, Middle Eastern, and Indian astronomers. An inconvenience of this system is that both circles are in motion in space.

In the 17th century, the need for navigation and timekeeping led to the establishment of national observatories in Paris (1667), Greenwich (1675), Berlin (1701), and St. Petersburg (1725). As astronomers sought improvements in knowledge and accuracy from observations, star catalogs were prepared, and discoveries were made. Edmond Halley showed in 1718 that the positions of the bright stars, Aldebaran, Sirius, and Arcturus, had changed by minutes of arc from their positions in antiquity. This angular motion perpendicular to the line of sight was called *proper motion*. In 1728, James Bradley discovered that the Earth's orbital motion caused a change of the direction of an object by approximately 30″ due to the finite speed of light, and this was called *stellar aberration*. Bradley detected the periodic motion of the Earth's celestial pole with respect to the stars in 1748. This motion, called *nutation*, can reach an amplitude of about 18″. In 1781, William Herschel discovered the planet Uranus from its motion and his systematic observations of the sky. He also discovered the solar motion toward the constellation Hercules from an analysis of proper motions. Friedrich Wilhelm Bessel, Wilhelm von Struve, and Thomas Henderson independently detected in 1838–1849 that star positions shifted as the observer moved in the Earth's orbit. This confirmation of the Copernican theory and measurement of stellar distances, called *parallaxes*, required observations with a precision of a few tenths of an arcsecond (Kovalevsky & Seidelmann, 2004).

The inclusion of these effects, along with refraction, has contributed to the improvements in accuracies over the years. Improved instrumentation, including photography, led to accuracy improvements in observations. Hence, there was a succession of improved star catalogs from Washington and Germany, including the FK5 catalog in the 1980s and the current standard, the Hipparcos Catalogue (Perryman et al., 1997). These catalogs were the basis for the reference systems defined in terms of the equinox and equator.

Newton's law of universal gravitation demonstrated that Kepler's three laws of planetary motion were the consequences of a gravitational central force. From that development, general theories of the motions of the Sun, Moon, and planets were produced by a number of scientists in different countries. Improved accuracies of these theories, along with comparisons with steadily improving observations, led to many mathematical developments, improvements in the knowledge of astronomical quantities, and the discovery of Neptune in 1849. However, there was a lack of international agreement on astronomical quantities or ephemerides. At the end of the 19th century, agreement was reached with international acceptance of Newcomb's constants, and solar system ephemerides, based on general theories by Newcomb and Hill, which were introduced in 1900.

The lunar theories presented a more challenging situation. Since the motion of the Moon is so rapid and the perturbations so large, the development of a theory for the motion of the Moon has presented continuing problems. Many scientists have worked on the theory, and a number of different methods were employed in attempting to develop lunar theories that would successfully represent observations and predict the future motion of the Moon. In the process, a great empirical term (Newcomb, 1878) was introduced, and tidal friction, variable rotation of the Earth, and the secular acceleration of the Moon were discovered. Still the understanding and accurate calculation of the ephemeris of the Moon remained a challenge.

1.5 Timekeeping

The Sun's position in the sky has always been an obvious means to keep track of time. The use of shadows cast by the Sun of sticks and sundials was a natural means of telling time. These tools could indicate the time of day by the direction of the shadow, and the time of year by the length of the shadow. The astrolabe first appeared in the third or second century BC and provided further improvement. Examples range from simple devices for measuring the angular separation between two directions, to quite sophisticated instruments. They could be used to measure the altitude of the Sun, Moon, planets, and stars, determine the hour of day or night, the latitude of the observer, and solve other astronomical problems without numerical calculations (Fraser, 1982; Dohrn-van Rossum, 1996).

The use of controlled flow of water produced alternative means of measuring time in Egypt, India, China, and Babylonia before 1500 BC. In the third century BC, water clocks were being used for scientific observations (Dohrn-van Rossum, 1996). In the 8th to 11th centuries AD, water was used to drive mechanical wheel clocks in China. Sand clocks were introduced in the late 14th century AD (Dohrn-van Rossum, 1996). In the late 13th or early 14th century, weight-driven mechanical clocks were developed, but their actual origin is uncertain. Initially, they were more decorative than accurate, and did not have minute hands. In the 17th century, Galileo recognized the value of the pendulum as a timekeeping device, but it was Huygens who built the first pendulum clock in 1656. It had an accuracy of 10 seconds per day, providing reasonably accurate, but not highly reliable timekeeping. It was the late 18th century when clocks were improved enough to provide reliable, accurate time (Jespersen, Fitz-Randolph, Robb, & Miner, 1999). Later, pendulum clocks were improved by various refinements to provide accurate sources of time for national timekeeping and for astronomy.

The challenge of safe navigation and the means of determining longitude at sea was a strong motivator for the development of robust, accurate mechanical clocks. This challenge was met by John Harrison with his H4 chronometer, made in 1759. In sea tests in 1762, it only lost five seconds in 81 days (Sobel, 1995).

In the 1800s, providing accurate time for civil purposes became an important function of local observatories. Typically, they made transit observations of the Sun by day and stars at night to determine local solar time for their location. Hence, each locality or region had its own time based on the location of the Sun in the sky.

1.6 Time Epochs

Throughout history there have been different choices for beginning the day. The ancient Egyptians began the day at dawn, but the Babylonians and Jews chose sunset. The ancient Romans switched to midnight after first using sunrise to mark the day's beginning. Sunrise was the most common choice in Western Europe before the general use of clock time. Time was counted in units of 12 hours of day and 12 hours of night. The ancient origin of 12-hour days and nights is lost, but they were transmitted from ancient Greece, Egypt, and Babylonia. A probable developmental explanation is given by Neugebauer (1957).

Since the periods of day and night varied during the year, the length of the hours differed from day to night and during the year. For astronomical purposes, these seasonal hours were replaced by a subdivision of the complete period of daylight and darkness into 24 equal and constant parts, known as *equinoctial hours*. Apparently Hipparchus was the first to adopt the equinoctial hours in place of the unequal and varying seasonal hours. The seasonal hours spread throughout the Greco-Roman world during the Hellenistic period, and until mechanical clocks became common during the Middle Ages, they remained in use for civil purposes. The equinoctial hours were used only in astronomical and calendrical works, or for other special purposes.

According to Pliny, Hipparchus reckoned the day from midnight, but Ptolemy reckoned it from noon at Alexandria, and in his *Handy Tables* (Neugebauer, 1975) divided the day and night into 24 equal hours, each hour subdivided into minutes and seconds. For astronomers and navigators observing the stars at night, it was convenient to avoid a change of days during the night, so they used days counted from noon to noon. This practice was established in astronomical tables and ephemerides within the Greek, Latin, and Arabic civilizations. For this reason the Julian day numbers, a continuous count of days from 4713 BC, continue to start each day at noon. Until 1925, the day in Greenwich Mean Time began at noon. In 1925, Greenwich Mean Civil Time was introduced as starting at midnight, and eventually Greenwich Mean Time came to be accepted as beginning at midnight. Thus, care must be taken when using observations from before 1930 as to what time system was used.

1.7 Time Transfer

Without widespread and comparatively rapid commerce, the distribution of a common time reference held little importance. Portable sundials were adequate tools to obtain the local time. In the early 14th century, European communities began to install weight-driven mechanical clocks in association with bells, either in town halls or in churches. These served as local time references and provided a means to notify people of the time for prayers, services, assemblies, and markets. Time balls on the top of buildings, which were visible by ships in ports, were used to signal the time on a daily basis for navigators. At the US Naval Observatory (USNO), for example, this service started in 1845. Well-calibrated portable mechanical clocks provided a means to distribute precise time. Local observatories made time available in various cities, based on their astronomical observations and mechanical clocks.

With the invention of the telegraph, time signals could be distributed over long distances. In 1865, the USNO started sending signals at 7:00 AM, noon, and 6:00

PM over the fire alarm system in Washington. That signal went to the State Department, which also had a Western Union telegraph signal, so by 1867, the time signal from the USNO was transferred there and sent throughout the country. In 1869, the signals were going across the country for the railroads and by 1871, the US Signal Service was distributing the signal to weather stations. In 1886, synchronized clocks in public offices were kept on USNO time via the signals. Telegraph signals, in conjunction with astronomical determinations of local time, were also used to measure the time difference between distant points and, thus, determine their relative positions (Bartky, 2000; Dick, 2003).

1.8 Rotation of the Earth

Solar time determined from astronomical observations was the independent argument used to calculate ephemerides of the solar system until the mid-20th century. This was done with the understanding that solar time provided a uniform measure of time. Discrepancies between observed and calculated positions of solar system objects appeared and were most evident for the Moon, because its motion is most rapid and complex. J. C. Adams showed that the observed secular acceleration of the Moon's mean motion could not be due to gravitational perturbations (Adams, 1853). That the tides exert a retarding action on the rotation of the Earth, along with a variation in the orbital velocity of the Moon, according to the conservation of momentum, was shown independently by W. Ferrel and C.-E. Delaunay (Ferrel, 1864; Delaunay, 1865). Newcomb considered an irregular rotation rate of the Earth as the explanation for lunar residuals, but he could not find corroboration from planetary observations (1878). The correlation of the irregularities of the motions of the inner planets and the Moon to prove the irregularity of the rotation of the Earth is described in detail in Chapter 4.

In 1765, Euler predicted that, if the axis of rotation were not coincident with the principal axis of inertia, the axis of rotation would have a circular motion with respect to the Earth's crust (1765). Using historical transit circle observations, Seth Chandler (1891a, 1891b, 1892) detected this effect, called *polar motion*, which can displace the direction of the rotation axis by angles of the order of 0.5″. However, the observed period of 433 days did not agree with Euler's theoretical prediction of 305 days. Newcomb explained the difference as due to the non-rigid Earth (1891). In the 1890s the International Latitude Service (ILS), with several stations at 39 degrees north latitude, was established to make optical observations to measure polar motion. That service continued until more accurate methods were developed in the 1970s. Actually, the axis of maximum moment of inertia moves around the axis of rotation with a complicated pattern made up largely of an annual component and a 14-month (Chandler) component.

1.9 Beginning the 20th Century

As the 20th century began, official time in each country was based on pendulum clock time standards. There was no international exchange of time. There were some accurate longitude measures by trans-oceanographic telegraph signals. The international monitoring of polar motion had just begun. Variations in the Earth rotation were suspected, but not proven. Mean solar time was based on Newcomb's theory of the Sun. Astronomical observations were improving based on photography, better instrumentation, recognition of the personal equation in observation timings, and the adoption of more accurate astronomical constants. The 19th century had experienced significant improvements in accuracy and knowledge, but the next century was to be more impressive.

References

Adams, J. C. (1853). On the Secular Variation of the Moon's Mean Motion. *Philosophical Transactions of the Royal Society of London Series I*, **143**, 397–406.

The Astronomical Phenomena. Washington, DC: US Government Printing Office.

Aveni, A. (2002). *Empires of Time*. Boulder, CO: University Press of Colorado.

Bartky, I. R. (2000). *Selling the True Time: Nineteenth-Century Timekeeping in America*. Stanford, CA: Stanford University Press.

Chandler, S. C. (1891a). On the Variation of Latitude, I. *Astron. J.*, **11**, 59–61.

Chandler, S. C. (1891b). On the Variation of Latitude, II. *Astron. J.*, **11**, 65–70.

Chandler, S. C. (1892). On the Variation of Latitude, VII. *Astron. J.*, **12**, 97–101.

Delaunay, C.-E. (1865). Sur l'existence d'une cause nouvelle ayant une action sensible sur la valeur de l'équation séculaire de la Lune. *Comptes rendus hebdomadaires des séances de l'Académie des sciences*, **61**, 1023–1032.

Dick, S. J. (2003). *Sky and Ocean Joined: The U.S. Naval Observatory, 1830–2000*. Cambridge: Cambridge University Press.

Dohrn-van Rossum, G. (1996). *History of the Hour: Clocks and Modern Temporal Orders*. Chicago, IL: University of Chicago Press.

Euler, L. (1765). Du mouvement de rotation des corps solides autour d'un axe variable. *Mémoires de l'académie des sciences de Berlin*, **14**, 154–193.

Explanatory Supplement to the Astronomical Ephemeris and the American Ephemeris and Nautical Almanac (1961). London: HM Stationery Office.

Ferrel, W. (1864). Note on the Influence of the Tides in Causing an Apparent Secular Acceleration of the Moon's Mean Motion. *Proc. American Acad. Arts and Sciences*, **6**, 379–383.

Fraser, J. T. (1982). *The Genesis and Evolution of Time: A Critique of Interpretation in Physics*. Amherst, MA: University of Massachusetts Press.

Jespersen, J., Fitz-Randolph, J., Robb, J., & Miner, D. (1999). *From Sundials to Atomic Clocks: Understanding Time and Frequency*. Washington, DC: US Dept. of Commerce, Technology Administration, National Institute of Standards and Technology.

Kovalevsky, J. & Seidelmann, P. K. (2004). *Fundamentals of Astrometry*. Cambridge: Cambridge University Press.

Neugebauer, O. (1957). *The Exact Sciences in Antiquity*. Providence, RI: Brown University Press.

Neugebauer, O. (1975). *A History of Ancient Mathematical Astronomy.* Berlin; New York, NY: Springer-Verlag.

Newcomb, S. (1878). *Researches on the Motion of the Moon.* Washington, DC.

Newcomb, S. (1891). On the Periodic Variation of Latitude and the Observations with the Washington Prime Vertical Transit. *Astron. J.*, **11**, 81–82.

Newton, I. & Motte, A. (1934). *Sir Isaac Newton's Mathematical Principles of Natural Philosophy and His System of the World.* Berkeley, CA: University of California Press.

Perryman, M. A. C., Lindegren, L., Kovalevsky, J., et al. (1997). The Hipparcos Catalogue. *Astron. Astrophys.*, **323**, L49–L52.

Richards, E. G. (1999). *Mapping Time: The Calendar and Its History.* New York, NY: Oxford University Press.

Seidelmann, P. K. (1992). *Explanatory Supplement to the Astronomical Almanac.* Mill Valley, CA: University Science Books.

Sobel, D. (1995). *Longitude: The True Story of a Lone Genius Who Solved the Greatest Scientific Problem of His Time.* New York, NY: Walker.

Urban, S. E. & Seidelmann, P. K. (2013). *Explanatory Supplement to the Astronomical Almanac.* Mill Valley, CA: University Science Books.

2

Time from the Earth's Rotation

2.1 Apparent Solar Time

From antiquity the direction of the Sun in the sky has been used as a way to measure the passage of time. Over the centuries this basic concept developed to the point that the angle measure, equivalent to the local hour angle of the Sun, became known as *apparent solar time*. This time depends on the longitude of the site in question. When the observer is on the Greenwich meridian we call it *Greenwich apparent solar time*, and for any other location it is called *local apparent solar time*. The length of a day, and, therefore, the second (determined by the fraction of 1/86,400 of the day), vary during the year, because the path of the Sun is inclined with respect to the equator, and because the eccentricity of the Earth's orbit affects the rate at which the Sun appears to move in the sky. Apparent solar time is the time given by sundials and was the argument in almanacs and national ephemerides until the early 19th century. Historically, the accuracy with which time can be determined using the Sun was limited by the accuracy of solar observations.

2.2 Mean Solar Time

Ancient astronomers were quite aware of the fact that measuring time using the daily motion of the Sun did not result in a uniform timescale. The solution of the problem goes back at least to the time of Ptolemy (AD 90–168), who was looking for a uniform time to construct tables of the motion of the Sun. *Mean solar time* uses the concept of a fictitious point on the equator of the celestial reference system defined so that, in one year, the Sun's motion along the ecliptic is equivalent to the motion of the fictitious point along the equator. The angle measure equivalent to the local hour angle of this point became known as mean solar time. The fiducial point is sometimes called the *fictitious mean Sun*, or just the *mean Sun*. As with apparent solar time, the mean solar time depends on the longitude of the observer's location.

It also depends on the mathematical description of the motion of the fictitious point. As mechanical clocks improved, apparent solar time became less useful, and mean solar time came into wider use. Mean solar time was introduced in almanacs in England in 1834 and in France in 1835. The concept of solar time, either mean or apparent for practical applications, relies on the uniformity of the Earth's rotation. The error of this assumption eventually led to the development of dynamical timescales (Chapter 9) and Universal Time (UT) (see Section 2.5).

The expression for the right ascension of the fictitious mean Sun depended on tables that were, in turn, based on the theory of the motion of the Sun that prevailed at the time. Tables of the difference between apparent and mean solar time (in the sense of apparent − mean solar time), called the *equation of time* (see Section 2.3), were prepared from the mathematical description of the motion of the fiducial point. In the 19th century there was no general agreement regarding tables of the Sun to be used to define mean solar time (see Chapter 3). However, in 1896 international agreement was reached to adopt Simon Newcomb's *Tables of the Sun* for mean solar time (Newcomb, 1898). The basis for mean solar time then became Newcomb's equation for the right ascension of the fictitious mean Sun:

$$R_U = 18h\ 38m\ 45.836s + 8640184.542s\ T + 0.0929s\ T^2, \tag{2.1}$$

where *T* is the number of centuries of 36,525 days elapsed since Greenwich mean noon on 1900 January 0. In his presentation of this expression Newcomb did not specify the timescale for what he called "T." His theory of the Sun was based on observations from the 18th and 19th centuries, and there are now modern solar theories and numerical integrations of the Earth's orbit that could provide a more accurate expression for mean solar time.

The equation of time reaches a minimum value of approximately −14 minutes around February 6 and a maximum value of about 16 minutes around November 3, as shown in Figure 2.2. The time-varying equation of time can also be plotted as a function of the solar declination, and that curve is called the *analemma* (Figure 2.1).

Because of the difficulty of observing the direction of the Sun with the required precision, the determination of mean solar time, in practice, came to rely on observations of the transit times of stars instead. The point-like images of stars make the angle measurements more precise. When using observations of stars instead of the Sun an observer would note the time (from a local clock) when a star with a known right ascension crossed the local meridian. Knowing the star's right ascension, the observer then knows the direction of the equinox at that instant. If that observer then also has a description of the direction to the fictitious mean Sun expressed in that conventional celestial reference system, it is possible to determine the mean solar time at that instant. The positions of stars in the

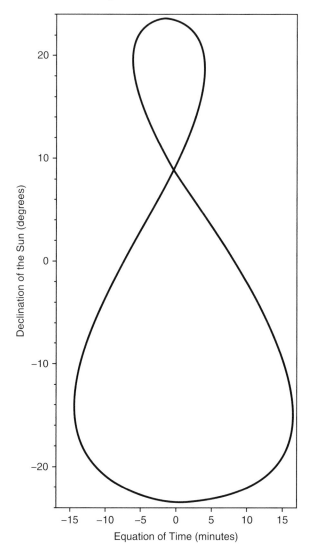

Figure 2.1 Analemma

reference frame are provided with respect to the equinox and equator of the system, as is the right ascension of the fiducial point. Consequently, in 1900 mean solar time was defined as:

> Greenwich Hour Angle of mean equinox of date − right ascension of
> fictitious mean Sun + 12h.   (2.2)

Prior to 1925 mean solar time was measured from noon, and the mean solar day beginning at noon was called the *astronomical day*. Mean solar time measured from noon on the Greenwich meridian was designated *Greenwich Mean Time* (GMT).

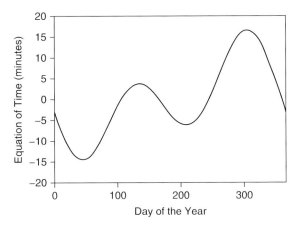

Figure 2.2 Equation of time

The use of the astronomical day was discontinued at the end of the year 1924, and replaced by the use of the civil day reckoned from midnight beginning with January 1, 1925. In 1925, January 1.0 was designated as the same instant as December 31.5 in the 1924 almanacs. The American Nautical Almanac introduced *Greenwich Civil Time* (GCT) and *local civil time* (LCT) as the name for the time reckoned from midnight. Starting in 1925, times measured from midnight were variously labeled Greenwich Civil Time or Greenwich Mean Time. The International Astronomical Union, at the Leiden General Assembly in 1928, recommended that the designation GMT not be used, because of the different meanings it had prior to 1925, and that GCT and UT, which are not ambiguous, be used instead. The term *Greenwich Mean Astronomical Time* (GMAT) was recommended when using a day beginning at noon. Finally, the name *Universal Time* was introduced for time measured from Greenwich midnight, even for times before 1925. The term *Greenwich Mean Time*, referring to time measured from midnight, continued to be used for navigation purposes and as civil time in the United Kingdom (*Trans. IAU*, 1928). Now GMT is only correctly used as the civil time in the United Kingdom (see Section 2.8), and the use of Greenwich mean solar time in practical applications has been superseded by Universal Time (see Section 2.5).

2.3 Sidereal Time

The rotation of the Earth can also be measured with respect to the stars without any reference to the Sun or a fictitious mean Sun. That time, equivalent to the hour angle of the equinox of the celestial frame, is called *sidereal time*. Since the Sun appears to move with respect to the stars in the sky by approximately one degree each day, sidereal time is quite different from solar time. Although it is a direct and observable measure of the rotation of the Earth in the celestial system, it is not a true measure of the Earth's rotation, because the equinox is continuously moving due to precession and nutation. Apparent sidereal time is measured with respect to the true equinox, the intersection of the true equator of date and the true ecliptic of date. The true equinox of date is affected by both precession and nutation, which introduces periodic variations into apparent sidereal time. Mean sidereal time is measured with respect to the mean equinox, which is only affected by the motion due to precession. Apparent sidereal time minus mean sidereal time is the equation of the equinoxes, formerly called nutation in right ascension.

The sidereal day is the period between two consecutive transits of the equinox. Due to precession the mean sidereal day of 24 hours of mean sidereal time is shorter than the period of rotation of the Earth by the amount of precession in right ascension in one day, which is approximately 0.0084 seconds. The sidereal day is about four minutes shorter than the solar day, because the continual motion of the Earth in its orbit during the day means that additional Earth rotation is necessary for the Sun to cross the local meridian again.

Local sidereal time depends on the longitude of the observer, and is related to Greenwich Sidereal Time by:

$$\text{local sidereal time} = \text{Greenwich Sidereal Time} + \text{east longitude}. \qquad (2.3)$$

Sidereal time is usually expressed in hours, minutes, and seconds. Longitude in degrees can be converted to units of time using a conversion of one hour being equivalent to 15 degrees.

Although no longer used in practice for precise timekeeping, optical and photographic observations of the diurnal motions of stars with respect to the local meridian do provide a measure of apparent sidereal time, since the stars' right ascensions are determined with respect to the true equinox. The equinox itself cannot be observed directly, and the stars are not fixed positions in the sky. Mean sidereal time is determined by applying the equation of the equinoxes. In practice these observations require corrections for refraction, aberration, parallax, and proper motion, and allowance must also be made for the variation in the position of the meridian due to the motion of the rotational poles over the Earth's surface. Until the mid-1980s, mean solar time was derived from apparent sidereal time using the conventional expression for the right ascension of the fiducial point that defined mean solar time.

2.4 Washington Conference of 1884

The growing desire for a worldwide conventional system of longitude measure led to the International Meridian Conference held in Washington, DC, in October 1884. Participants agreed that:

1. The Greenwich meridian would be the initial meridian for longitudes.
2. Longitudes would be measured in two directions up to 180 degrees, east longitude being positive and west longitude being negative.
3. A universal day would be adopted for all purposes for which it may be found convenient.
4. This universal day would be a mean solar day, to begin for the entire world at the moment of mean midnight of the initial meridian, coinciding with the beginning of the civil day and date of that meridian, and would be counted from zero up to 24 hours.

The conference expressed the hope that as soon as practicable the astronomical and nautical days would be arranged everywhere to begin at mean midnight (*Expl. Supp.*, 1961). The Greenwich meridian was adopted as the zero meridian with some countries continuing to use a secondary meridian internally. In some cases, this was because the longitude difference between Greenwich and a reference national meridian was much less accurate than local longitude differences. The astronomical and nautical day change was not made until 1925. The adoption of east positive for longitudes was not generally accepted in the West until the 1980s, when computers required the use of plus and minus numbers, rather than the letters W and E.

2.5 Universal Time

The term *Universal Time* was first recommended by the International Astronomical Union (IAU) (*Trans. IAU*, 1935) to designate mean time on the meridian of Greenwich reckoned from midnight. It now refers to a set of timescales related to the mean diurnal motion of the Sun. There has been a progression in definitions and means of determining UT, because of improvements in knowledge and accuracy and increasing demands for accuracy. UT1 is the timescale derived from direct observations of the Earth's rotation angle in space. Astronomical observations are made to determine this angle with respect to a celestial reference system, and then it can be related to mean solar time by using a mathematical expression adopted for this purpose.

Before 1984 Universal Time was defined as 12 hours plus the Greenwich hour angle of a point on the equator whose right ascension, measured from the mean equinox of date, is given by Equation 2.4. The measure of Universal Time at T_U, in hours, minutes, and seconds, was then:

$$12^h + \text{Greenwich hour angle of the mean equinox of date} - R_U. \qquad (2.4)$$

As with sidereal time there are local mean solar times related to Universal Time by:

$$\text{Local mean solar time} = \text{Universal Time} + \text{east longitude (in time units).} \quad (2.5)$$

Local observations of the Earth's rotation angle made by timing star transits are affected by the motion of the pole of the Earth's rotation axis over the surface of the planet. Consequently these local observations provide a local measure of Universal Time known as UT0, which is a designation no longer in common use because modern measures of the Earth's rotation angle no longer depend on the location of the local meridian. UT1 was obtained by correcting the UT0 observations for the motion of the pole at the observing site.

UT2 is a version of Universal Time that is also rarely used in practice today. UT2 corrects UT1 for the annual seasonal variation in the Earth's rotational speed by applying a conventionally adopted mathematical expression to UT1. The correction accounts for the fact that the Earth rotates slower in the spring and faster in the fall. That expression is (McCarthy, 1991):

$$UT2 = UT1 + 0.022\text{s}\ \sin 2\pi\ t_B - 0.012\text{s}\ \cos 2\pi\ t_B$$
$$-0.006\text{s}\ \sin 4\pi\ t_B + 0.007\text{s}\ \cos 4\pi\ t_B \qquad (2.6)$$

where t_B is the fraction of the Besselian year (see Section 3.11) and given by:

$$t_B = \text{mod}\left[2000 + \frac{(\text{Julian Date} - 2\ 451\ 544.533)}{365.2422},\ 1\right]. \qquad (2.7)$$

At 12^h UT, the Greenwich hour angle of the mean Sun is zero, which implies that the Greenwich Mean Sidereal Time (GMST) at that instant is equivalent to the right ascension of the mean Sun, R_U (Equation 2.1). This led to the expression used from 1900 to 1984:

$$0^h\ \text{UT} = 6^h\ 38^m\ 45.836\text{s} + 8640184.542\text{s}\ T_U + 0.0929\text{s}\ T_U^2, \qquad (2.8)$$

where T_U takes successive values at a uniform interval of 1/36525 from 1900.0 (*Expl. Supp.*, 1961).

The formula was revised when improved values of astronomical constants were introduced in 1984, and the expression by Aoki et al. (1982) was introduced. Greenwich Mean Sidereal Time was then related to UT1 by:

$$\text{GMST1 of 0h UT1} = 24110.54841\text{s} + 8640184.812866\text{s } T_U +$$
$$0.093104\text{s } T_U^2 - 6.2 \times 10^{-6}\text{ s } T_U^3, \tag{2.9}$$

where $T_U = d_U/36525$, d_U is the number of days of Universal Time since JD 2451545.0 UT1 (2000 January 1, 12h UT1), taking on the values of ±0.5, ±1.5, ±2.5, ±3.5 ... The equation is consistent with the position and motion of the equinox specified by the IAU 1976 System of Astronomical Constants, the 1980 IAU Theory of Nutation, and the positions and proper motions of stars in the FK5 catalog. This equation, which is dependent on the value of the constant of precession and on variations introduced through the timescale, was considered to be the definition of UT1 until 2003.

In 2000 the XXIVth General Assembly of the International Astronomical Union recommended the use of the "non-rotating origin" (Guinot, 1979) both in the *Geocentric Celestial Reference System* (GCRS) and in the *International Terrestrial Reference System* (ITRS). These origins are now called the *Celestial Intermediate Origin* (CIO) and the *Terrestrial Intermediate Origin* (TIO), respectively. The *Earth Rotation Angle* (ERA) is then the angle measured along the equator of the *Celestial Intermediate Pole* (CIP) between the CIO and the TIO. The IAU also recommended that UT1 be linearly proportional to the ERA. As a result, the Earth Rotation Angle θ, is linked to UT1 by the conventional relationship:

$$\theta(t_u) = 2\pi(0.7790572732640 + 1.00273781191135448 t_u), \tag{2.10}$$

with t_u = the Julian UT1 date $- 2\ 451\ 545.0$, the Julian Date being the interval in days and fractions of a day since 4713 BCE January 1, Greenwich noon.

Equation (2.10) is strictly based on the Earth's rotation, since the ICRS is a fixed reference system (see Chapters 5 and 7), and represents the linear relationship between the Earth Rotation Angle (θ) and UT1, which continues the previously defined phase and rate of UT1 (Capitaine, Guinot, & McCarthy, 2000). Although Universal Time is no longer 12^h + the Greenwich hour angle of the fictitious mean Sun, it is sufficiently close compared to the difference between the mean and true Sun to continue to refer to it as a substitute for mean solar time, as in the past. However, Universal Time is not precisely mean solar time.

Today UT1 is determined using *Very Long Baseline Interferometry* (VLBI) measurements of selected radio point sources, mostly quasars, and interpolated by the tracking of Global Positioning System satellites. Strictly speaking, because of the motion of satellite orbital nodes in space, VLBI provides the only rigorous determination of UT1.

2.6 UT1 as Mean Solar Time

Both UT1 and mean solar time are measures of the Earth's rotation. How are they related? Mean solar time at Greenwich can be given by:

$$\text{Mean solar time at Greenwich} = \text{Greenwich Mean Sidereal Time}$$
$$- \alpha_\odot + 12h, \tag{2.11}$$

where $\alpha_\odot$ is the right ascension of the fictitious mean Sun, Greenwich Mean Sidereal Time is a function of two timescales: t_u, the Julian UT1 date $-$ 2 451 545.0, and t, the Julian centuries of 36,525 days of Terrestrial Time (TT) since 2 451 545.0 TT. It can be expressed (Urban & Seidelmann, 2012) as:

$$\text{GMST}(t_u, t) = \theta(t_u) - E_{prec}(t), \tag{2.12}$$

where $\theta(t_u)$ is the Earth Rotation Angle and $E_{prec}(t)$ is the precession portion of the equation of the origins. Those quantities are given by

$$\theta(t) = 2\pi(0.779\ 057\ 273\ 2640 + 1.002\ 737\ 811\ 911\ 354\ 48 t_u), \text{ and} \tag{2.13}$$

$$E_{prec}(t) = -0.014\ 506'' - 4\ 612.156\ 534'' t - 1.391\ 5817'' t^2$$
$$+ 0.000\ 000\ 44'' t^3 + 0.000\ 029\ 956'' t^4. \tag{2.14}$$

Newcomb's expression for mean solar time (Equation 2.1) was based on the constants and his theory of 1900 and on the epoch 1900.0. That can be converted into modern constants and the epoch 2000.0 (Seago & Seidelmann, 2016):

$$R_\odot = 1009658.2260'' + 1296027721.93'' \ T_E$$
$$+ 139.4'' \ T_E^2 \ [\text{epoch 2000}], \tag{2.15}$$

where T_E is in units of Julian millennia from epoch J2000.0 measured in Barycentric Dynamical Time (see Chapter 9).

Using the most contemporary expressions for the mean elements of the Earth in the VSOP 2013 analytical theory (Simon, Francou, Fienga, & Manche, 2013), the mathematical expression for the right ascension of the fictitious mean Sun ($\alpha_\odot$) has also been developed (Seago & Seidelmann, 2016):

$$\alpha_\odot = 1\ 009\ 658.636\ 84'' + 129\ 602\ 771.387\ 492'' t$$
$$+ 1.396\ 56'' t^2 - 0.000\ 0927'' t^3. \tag{2.16}$$

This leads to two different expressions for the mean solar time at Greenwich, mst_G, depending on the representation of the right ascension of the fictitious mean Sun (either Equation 2.15 or 2.16).

For any interval, τ, in TT days since 2 451 545.0TT, we note that $t_u = \tau - \Delta T(\tau)/86\ 400$, that the number of UT1 seconds in that interval is $86\ 400 t_u$, and that:

$$\theta(\tau - \Delta T(\tau)/86400) = 67\,310.548\,410s + 86\,636.546\,9491s\,\tau$$
$$-1.002\,737\,811\,911\,35\Delta T(\tau). \tag{2.17}$$

The difference between mean solar time and UT1 can be estimated in terms of τ as:

$$mst_G - UT1(\tau) = \theta(\tau - \Delta T(\tau)//86400) - E_{prec}(\tau) - \alpha_\odot(\tau) - 86400\tau + \Delta T(\tau). \tag{2.18}$$

Two estimates of the difference between mst_G and UT1, corresponding to the two different expressions for the right ascension of the fictitious mean Sun, then result from this expression. Using the "updated" Newcomb expression:

$$mst_G - UT1(\tau) = 0.000977 - 0.002\,738\Delta T(\tau) - 0.184\,03s\,\tau - 1.61220s\,\tau^2,$$

and using the VSOP expression:

$$mst_G - UT1(\tau) = -0.026\,412s - 0.0027\,738\Delta T(\tau) + 0.35297s\,\tau$$
$$- 0.033\,189s\,\tau^2, \tag{2.19}$$

where τ is now given in millennia (365.250 days).

These results are illustrated, along with the observed UTC-UT1, in Figure 2.3. The discontinuities in the plot of UTC result from the introduction of leap seconds,

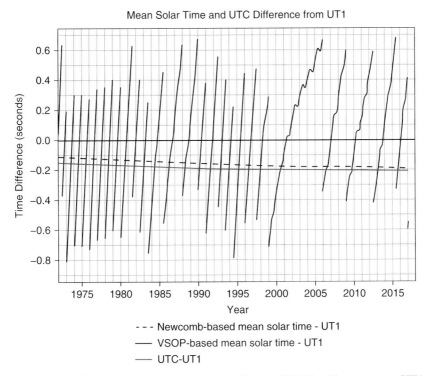

Figure 2.3 Difference between mean solar time and UTC with respect to UT1 from Seago and Seidelmann, 2016

which maintain UTC's proximity to UT1. The time expressed by either mean Sun differs from Universal Time by an amount that is, currently, much less than the variation between UTC and UT1.

2.7 Coordinated Universal Time

In the mid-1940s electronic clocks began to be used to broadcast radio time signals. They kept time with a uniform rate and were adjusted as needed to keep pace with time determined astronomically. Most time signals were adjusted in rate and offsets to match the time that was determined from star transits, and each country or organization broadcasted its own timescale.

On January 1, 1960, the United Kingdom and the United States started to coordinate adjustments made to their timescales. The resulting timescale began to be called *Coordinated Universal Time*, referring to the coordinated actions between the two institutions. Timing laboratories from other countries also started to participate over time, and in 1961 the Bureau International de l'Heure at Paris Observatory began to coordinate the process internationally. In 1965 the IAU officially approved the name *Coordinated Universal Time* with the abbreviation UTC.

The current system of Coordinated Universal Time can be traced back to a meeting of the International Union of Radio Science (URSI) in 1966 when participants noted the need for a uniform atomic frequency. At the 1967 meeting of the URSI, participants agreed that all adjustments to atomic time should be eliminated, and that UT2 information could be distributed in tables or in radio transmissions. In May 1968, the idea of the current practice of introducing one-second adjustments in the UTC timescale was introduced independently by Louis Essen and Gernot Winkler at a meeting of a commission organized by the International Committee for Weights and Measures (CIPM) to discuss the issue. In that same year, Study Group 7 of the International Radio Consultative Committee, abbreviated CCIR, meeting in Boulder, Colorado, discussed possible changes in the definition of UTC. They formed an "Interim Working Party" to provide proposals for a possible new definition of UTC. The options considered were (1) steps in UTC of 0.1 or 0.2 seconds to keep UTC close to UT2, (2) replacing UTC with a timescale with no adjustments, and (3) one-second adjustments.

In 1970 the CCIR approved proposals at its XIIth Plenary Assembly in New Delhi that provided the current definition of UTC. It specified that (a) radio carrier frequencies and time intervals should correspond to the atomic second based on the caesium atom; (b) step adjustments should be exactly one second to maintain approximate agreement with UT; and (c) standard time signals should contain information on the difference between UTC and UT. The new system began on January 1, 1972. The difference between UT1 and UTC is provided routinely by the International Earth Rotation and Reference Systems Service (IERS) based on astronomical observations of quasars using Very Long Baseline Interferometry.

2.8 Greenwich Mean Time (GMT)

The use of the term *Greenwich Mean Time* or its abbreviation GMT remains a source of confusion today. As discussed in Section 2.2, the mean solar time on the Greenwich meridian reckoned from noon was designated originally as *Greenwich Mean Time*, and the mean solar day beginning at noon, twelve hours after midnight at the beginning of the same civil day, was known as the *astronomical day*. From 1780 to 1833 GMT in the British Nautical Almanac was based on apparent solar time. Starting in 1834 mean solar time was the basis of the tabulations in the Nautical Almanac. The International Meridian Conference in Washington in 1884 gave special significance to GMT, as the meridian of Greenwich was designated as the zero meridian and the origin of time zones around the world.

In 1928 the IAU recommended that GMT not be used and that GCT and UT be used instead. The name *Greenwich Mean Astronomical Time* was recommended when reckoning time from noon for any dates after 1925. This confusion was modified when, in 1935 in Paris, the IAU recommended that the use of GCT be discontinued, and that Universal Time (Temps Universel, in French, or Weltzeit, in German) be adopted for international use. A change could not be made in the British publications, and none was made in the American Nautical Almanac, but beginning with the 1939 volume of the American Ephemeris, the double designation *Universal Time or Greenwich Civil Time* was introduced. It was still called *Greenwich Mean Time* in the navigation publications in English-speaking countries. Greenwich Mean Astronomical Time was introduced in some cases for time reckoned from noon. The astronomical almanacs ceased to use GMT after 1960; however, the navigational almanacs continued to use it as UT1. Since 1976 the IAU urged that the term GMT should be replaced by UT0, UT1, UT2, or UTC, as appropriate.

GMT was always compulsory in official British publications. The CCIR referred to GMT for radio broadcasts of time signals, and through the 1950s GMT was used by radio time signals as equivalent to UT2. GMT is still used as the official timescale of the United Kingdom and in some communications systems as UTC. During the summer, GMT can be used in the United Kingdom as the name for daylight savings time, so it is UTC plus one hour. So, since 1925 the use of GMT has had different meanings and its use can be very confusing. Thus, it is recommended that for any precise timekeeping, the name GMT should not be used.

2.9 Time Zones

A worldwide system of standard time zones, based on increments of 15 degrees in longitude, provides the basis for local civil times that are related loosely to solar time. Time zones were first proposed in 1870 by Charles F. Dowd (1930) to regulate time for railroads in the United States. For political and geographical reasons, the time zones are not necessarily uniform longitude strips, 15 degrees wide, running from the north to the south pole. Rather the zone boundaries are set by individual countries and usually follow country, state, or province boundaries. Some countries also introduce time zones on fractions of an hour. The zones are designated by letters with Z specifying the Greenwich-centered zone, and then the zone to the east being A and progressing to the east to zone M, which specifies the 180-degree east zone. The zone for 15 west is designated N, and then the designations continue alphabetically until 180 west is designated Y. A map of the time zones is given in Figure 2.4.

2.10 Daylight Savings Time

Benjamin Franklin first suggested the idea of advancing the standard time by one hour to increase the number of daylight hours in the evening as a means of reducing energy consumption, but it took until the First World War for the idea to be adopted. Since then there has been an inconsistent history of adoption by states and countries, of the dates for the beginning and ending times, and of the political support or opposition, by locations, both in the United States and in other countries. There is also an inconsistency as to how it is designated. In some cases, the names do not change, only the hours have shifted by an hour. In other cases, the word "Standard" is replaced by "Daylight," so for example, in the United States, it is called informally *Eastern Daylight Time* in the summer and *Eastern Standard Time* in the winter. Daylight savings time is sometimes called *fast, advanced*, or *summer time*. According to the US code, however, both times are referred to formally as *standard time*.

The issues concerning daylight savings time continue to be energy consumption, safety, airline schedule changes, agriculture, and personal living styles. One current problem, with different countries changing times on different dates, is that international airlines have to change schedules each time there is a time change to satisfy airport curfew rules.

Starting in 2007 the United States has changed the dates of daylight savings time from the second Sunday in March to the first Sunday in November, with the changes taking place at 2:00 AM. Most of Europe goes on advanced time on the last Sunday in March at 2:00 AM local time. They end advanced time on the last Sunday in October.

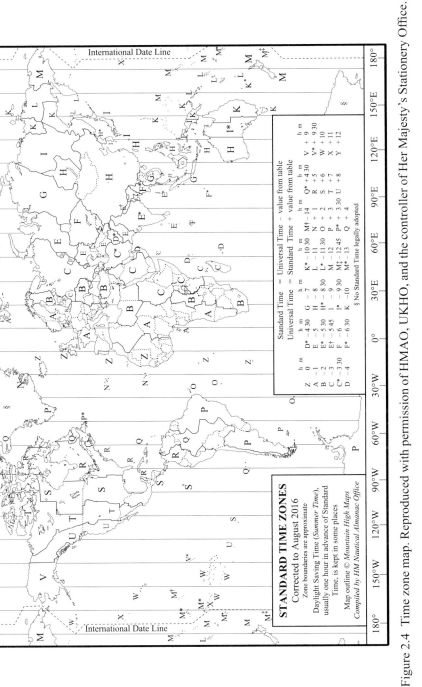

Figure 2.4 Time zone map. Reproduced with permission of HMAO, UKHO, and the controller of Her Majesty's Stationery Office.

References

Aoki, S., Guinot, B., Kaplan, G. H., Kinoshita, H., McCarthy, D. D., & Seidelmann, P. K. (1982). The New Definition of Universal Time. *Astron. Astrophys.*, **105**, 359–361.

Capitaine, N., Guinot, B., & McCarthy, D. D. (2000). Definition of the Celestial Ephemeris Origin and of UT1 in the International Celestial Reference Frame. *Astron. Astrophys.*, **355**, 398–405.

Dowd, C. N. (1930). *Dowd, C.F. A.M., PhD, a Narrative of His Services in Originating and Promoting the System of Standard Time*. New York, NY: Knickerbocker Press.

Explanatory Supplement to The Astronomical Ephemeris and The American Ephemeris and Nautical Almanac. (1961). London: Her Majesty's Stationery Office.

Guinot, B. (1979). Basic Problems in the Kinematics of the Rotation of the Earth. In D. D. McCarthy & J. D. H. Pilkington, eds., *Time and the Earth's Rotation, IAU Symp 82*. Dordrecht: D. Reidel Publishing Company, p. 7.

McCarthy, D. D. (1991). Astronomical Time. *IEEE, Proceedings*, **79**, July 915–920.

Newcomb, S. (1898). Tables of the Motion of the Earth on Its Axis around the Sun. *Astronomical Papers of the American Ephemeris and Nautical Almanac*. Washington, DC: US Government Printing Office.

Seago, J. H. & Seidelmann, P. K. (2016). Mean Solar Time and Its Connection to Universal Time. In E. F. Arias, L. Combrinck, P. Gabor, C. Hohenkerk, & P. K. Seidelmann, eds., *The Science of Time 2016*. Springer, 205–226.

Simon, J.-L., Bretagnon, P., Chapront, J., Chapront-Toouze, M., Francou, G., & Laskar, J. (1994). Numerical Expressions for Precession Formulae and Mean Elements for the Moon and Planets. *Astron. Astrophys.*, **281**, 2, 563–683.

Simon, J.-L., Francou, G., Fienga, A., & Manche, H. (2013). New Analytical Planetary Theories VSOP2013 and TOP20123. *Astron. Astrophys.*, **557**, A49.

Trans. IAU, 1928, **III**, pp. 224, 300.

Trans. IAU 1935, **V**, pp. 29–30, 286, 369.

Urban, S. E. & Seidelmann, P. K. (2012). *Explanatory Supplement to the Astronomical Almanac*. Mill Valley, CA: University Science Books.

3

Ephemerides

3.1 Ephemerides and Time

The term *ephemerides*, the plural of the word *ephemeris*, refers to tables of the positions of celestial bodies, either angular or three-dimensional, listed at regular time intervals. Historically, these were made available in various formats. Some would even argue that prehistoric stone circles indicate primitive knowledge of the motions of the solar system bodies. From early times, numerical tables were prepared as a way of determining the directions of the Sun, Moon, and planets. By entering these tables with a given date and performing appropriate arithmetical operations, the location of a body in the sky could be determined for the specified epoch. The primary uses of the tables were for astrology or determining astronomical positions for specific times. Thus, it did not make sense to compute a lot of positions when only a few were needed. This lack of early ephemerides may explain the lack of improvements of astronomical positions over the centuries. It was not until the availability of the printing press that ephemerides were computed and printed (Gingerich, 2007).

With improved mathematical capabilities, theories of motions were developed, so the positions of the solar system bodies could be determined from algebraic and trigonometric expressions. Later, numerical integrations were used to provide ephemerides. Currently, ephemerides are generally available in printed or electronic formats.

The timescales for the tables, theories, and ephemerides are the independent variables, and it is desirable that they be uniform in rate. Through the years they have changed in order to meet the prevailing understanding of time, the improving precision in timekeeping, and the accuracy of the observations of the solar system positions.

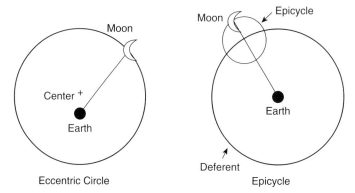

Figure 3.1 Geometric models of planetary motion

3.2 Before Kepler and Newton

In the earliest times, observers noticed, in addition to the Sun and Moon, the bright planets, Mercury, Venus, Mars, Jupiter, and Saturn, moving among the stars. These were called wanderers, which in Greek is the derivation of the word *planet*. The earliest observers of the sky most likely concluded that the Earth was stationary and at the center of the universe, and this was sufficient to enable predictions of eclipses and seasons. In about the fifth century BC, it appears that Babylonian astronomers were developing arithmetic tables for the Sun and Moon (Neugebauer, 1969). A century later, the world system of Aristotle (384–322 BC) incorporated a series of geocentric spheres borrowed from the speculations of earlier Greek philosophers to describe the motions of celestial bodies (Dreyer, 1953). The geometric basis was a Greek innovation. Aristarchus of Samos in about 250 BC developed an earlier Pythagorean philosophy and proposed a heliocentric view of the solar system, but this was an idea without a theory, tables, or predicted positions. Although he left no astronomical writing, we know from other writers that Apollonius around 200 BC appears to have developed a geocentric model of the solar system in which the planetary motions were described using either eccentric circular motions or epicycles (Figure 3.1). In the former the planet moves uniformly along a circle, but the Earth is eccentric (i.e., not at the center), so the apparent angular speed of the Moon or planet varies. In the latter, the planet moves uniformly on a circular orbit, called an *epicycle*, whose center moves uniformly over a circular orbit, called a *deferent*, around the Earth.

Hipparchus (ca. 190–120 BC) from Nicaea (Turkey) and Rhodes developed accurate, geocentric models for the motion of the Sun and Moon using Babylonian and Alexandrian observations and knowledge accumulated over the centuries. He also compiled trigonometric tables and the first catalog of star positions, and discovered precession. Claudius Ptolemaeus (Ptolemy) of uncertain ancestry, who lived in Alexandria, wrote the *Syntaxis Mathematica* about AD 150 (Toomer, 1998). Commonly known as the *Almagest* after the Latin form of the title of the Arab translation, it described his geocentric model and computational methods, and it included tables for determining solar system positions. The *Almagest* also contained a star catalog, probably an updated version of Hipparchus's catalog, and a list of 48 constellations covering the sky visible to Ptolemy. He later produced the so-called *Handy Tables* that provided the data to compute the positions of the Sun, Moon, and planets and the rising and setting of the stars, and to predict eclipses of the Sun and Moon. Later astronomical tables or *zijes* were modeled after these tables. *Zij* is a generic name for Arabic astronomical books that include tabular parameters for calculating positions of astronomical bodies. Ptolemy's solar tables were based on an Earth-centered solar system using circular motions and epicycles to represent the motions. The *Almagest*, which followed Aristotle's geocentric cosmology, was generally accepted in Western culture until the 16th century.

Aryabhata (AD 476–550), an Indian astronomer-mathematician, explicitly proposed a rotating spherical Earth to explain the diurnal motion of the stars. His work, *Āryabhatiya*, written in 499, explains a geocentric system with planetary motions being described by epicyclic models (Sharma, 1977). In the Islamic period, the *Handy Tables* became the basis of the many *zijes*. Then followed the *Toledan Tables* and the relatively common *Alfonsine Tables*. From the 12th century, the Arabic computational "Tables of Toledo" met the accuracy needs of astronomers and navigators until the time of the great explorations. In the Islamic world, Al-Biruni discussed the Indian heliocentric theories of Aryabhata and others in his *Chronicles of India* in AD 1030. Arab scholars were also familiar with ancient Greek sources (Khan, 1977).

In the 13th century, the *Alfonsine Tables* became a widely accepted means to obtain planetary positions. They were named in honor of Alfonso X of Leon and Castile who encouraged the translation of a number of older Arabic publications. The *Alfonsine Tables* also contained a set of instructions written by John of Saxony that explained their use. Nevertheless, a set of "Resolved Tables" followed that were easier to use and largely for astrological purposes. They were a different means of getting mean motions, but the equations for eccentricity and epicycle were the same as the *Alfonsine Tables*'s. The 15th-century astronomer Johannes Müller von Königsberg (1436–1476), who often used the name Regiomontanus, provided, for the first time, daily positions of planets in his *Ephemerides* published for 1474–1506. He had expressed some concerns about previous tables, which he sought to improve by constructing his own, but his ephemerides closely matched the positions from the *Alfonsine Tables* (Chabas & Goldstein, 2003).

The developments from the 15th century onward can, in most cases, be attributed to specific scientists, who made major contributions to our knowledge of solar system motions. In the 15th century, Nicolaus Copernicus (1473–1543), a mathematician, astronomer, physician, classical scholar, Roman Catholic cleric, administrator, diplomat, and economist, proposed in *De revolutionibus orbium coelestium* a heliocentric system in which he may have drawn on the Greek and Islamic traditions of mathematics and astronomy, including the works of Nasis al-Din Tusi, Mu'ayyad al-Din al-'Urdi, and ibu al-Shatir (Saliba, 2002). The detailed models generally agree due to the geometry, which had to do with the aesthetics of uniform circular motion, but did not result in improved accuracy of positions. Copernicus did provide planetary tables in his publication, but they were limited. Erasmus Reinhold (1511–1553) sought to correct that situation by publishing the *Prutenic Tables* in 1551 (Gingerich, 1973).

3.3 Kepler and Newton

Modern work on ephemerides begins in the 16th century with Tycho Brahe (1546–1601) making observations of solar system bodies with improved accuracy. Johannes Kepler (1571–1630) used these observations, primarily observations of Mars, which were the only ones accurate enough, to develop his three laws of planetary motion. The first two were published in *Astronomia Nova* in 1609 and the third appeared in 1619 in his work *Harmonices Mundi*. Specifically, Kepler's laws are:

1. Planets move around the Sun in elliptic orbits with the Sun located at one focus of the ellipse.
2. As a planet moves in its orbit around the Sun, equal areas, as measured from the focus, are swept out in equal times.
3. The square of the period of the orbit is proportional to the cube of the semi-major axis of the elliptic orbit.

Kepler's *Epitome Astronomiae Copernicanae* (1618–1621 in three parts) was a widely read work on theoretical astronomy. He went on to publish a star catalog and tables providing planetary positions in 1627. The *Tabulae Rudolphinae* or *Rudolphine Tables* were based on the observations of Tycho Brahe and Kepler's first two laws, and were intended to complete his work. The *Rudolphine Tables* were difficult to use due to their basis on ellipses and their use of logarithms. The tables of Phillip Landsberg (1631) appeared a few years later and were easier to use as they were based on circular orbits, but they were much less accurate. Kepler's laws and tables became the basis for further developments in Europe in the first half of the 17th century (Russell, 1964).

Isaac Newton (1642–1727) developed his laws of dynamics and gravitation based, in part, on Kepler's third law. His laws of dynamics can be given as:

1. Every body perseveres in its state of rest or uniform straight-line motion, unless it is compelled by some impressed force to change that state.
2. The change of motion is proportional to the motive force impressed and takes place in the same direction as the force.
3. Action is always contrary and equal to reaction.

Newton's law of gravitation, as given in his *Philosophiae Naturalis Principia Mathematica*, published in 1687, became the basis for Newtonian mechanics. It states that the gravitational force, *F*, acting between two bodies of mass, *m* and *M*, is proportional to the product of the masses and inversely proportional to the square of the distance, *r*, between them (Newton, 1999). This can be written as an equation in vector form:

$$\vec{\mathbf{F}} = \frac{GMm}{|r|^3}\vec{\mathbf{r}}$$

(3.1)

where *G* is the gravitational constant. From this law of gravitation, ephemerides of the Sun, Moon, and planets could be calculated using the available computational capabilities (Szebehely & Mark, 1998). Newton's *Principia* did not lead directly to any ephemerides; however, it did influence Mayer's lunar tables (1753) and Edmond Halley (Cook, 1998). The Copernican system was establishing itself; the telescope was bringing new knowledge of the skies, and the ellipse was replacing the epicycle and eccentric circle.

3.4 Tables, General Theories, and Ephemerides

In the days of computation by paper and pencil, tables of logarithms, and mental arithmetic, the calculation of astronomical positions and ephemerides was done from general theories and tables. The general theories used mean elements for the unperturbed, or Keplerian, motion of the planets, and trigonometric expressions for the periodic perturbations of one body by another. The tables gave evaluations of the trigonometric terms tabulated at time intervals small enough for the accuracy sought for the ephemerides, so the trigonometric functions did not have to be evaluated repeatedly. General theories and tables were prepared for the Sun, Moon, and planets using this process. The accuracy of the tables was limited by the accuracy of the orbital elements and the number of trigonometric terms included. The observations were then compared to the predicted positions of the ephemerides to determine improvements to the general theories and tables and to the values of the orbital elements. In this process the masses of the planets and values of astronomical constants were also determined.

In the 17th and 18th centuries, ephemerides were computed by a small number of astronomers or astrologers. Kepler began one series of ephemerides in 1617 and continued until 1637, when Lorenz Eichstadt (1596–1660) continued the series in Gdansk. Johann Hecker computed ephemerides from 1666 through 1680. Then the astronomers at Paris Observatory initiated the *Connaissance des Temps* in 1679 to continue the tabular data annually. There was a competitive activity concerning the French publication, but it continued over the years with varying accuracies (Gingerich & Welther, 1983). Other contributions were made by Andrea Argoli (1570–1657) and Francesco Montebruni and his successors. Ephemerides were also compiled by Eustachio Manfredi (1674–1739) in Bologna (Gingerich, 1997).

Thomas Streete (1621–1689) wrote *Astronomia Carolina: a New Theorie of Coelestial Motions* (1661) on computational astronomy, including tables of planetary positions and motions. This was the standard textbook into the 18th century and was consulted by Newton, Flamsteed, and Halley (Curry, 2004). Edmond Halley (1656–1742), who was personally responsible for publishing Newton's *Philosophiae Naturalis Principia Mathematica*, prepared tables of planetary positions, *Tabulae Astronomicae* (Halley, 1749), which were eventually published posthumously. He wrote an appendix for the 1710 edition of Thomas Streete's *Astronomia Carolina*, which was probably used more than Halley's tables. Halley was also involved in improvements to the lunar tables, but the basis of the knowledge of the lunar orbit is uncertain. In 1767, the first *Nautical Almanac* was published in Great Britain. It continued to make use of Halley's work through the 1779 edition.

Early accurate tables of the Sun and the Moon, *Novae Tabulae Motuum Solis et Lunae 1753*, were determined by Tobias Mayer (1723–1762) (1753), and were used for the British *Nautical Almanac* until 1804 (Forbes, 1980). Jérôme Lalande's (1732–1807) planetary tables (1792), which included mutual perturbations, represented the state of the art at the end of the 18th century. The *Nautical Almanac* editions from 1780 through 1804 used his tables. After that, a number of different theories were used until the 1860s (Dunkin 1898a, 1898b; Lewis 1898).

William Herschel announced the discovery of the planet finally named Uranus on March 13, 1781. This was the first planet discovered with a telescope and expanded the boundaries of the solar system. When the first minor planet was discovered on January 1, 1801, only a few observations were made before it was lost in the glare of the Sun, so there was a problem trying to locate it again. Calculating a prediction ephemeris from the few observations to recover the minor planet at later dates was a new problem that Carl Friedrich Gauss

(1777–1855) solved in three months. In the process, he improved the mathematics of orbit prediction, which he published in *Theory of Celestial Movement* (1809). This introduced the *Gaussian Gravitational Constant* and the *method of least-squares*, which is used to this day to minimize the effects of measurement errors. Gauss also investigated the Earth's magnetic field and developed a method to measure its horizontal intensity. He also worked out the mathematical theory for separating the inner (core and crust) and outer (magnetosphere) sources of the Earth's magnetic field (Dunnington, Gray, & Dohse, 2004).

Urbain Jean Joseph LeVerrier (1811–1877) developed planetary theories, which were used for a long period in the *Connaissance des Temps*. In the process, he predicted the existence of a planet that was causing differences between observations of the position of Uranus and the positions predicted by ephemerides. His prediction for the position of the perturbing planet, which was eventually named Neptune, was independent of that of John Couch Adams, an English astronomer who was also investigating the problem. LeVerrier sent his prediction in 1846 to German astronomer Johannes Gottfried Galle (1812–1910), who, along with his Danish assistant Heinrich Louis d'Arrest (1822–1875), found the planet on the first night of observing, in part because they had a new, sufficiently faint, star catalog. Although there was no ill will between LeVerrier and Adams, an international disagreement arose concerning whether LeVerrier or Adams should receive credit for the predicted position. In 1855, LeVerrier discovered the advance of the perihelion of Mercury, which exceeded that predicted by Newton's theory and which eventually became evidence for Einstein's theory of relativity. LeVerrier attributed the perihelion advance to a planet called Vulcan, or a second asteroid belt, which was closer to the Sun than Mercury (Lequeux, 2013).

The British and US Nautical Almanac Offices used the tables of LeVerrier until 1900, when the tables of Newcomb and Hill were introduced for the Sun and the planets. A list of the tables used in the British *Nautical Almanac* and the *American Ephemeris and Nautical Almanac* prior to 1900 is given in the *Explanatory Supplement* (1961). An excellent description of the details and accuracies, including figures, showing the sources of the solar system ephemerides and the progression of accuracies from 1600 to 1800, is given by Gingerich and Welther (1983).

Many scientific developments during the 20th century both permitted and required improvements in the ephemerides of the solar system. The new developments included Einstein's theory of relativity; the variable rotation of the Earth; the discovery of Pluto; the need for a new uniform timescale; improved observational accuracies, including radar, laser ranging, and spacecraft observations; and improved astronomical constants. The lunar theory was improved (see Section 3.5), and the ephemerides of the five outer planets were improved with numerical integration (Eckert, Brouwer, & Clemence, 1951), but the ephemerides of the other planets continued to rely on older theories until the 1980s. By 1980, the planetary ephemerides were in error by about 0.10 seconds in right ascension and $0''.3$ in declination, except for Neptune and Pluto whose errors were significantly larger. In 1984, numerically integrated ephemerides by the Jet Propulsion Laboratory, designated DE200, were introduced with significantly improved accuracies. Astronomers have made many improvements in the planetary and lunar ephemerides over recent years. Details concerning the numerical integrations, the equations of motion, and observational data are given by Standish and Williams (2009). See Section 3.9 for the modern ephemerides available today.

3.5 Lunar Theories

The Moon is the most rapidly moving and closest celestial body to the Earth, so representing the motion of the Moon is the most complicated. Together with the Sun and the Earth, this is an example of the classical celestial mechanics three-body problem. In addition to the challenge of determining the periodic motions due to the perturbations by the Sun and the other planets, complicated interactions take place between the Earth and the Moon. Further, all the kinematics of the Earth's motion are involved in determining the lunar ephemeris.

Hipparchus studied the motion of the Moon and confirmed the values of the periods of its motion as determined by Babylonian astronomers. The error in their value of the synodic month was less than 0.2 second in the fourth century BC, and less than 0.1 second in Hipparchus's time. In this work, Hipparchus used both the eccentric circle and epicycle models to determine relative proportions and actual sizes of the orbits (Dreyer, 1953).

With the founding of the Royal Greenwich Observatory in 1675, John Flamsteed (1646–1719) became Charles II's "astronomical observator." Now recognized as the first Astronomer Royal, he began his work at the observatory by demonstrating that the rotational speed of the Earth did not vary. Of course, the rotational speed does vary, but his clocks were not accurate enough to detect this fact. He then set about making systematic observations of the Moon to obtain the data necessary to implement the lunar distance method for determining longitude at sea (Kollerstrom & Yallop, 1995). In the course of this work Flamsteed compiled his tables of the Moon's position from 1689 to 1704.

Using Flamsteed's observations, Edmund Halley (1695) noted that the Moon's motion appeared to be gradually speeding up. This "secular acceleration" could not be explained by gravitational theory. Although Newton had developed a lunar theory in 1702 (Gregory, 1715), Alexis Clairaut (1713–1765) computed the first full analytical theory of the Moon in 1750. Leonhard Euler (1707–1783) published lunar tables in 1746, and in 1753 provided his theory of lunar motion (Euler, 1753), which Tobias Mayer used in his tables. Mayer's work resulted in an award of £3,000 from the British government to his widow for its contribution to the problem of longitude determination. Euler also received £300 for his theoretical work. Mayer did the first careful investigation of the librations of the Moon in 1750. In 1755, James Bradley (1693–1762), England's Astronomer Royal from 1742 until his death in 1762, compared Mayer's lunar tables to Greenwich observations and found them capable of providing the Moon's position with an accuracy of 5″ (Goldstine, 1993).

In the 18th century, significant contributions to the lunar theory were also made by Jean le Rond d'Alembert (1717–1783) and Pierre-Simon, Marquis de Laplace (1749–1827), among others. Laplace (1786) addressed the issue of the observed secular acceleration in the Moon's motion and theorized that this was due to a secular change in the eccentricity of the Earth's orbit. In fact, this accounts for only a portion of the observed value, the rest being caused by the effect of the Earth's secular rotational deceleration due chiefly to tidal friction. In 1818, Laplace proposed that the Académie des Sciences in Paris award a prize to whoever succeeded in constructing lunar tables based solely on the law of universal gravity. In 1820, that prize was awarded to Francesco Carlini, Giovanni Plana, and Marie-Charles-Théodore de Damoiseau (Tagliaferri & Tucci, 2003).

In 1846, Charles-Eugène Delaunay (1816–1872) published his work on the lunar theory, containing what became known as *Delaunay's method* (Delaunay, 1846), and in 1865, he pointed out that problems with the lunar orbit might be attributed to the slowing of the Earth's rotation due to tidal friction (Delaunay, 1866). His work on the lunar theory is summarized in two volumes (1860, 1867). Other lunar theories were developed by Gustave de Pontecoulant (1840) and Peter Hansen (Brown, 1896, 1960; Cook, 1988). Hansen's tables (1857) were used generally in the almanacs after 1882. However, soon after that, corrections to Hansen's tables, determined by Simon Newcomb (1876), were gradually introduced in the published lunar ephemerides.

G. W. Hill (1838–1914) took a new approach to the lunar theory. Instead of polar coordinates referred to fixed axes, he used rectangular coordinates referred to moving axes. This method led to the use of a *variational curve*, which contains an important part of the solar perturbation, known as the *variation*. Thus, Hill obtained differential equations in a simple algebraic form, suitable for solution in infinite series. Ernest Brown (1866–1938) used Hill's method to develop his lunar tables (1896). The *Theory of the Motion of the Moon* was published in five parts in the memoirs of the Royal Astronomical Society, 1897–1908. Brown's Lunar Tables (Brown, 1897–1908) were used in the almanacs from 1923 to 1960, when the Improved Lunar Ephemeris (ILE) (1954), based on Brown's theory, was introduced. Observations showed that Brown's tables were better that those of

Hansen, which had been used since 1857, but there was still a 10″ fluctuation in the Moon's mean longitude. So, a "great empirical term" of magnitude 10.71″ and period 257 years was introduced to correct for the fluctuation. Simon Newcomb suggested that it was due to a gradual deceleration of the Earth's rate of rotation, due to friction generated by the tides. In other words, the Moon was not speeding up, but time, as measured by the Earth's increasingly longer day, appeared to be slowing down. Brown concluded that the Earth's rate of rotation was slowing, and that there were random, unpredictable fluctuations in the Earth's rotation with periods of 60 to 70 years. In practice, this is one of the complications in separating the lunar secular acceleration, due to tidal interaction, and secular change in the eccentricity of the Earth's orbit, from variations in the Earth's rotation.

The lunar theory presented many challenges to achieving accuracies matching those of the observations. In addition to the problems of developing a fully accurate gravitational theory including the Sun, Earth, and Moon interactions, the planetary perturbations, both direct and indirect effects, and the figure of Earth and Moon, all of which required extensive lists of periodic terms, there were the problems of the secular acceleration, the tidal interactions, the librational motions, limb variations, lunisolar precession and nutation, the location of the equinox, and variable Earth rotation.

Beginning in 1960, the British and US almanacs used the ILE calculated directly from Brown's lunar theory instead of from his tables. In order to obtain a gravitational ephemeris on the same measure of time as Newcomb's Tables of the Sun, the empirical term was removed from the orbital elements and the following correction was applied to the mean longitude:

$$\Delta l = -8''.72 - 26''.74T - 11''.22T^2, \tag{3.2}$$

where T is measured in Julian centuries from 1900 January 0.5 (*Explanatory Supplement*, 1961, p. 106).

Brown's theory represented the lunar main problem with accuracies of 0″.01 in longitude and latitude and 0″.000 1 in parallax (Henrard, 1973). The perturbations due to Earth flattening were less precise, about 0″.07 in longitude and latitude. However, the planetary perturbations were shown to produce differences from numerical integrations in position by up to 500 meters (Mulholland, 1969). Inaccuracies in fundamental constants could increase this error, particularly in secular and long-period terms. A probable secular error in longitude of 8″ per century was quoted by Mulholland (1972) (Kovalevsky, 1977). Observations of occultations, where the Moon is observed passing in front of a star, accurately measured the Moon's position with respect to the star's position in a star catalog reference system. These indicated lunar ephemeris errors of as much as 0″.5, or 5 km, in the 1970s.

The ILE was replaced in 1984 by a numerically integrated lunar ephemeris, LE200. However, the long-term trend of the lunar motion in the numerical integration was fit to that of the ILE. See Section 3.9 for the modern lunar ephemerides available today.

3.6 The Advent of Computers

Following early attempts to ease computational burdens, including the adding machine of Blaise Pascal (1623–1662) in 1645, and the desk calculator "step reckoner" of Gottfried Leibniz (1646–1716) in 1671, Charles Babbage (1791–1871) in the early 1800s conceived an "Analytical Engine" to perform calculations (Dubbey, 1978). Due to lack of funds, the machine was never completed, but Georg and Edvard Scheutz used the description to construct such a machine for display in London and Paris in 1854 and 1855. B. A. Gould purchased the machine for Dudley Observatory, where it was used to calculate tables of the Sun's longitude, the radius vectors of Venus and Earth, and the geocentric distance to Mars (Seidelmann, 1976).

With the introduction of punched card equipment in the 1920s, L. J. Comrie brought to Britain the computation of ephemerides from the tables by use of punched card equipment, for the determination of the lunar ephemeris (Comrie, 1925). In the United States, Wallace Eckert was hired as the director of the Nautical Almanac Office of the US Naval Observatory at the beginning of World War II. He introduced punched card equipment for computations and a punch card–operated typewriter to print camera-ready pages of the almanacs (Dick, 1999).

In 1948, the IBM Selective Sequence Electronic Calculator (SSEC) computed the ephemeris of the Moon directly from Brown's theory for 20 years to an accuracy of $0''.01$. The SSEC was used for a numerical integration of the Sun and five outer planets from 1753 to 2060 at 40-day intervals with an accuracy of 16 digits (Eckert et al., 1951). These became the basis for publication of the outer planet positions in the *American Ephemeris*.

With modern computer capabilities, more accurate ephemerides could be determined by numerical integrations than from general theories. So, when new astronomical constants, a new fundamental star catalog, and a new dynamical reference frame were introduced in 1984, ephemerides based on numerical integrations were also introduced (Seidelmann, 1976).

3.7 Numerical Integrations

Numerical integration of differential equations requires the use of initial conditions, specifically in the case of ephemerides, a position and velocity vector for each body. The standard method used is based on the formulation by P. H. Cowell (1870–1949) and uses an equation of motion that can incorporate all the appropriate forces along with terms for relativistic effects. The perturbations from all bodies can be included (*Explanatory Supplement*, 1992, 2012). The derivatives and positions are determined from the differential equations of motion. Then difference tables, evaluated for successive time steps, are used to determine the successive positions and velocities of the bodies. Once an ephemeris is determined, it is compared to observations, a least-squares solution is made, and a differential correction is made to determine new initial conditions, and the next numerical integration is started. By an iterative process, the numerical integration is fit to the observations as well as possible. This process is computationally intensive, so that it is time consuming if done by hand, but easily carried out on a computer.

3.8 Observational Data

Solar system observational data have increased in quantity and accuracy over the years. Until the 1960s, the data were restricted to optical observations, primarily transit circle observations. The solar and lunar observations were particularly subject to systematic errors, since they differed significantly from stellar observations and involved observing relatively bright limbs of the bodies. In addition, solar observations were affected by instrumental heating. The inner planets were subject to phase effects based on their positions relative to the Sun. Lunar occultation observations provided a different type of observation. Systematic discussions of the different planetary observations were undertaken by a number of individuals, resulting in improved ephemerides. Examples are the investigations of Mercury by Clemence (1943), of Venus by Duncombe (1958), of Mars by Ross (1917) and Laubscher (1981), and of Neptune by Jackson (1974).

There were significant improvements in the observational data due to improved methods, instruments, and scientific knowledge, particularly about 1830 and again about 1900. Hence, for those planets with shorter periods of revolution about the Sun, observations before 1900 were not used. For those with longer periods, like Uranus and Neptune, all available observations had to be used.

3.8.1 Radar Observations

Radar observations of planets with timing by atomic clocks began in 1966. The observed delay in receiving the reflected signal depends not only on the distance of the target but also on relativistic effects, the solar corona, and the troposphere. Variations in planetary topography are responsible for the largest uncertainties in the radar measurements. For the inner solar system, the radar observations dominate the determination of the ephemerides and the dynamical reference system (*Explanatory Supplement*, 1992, 2012) (see Section 16.6.2).

3.8.2 Lunar Laser Ranging

The *Apollo* 11, 14, and 15 missions placed retro-reflectors on the Moon at their landing sites starting in 1969. Also, there are retro reflectors on the *Lunakhod* vehicles. Since the lunar laser ranging measures are from an observatory fixed on the surface of the Earth to a retro-reflector fixed on the surface of the Moon, the observations of the length of time for the laser signal to be reflected from the target are dependent on the distance between the Earth and the Moon, the librations of the Moon, and all the kinematics of the Earth. So the measurements can be used to determine the orbit and the librations of the Moon, the rotation of the Earth, polar motion, precession, nutation, crustal motion, and relativistic effects.

The lunar laser ranging measurement accuracies have proceeded from the centimeter level to the millimeter level. At this accuracy level, the Weak and Strong Equivalence Principles can be tested with sensitivity approaching 10^{-14}. In addition to improving values of the post-Newtonian (PPN) parameters, effects of the next post-Newtonian orders (c^4) of light deflection and other relativistic effects can be investigated (Turyshev et al., 2004) (see Section 16.6.3).

3.8.3 Spacecraft Observations

With spacecraft missions to planets, round-trip range and Doppler measurements became available. These include missions making close passes by the planets, those in orbits around the planets, and landers on the surfaces of planets. Resulting data provide distances to the planets, planet masses and gravity fields, and planetary satellite information (*Explanatory Supplement*, 1992, 2012).

3.9 Modern Ephemerides

Three organizations currently generate planetary and lunar ephemerides. They are the DE430/LE430 by the US Jet Propulsion Laboratory (JPL) since 2013 (Folkner et al., 2014), EPM 2011 by the Russian Institute of Applied Astronomy (IAA) since 2012 (Pitjeva & Pitjeva, 2013), and the INPOP10e by the French Institut de Mécanique Celeste et Calcul des Ephemerides (IMCCE) since 2012 (Fienga et al., 2015). They are in general agreement with similarities, differences, strengths, and weaknesses. Comparisons are available from the International Astronomical Union (IAU) Division I commission 4 Ephemerides website www.iaucom4.org

(click Ephemerides, then Comparison of Ephemerides). The commission is now named X2, Solar System Ephemerides.

These ephemerides are based largely on radiometric ranging observations. For the outer planets, the observational data are still primarily optical, so the accuracies are much poorer. The plane-of-the-sky directions are accurate to a few hundredths of an arcsecond at present, but they deteriorate significantly with time, particularly for the outermost planets. The primary limitation on the accuracy of current planetary ephemerides is the uncertainty of minor planet masses and their effect on the orbit of Mars.

3.10 Reference System

Astronomical observation of time depends on definitions of terrestrial and celestial reference systems. Newcomb's Tables of the Sun (1895), astronomical constants, the definitions of mean solar time, tropical year, the Besselian year, and the star catalog equinox provided the reference systems from 1900 to 1984. When new constants and ephemerides were adopted in 1976, and introduced in 1984, Newcomb's values were replaced, and the values of the tropical and Besselian year were improved with new ephemerides. The Julian year of 365.25 days was adopted to define standard epochs. The notations of B for Besselian and J for Julian were introduced to differentiate between the two types of epochs. The standard epoch 2000.0 was introduced as a Julian epoch and designated as J2000.0. The different epochs are specified in the *Explanatory Supplements* (1961, 1992, 2012).

The reference frames that realize the celestial reference systems were based on fundamental star catalogs of nearby bright stars. These catalogs were formed from observations, primarily made with transit circles, and included an American Series of Fundamental Catalogs starting with Newcomb and continuing to the N30 catalog, as well as a series of German fundamental catalogs named FK, NFK, FK3, FK4, and FK5 (Eichhorn, 1974). The IAU recommended the use of the FK3 in 1938. It was replaced by the FK4 when it became available in 1963 and by the FK5 in 1985.

Differences in the reference systems implied by the solar, lunar, and planetary ephemerides, and those of the star catalogs, were evident. In the 1950s, discrepancies between the origins of the solar and lunar ephemerides were recognized and corrections were introduced. By the 1970s, it was recognized that the equinox defined by the solar system ephemerides, called the *dynamical equinox*, differed from the origin of the fundamental star catalog right ascensions, called the *catalog equinox*, and that the star catalog origin of right ascensions varied systematically with declination due to systematic errors in the star catalogs. Also, it was recognized that the presence of systematic errors in the proper motions of stars could cause an apparent rotation of the equinox of the star catalogs. Thus, an equinox motion correction was introduced between the FK4 (Fricke et al., 1963) and FK5 (Fricke et al., 1988) star catalogs.

Prior to the 1960s, the ephemerides were determined using optical observations of the positions of the Sun, Moon, and planets. With the availability of radar, lunar laser, and spacecraft observations, the ephemerides for the inner planets and the Moon began to make use of new types of observations, and the reference frame of the ephemerides grew to become independent of the star catalogs. Each new ephemeris defined its own version of the equinox by its origin of right ascension.

It was also recognized that two methods of determining the position of the dynamical equinox were being used. One method is based on a "rotating ecliptic" related to the geometrical path of the Earth-Moon barycenter. This has been used historically since Newcomb. The other method is based on an ecliptic pole defined by the mean orbital angular momentum vector of the Earth-Moon barycenter in a Barycentric Celestial Reference System. This is an "inertial" ecliptic, related to the Earth-Moon barycentric orbital angular momentum vector. The difference between these two methods of determining the equinox is about $0''.1$ (Hilton et al., 2006).

The International Celestial Reference System (ICRS), introduced in 1992, is now maintained by the International Earth Rotation and Reference systems Service (IERS), and realized in practice by the directions to a set of quasars in the radio wavelengths and the directions to stars in the Hipparcos Catalogue in the visual wavelengths. The *Gaia* spacecraft is providing new data at optical wavelengths for positions of optical sources and ICRF radio sources (Mignard et al., 2016). Similarly, the IERS also maintains the International Terrestrial Reference System (ITRS) that is realized by the adopted positions of a large number of observing sites on the Earth's surface (see www.iers.org).

3.11 Besselian Year

The Besselian year is measured from the instant when the right ascension of the fictitious mean sun, affected by aberration and measured from the mean equinox, is 18h 40m. This instant occurs near the beginning of the calendar year. The Besselian year is shorter than the tropical year by $0.148T$ seconds, where T is measured in centuries after 1900. This difference is due to the excess of the acceleration of the right ascension of the fictitious mean Sun over the mean longitude of the Sun (*Explanatory Supplement*, 1961). The Besselian year was used as the time measure for standard star catalogs prior to 1984. The epochs 1900.0 and

1950.0, for example, were on the Besselian year and now are designated B1900.0 and B1950.0.

3.12 Time Arguments

The independent variable in ephemerides is assumed to be uniform time. From the time of the Hellenistic astronomer Ptolemy, the concept of mean solar time was used for this independent variable. Following the recognition of the variability of the Earth's rotation, a new timescale, Ephemeris Time, based on the Earth's orbital motion was introduced (see Chapter 5). Problems with real-time realization of Ephemeris Time soon became apparent, and atomic time became available as an alternative source of uniform time. With improvements in accuracies of observations and ephemerides, it became necessary to include relativistic theories in the definitions of timescales, so a category of dynamical times was introduced in 1984. These timescales were refined in definitions and names in the 1990s, so there are dynamical timescales for ephemerides that are precisely related to an atomic timescale. Details concerning these developments are described in the following chapters.

3.13 Astronomical Constants

Each ephemeris requires a set of astronomical constants, either explicitly adopted for that purpose or developed with the ephemeris. These constants might include planetary masses, the astronomical unit, the precession constant, nutation theory, aberration constant, the gravitational constant, etc. Newcomb became the first to undertake a completely new set of consistent ephemerides for the Sun, Moon, and planets, along with the determination of a complete set of astronomical constants. This collected work was adopted by the international community in 1900, including an expression for mean solar time based on his solar theory (Newcomb, 1895, 1898).

Improving observational accuracy drove further improvements in the ephemerides and the astronomical constants. Newcomb's constants were partially updated by the IAU in 1968 (*Supplement to the A. E., 1968*, 1966). With the introduction of a new reference frame, new ephemerides based on numerical integration, new timescales, and new astronomical constants in 1984, Newcomb's system was finally replaced to meet the requirements for better accuracies (*Explanatory Supplement*, 1992, 2012).

In current practice, the values of some of these constants may vary at the highest levels of precision depending on the application. Relativistic scaling of the astronomical quantities must be considered in a system of astronomical constants

(Klioner, 2008). The International Union of Geodesy and Geophysics (IUGG), the IERS, and the IAU are sources for conventional values that are updated as required to be consistent with modern applications. *The IERS Conventions* (2010) is one source of the most current constants and procedures. Producers of modern ephemerides also generally publish the numerical values of the constants employed in the production of their ephemerides, and tables of constants can be found in *The Astronomical Almanac*.

3.14 Redefinition of the Astronomical Unit (au)

Historically, the astronomical unit has been considered to be the mean distance of the Earth from the Sun and regarded as the length of the semi-major axis of the Earth's orbit. Since the 19th century, and officially from 1938 to 2012, the value of the Gaussian gravitational constant, k (Gauss, 1809), was adopted as a fixed constant used to define the astronomical unit. That definition of the astronomical unit, specified in the IAU 1976 system of constants, was the length such that the Newtonian gravitational constant, G, is the square of the Gaussian constant, k,

$$G = k^2 = 0.000\ 295\ 912\ 208\ 285\ 591\ 102\ 5, \tag{3.3}$$

with the mass of the Sun and the day (864000 SI seconds) as the units of mass and time, respectively.

The Gaussian constant, $k = 0.017\ 202\ 098\ 95$, with units being (astronomical unit)$^{3/2}$ (day)$^{-1}$ (solar mass)$^{-1/2}$ and thus having dimensions $L^{3/2}\ M^{-1/2}T^{-1}$ was considered to be a defining constant used to derive the astronomical unit. This definition was adopted because of the lack of solar system distance measurements with adequate precision. The scale of the solar system, determined by the value of the astronomical unit in SI meters, was derived by fitting a planetary ephemeris, which depended on the theory of motion and observations being used. Using this process, the astronomical unit was derived to be $1.495\ 978\ 70 \times 10^{11}$ m in 1976 and $1.495\ 978\ 707\ 00 \times 10^{11}\ \pm 3$ m in 2009 (Pitjeva & Standish, 2009).

The value of the solar mass parameter GM_S could be calculated using k and the observationally determined astronomical unit, A, from

$$GM_S = A^3\ k^2\ /D^2 \tag{3.4}$$

In 2012 the IAU, in consideration of the need for a self-consistent set of units and numerical standards for dynamical astronomy in the framework of General Relativity, and that GM_S and its expected time variability can be determined directly in SI units from modern planetary ephemerides, redefined the astronomical unit to be 149 597 870 700 m exactly, in agreement with the observationally derived value of 2009. This definition of the astronomical unit (to be abbreviated "au" uniquely after 2012) can be used with all timescales, TCB, TDB, TCG, and TT. The Gaussian gravitational constant, k, could be deleted from the system of astronomical constants (*Trans. IAU*, 2015). This action provided a self-consistent set of units in the relativistic framework and permitted direct determination of possible time variation in the solar mass parameter in SI units (Capitaine, Guinot, & Klioner, 2010; Brumfiel, 2012; Capitaine, 2012; Capitaine, Klioner, & McCarthy, 2012).

3.15 Artificial Satellite Theories

Following the launch of *Sputnik* on October 4, 1957, the computation of orbits for artificial Earth satellites created the field of astrodynamics. This topic is essentially celestial mechanics related to objects whose propulsion systems can be used to change their orbits (Gurfil & Seidelmann, 2016). Artificial satellites could be launched for a variety of purposes, including communication, navigation, reconnaissance, and scientific observations. The use of artificial satellites for these purposes introduced new accuracy requirements for orbit computations and for timekeeping. Analysts can choose among geocentric rotating or non-rotating, or barycentric reference frames depending on their needs. The relativistic equations must be formulated according to the reference frame chosen. The use of satellites for time transfer improved the international system of timekeeping, and the use of accurate clocks on satellites introduced a new means of accurate navigation, specifically the Global Positioning System (GPS).

3.16 Theory of Relativity

Although Einstein published his theory of relativity in 1905, its incorporation into ephemerides, timescales, and astrometry was delayed because the accuracies of observations and theories did not require the inclusion of relativity in celestial mechanics, except for the advance of the perihelion of Mercury. The introduction of Ephemeris Time in 1952 did not consider the theory of relativity. The ephemerides of the Sun, Moon, and planets were not corrected for relativity until the numerically integrated ephemerides were introduced in 1984. In this case, the parameterized post-Newtonian n-body metric equations of motion were introduced (Will, 1974). Planetary observations were not corrected for relativistic effects until after 1984. The dynamical timescales were introduced in 1984 to recognize relativistic effects. Atomic timescales and time transfer methods began to be specified in conformity with relativity starting in the 1970s. Subsequent chapters provide the details of these changes.

References

Brown, E. W. (1896). *An Introductory Treatise on the Lunar Theory*. Cambridge, UK: Cambridge University Press, and (1960) New York, NY: Dover Publications.

Brown, E. W. (1897–1908). Theory of the Motion of the Moon: Containing a New Calculation of the Expressions for the Coordinates of the Moon in Terms of the Time. *Memoirs of the Royal Astronomical Society*, **53**, 39–116, 163–202; **54**, 1–63; **57**, 51–145; **59**, 1–103.

Brown, E. W. (1914). Cosmic Physics. *Science*, **40**, 389–401.

Brumfiel, G. (2012). The Astronomical Unit Gets Fixed. *Nature News*, Macmillan Publishers Lmt., September 14.

Capitaine, N. (2012). Toward an IAU Resolution for the Re-definition of the Astronomical Unit of Length. In H. Schuh, S. Boehm, T. Nilsson, & N. Capitaine, eds., *Proceedings of the Journees 2011 Systemes de reference spatio-temporels*. Vienna: Vienna University of Technology.

Capitaine, N., Guinot, B., & Klioner, S. A. (2010). Proposal for the Re-Definition of the Astronomical Unit of Length through a Fixed Relation to the SI Metre. In N. Capitaine, ed., *Proceedings of the Journees 2010 Systemes de reference spatio-temporels*. Paris: Observatoire de Paris, pp. 20–23.

Capitaine, N., Klioner, S., & McCarthy, D. D. (2012). The Re-Definition of the Astronomical Unit of Length: Reasons and Consequences. *IAU Joint Discussion 7: Space-Time Reference Systems for Future Research at IAU General Assembly-Beijing*. Online at http://referencesystems.info/iau-joint-discussion-7.html.

Chabas, J. & Goldstein, B. R. (2003). *The Alfonsine Tables of Toledo*. Dordrecht: Springer.

Clemence, G. M. (1943). The Motion of Mercury, 1765–1937. *Astronomical Papers of the American Ephemeris and Nautical Almanac*, vol. XI. Washington, DC: US Government Printing Office.

Comrie, L. J. (1925). The Application of Calculating Machines to Astronomical Computing. *Popular Astronomy*, **33**, 1–4.

Cook, A. (1988). *The Motion of the Moon*. Bristol, UK, and Philadelphia, PA: Adam Hilger.

Cook, A. (1998). *Edmond Halley, Charting the Heavens and the Seas*. Oxford: Clarendon Press.

Curry, P. (2004). Streete, Thomas (1621–1689). In *Oxford Dictionary of National Biography*. Oxford: Oxford University.

Delaunay, C.-E. (1846). Mémoire sur une Méthode nouvelle pour la détermination du mouvement de la Lune. *Comptes rendus hebdomadaires des séances de l'Académie des sciences*, **22**, 32–37.

Delaunay, C.-E. (1860). Théorie du mouvement de la Lune, premier volume. *Mémoires de l'Académie des Sciences*, **28**.

Delaunay, C.-E. (1866). *Conference sur l'astronomie et en particulier sur le ralentissement du mouvement de rotation de la terre*. Paris: G. Bailliere.

Delaunay, C.-E. (1867). Théorie du mouvement de la Lune, deuxième volume. *Mémoires de l'Académie des Sciences*, **29**.

Dick, S. (1999). History of the American Nautical Almanac Office, The Eckert and Clemence Years, 1940–1958. In A. D. Fiala and S. J. Dick, eds., *Proceedings, Nautical Almanac Office Sesquicentennial Symposium*. Washington, DC: US Naval Observatory, pp. 35–46.

Dreyer, J. L. E. (1953). *The History of Astronomy from Thales to Kepler*. Dover Publications.

Dubbey, J. M. (1978). *The Mathematical Work of Charles Babbage*. Cambridge: Cambridge University Press.

Duncombe, R. L. (1958). The Motion of Venus, 1750–1949. *Astronomical Papers of the American Ephemeris and Nautical Almanac*, vol. XVI. Washington, DC: US Government Printing Office.

Dunkin, E. (1898a). Notes on some Points connected with the Early History of the 'Nautical Almanac', *The Observatory*, **21**, 49–53, 123–127.

Dunkin, E. (1898b). Some further Notes on some Points connected with the Early History of the 'Nautical Almanac', *The Observatory*, **21**, 165–168.

Dunnington, G. W., Gray, J., & Dohse F.-E. (2004). *Carl Friedrich Gauss, Titan of Science*. Washington, DC: American Mathematical Association.

Eckert, W. J., Brouwer, D., & Clemence, G. M. (1951). Coordinates of the Five Outer Planets. *Astronomical Papers of the American Ephemeris and Nautical Almanac*, vol. XII. Washington, DC: US Government Printing Office.

Eichhorn, H. (1974). *Astronomy of Star Positions*. New York, NY: Frederick Ungar Publishing Company.

Euler, L. (1753). *Theoria Motus Lunae exhibens omnes eius inaequalitates*. St. Petersburg: Academiae Imperialis Scientiarum.

Explanatory Supplement to The Astronomical Ephemeris and The American Ephemeris and Nautical Almanac (1961). London: Her Majesty's Stationery Office.

Explanatory Supplement to the Astronomical Almanac (1992). P. K. Seidelmann, ed. Mill Valley, CA: University Science Books.

Explanatory Supplement to the Astronomical Almanac (2012). S. E. Urban & P. K. Seidelmann, eds. Mill Valley, CA: University Science Books.

Fienga, A., Manche, H., Laskar, J., Gastineau, M., & Verma, A. (2015). *INPOP new Release INPOP13b*, reprint arXiv:1405.0484v2.

Folkner, W. M., Williams, J. G., Boggs, D. H., Park, R. S., & Kuchynka, P. (2014). *The Planetary and Lunar Ephemerides DE430 and DE431. The Interplanetary Network Progress Report*. 42–196, pp. 1–81.

Forbes, E. G. (1980). *Tobias Mayer (1723–1762), Pioneer of Enlightened Science in Germany*. Gottingen: Vandenhoech & Ruprecht.

Fricke, W., Kopff, A., Gliese, W., et al. (1963). Fourth Fundamental Catalogue and Supplement, *Veröff. Astron. Rechen-Inst., Heidelberg*, **10**, 11.

Fricke, W., Schwan, H., Lederle, T., et al. (1988). Fifth Fundamental Catalogue (FK5). Part I. The Basic Fundamental Stars, *Veröff. Astron. Rechen-Inst. Heidelberg*, **32**, 1–106.

Gauss, C. F. (1809). *Theoria motus corporum coelestium in sectionibus conicis solem ambientium* [Theory of the Motion of the Heavenly Bodies Moving about the Sun in Conic Sections]. Dover Publications, 2004.

Gingerich, O. (1973). The Role of Erasmus Reinhold and the Prutenic Tables in the Dissemination of Copernican Theory. *Studia Copernicana*, **6**, 43–62.

Gingerich, O. (1997). Astronomical Tables and Ephemerides. In J. Lankford, ed., *History of Astronomy: An Encyclopedia*, pp. 505–508.

Gingerich, O. (2007). Gutenburg's Gift, Library and Information Services in Astronomy V, Common Challenges, Uncommon Solutions, *ASP Conference Series*, S. Ricketts, C. Birdie and E. Isaksson, eds., **377**, 319–328.

Gingerich, O. & Welther, B. L. (1983). Planetary, Lunar, and Solar Positions, New and Full Moons, A.D. 1650–1805. *American Philosophical Society, Memoirs Series*, **59**.

Goldstine, H. H. (1993). *The Computer from Pascal to von Neumann*. Princeton, NJ: Princeton University Press, pp. 27–28.

Gregory, D. (1972). *Astronomiae Physicae at Geometricae Elementa 1702*, with "*Lunae Theoria Newtoniana*" on pp. 332–336; reprinted 1726; translated as *The Elements of Physical and Geometrical Astronomy* 1715, 2nd Edition 1726, 2 vols., with TMM on pp. 563–571; facsimile reprint (Sources of Science, no. 119), New York and London: Johnson. Reprint.

Gurfil, P. & Seidelmann, P. K. (2016). *Celestial Mechanics and Astrodynamics: Theory and Practice*. Springer.

Halley, E. (1693). Emendationes ac Notae in vetustas Albatenii Observationes Astronomicas cum restitutione Tabularum Lunisolarium ejusdem Authoris. *Phil. Trans. Roy. Soc.*, **17**, 913.

Halley, E. (1749). *Tabulae Astronomicae*. J. Bevis, ed. London.

Hansen, P. (1857). *Tables De La Lune, Construites D'après La Principe Newtonien De La Gravitation Universelle*. London: G. E. Eyre and G. Spottiswoode.

Henrard, J. (1973). L'éphéméride analytique lunaire – ALE, *Ciel et Terre*, **89**, 1.

Hilton, J. L., Capitaine, N., Chapront, J., et al. (2006). Report of the International Astronomical Union Division I Working Group on Precession and the Ecliptic. *Celest. Mech. & Dyn. Astron.*, **94**, 351–367.

IERS Conventions 2010 (2010). G. Petit and B. J. Luzum, eds. Frankfurt am Main: Verlag des Bundesamts für Kartographie und Geodäsie.

Improved Lunar Ephemeris, 1952–1959: A Joint Supplement to the American Ephemeris and the (British) Nautical Almanac (1954). Washington, DC: US Government Printing Office.

Jackson, E. S. (1974). A Discussion of the Observations of Neptune 1846–1970. *Astronomical Papers of the American Ephemeris and Nautical Almanac*, vol. XXII, no. II. Washington, DC: US Government Printing Office.

Khan, M. S. (1977). Āryabhata I and Al-Bīrunī. *Indian J. Hist. Sci.*, **12**, 237.

Klioner, S. A. (2008). Relativistic Scaling of Astronomical Quantities and the System of Astronomical Units. *Astronomy & Astrophysics*, **478**, 951–958.

Kollerstrom, N. & Yallop, B. (1995). Flamsteed's Lunar Data 1692–5, Sent to Newton. *Journal for the History of Astronomy*, **26**, 237–246.

Kovalevsky, J. (1977). Lunar Orbital Theory. *Phil. Trans. Roy. Soc. London. A.* **284**, 565–571.

Lalande, J.(1792). *Traité d'astronomie*, vol. 1. Paris: Desaint.

Laplace, P.-S. (1786). Sur l'équation séculaire de la Lune. *Mém Acad Roy Sci*, 235.

Laubscher, R. E. (1981). The Motion of Mars 1751–1969. *Astronomical Papers of the American Ephemeris and Nautical Almanac*, vol. XXII, no. IV. Washington, DC: US Government Printing Office.

Lequeux, J. (2013). *Le Verrier: Magnificent and Detestable Astronomer,* W. Sheehan translator. Springer.

Lewis, T. (1898). Almanacs, *The Observatory*, **21**, 299–305, 327–334.

Mayer, T. (1753). Novae Tabulae Motuum Solis et Lunae. In *Commentarii Societatis Regiae Scientiarum Gottingensis*, vol. II. Göttingen.

Mignard, F., Klioner, S., Lindegren, L. et al. (2016). Gaia Data Release 1: The Reference Frame and the Optical Properties of ICRF Sources. *Astronomy & Astrophysics*, **595** Eprint: arXiv:1609.07255.

Mulholland, J. D. (1969). Numerical Studies of Lunar Motion. *Nature*, **223**, 247–249.

Mulholland, J. D. (1972). Numerical Isolation of Flaws in the Lunar Theory. *Celestial Mechanics*, **6**, 242–246.

Neugebauer, O. (1969). *The Exact Sciences in Antiquity.* Mineola, NY: Dover Publications.

Newcomb, S. (1876). *Investigation of Corrections to Hansen's Tables of the Moon, with Tables for Their Application*. Washington, DC: US Government Printing Office.

Newcomb, S. (1895). *The Elements of the Four Inner Planets and the Fundamental Constants of Astronomy*. Washington, DC: US Government Printing Office.

Newcomb, S. (1898). Tables of the Motion of the Earth on Its Axis and around the Sun. *Astronomical Papers of the American Ephemeris and Nautical Almanac*, vol. VI, no. I. Washington, DC: US Government Printing Office.

Newton, I. (1999). *The Principia: Mathematical Principles of Natural Philosophy*, Cohen, I. B., Whitman, A. translators. University of California Press.

Pitjeva, E. V. & Pitjev, N. P. (2013). Relativistic Effects and Dark Matter in the Solar System from Observations of Planets and Spacecraft, *Monthly Not. Roy. Astron. Soc.*, **432**(4), 3431–3437.

Pitjeva, E. V. & Standish, E. M. (2009). Proposals for the Masses of the Three Largest Asteroids, the Moon-Earth Mass Ratio and the Astronomical Unit. *Celest. Mech. & Dyn. Astr*, **103**, 365–372.

Pontecoulant, G. de (1840). *Traité élémentaire de physique céleste, ou précis d'astronomie théorique et pratique, servant d'introduction à l'étude de cette science.* Paris: Carillan-Goeury.

Ross, F. E. (1917). New Elements of Mars and Tables for Correcting the Heliocentric Positions Derived from Astronomical Papers, Vol VI, Part IV. *Astronomical Papers of the American Ephemeris and Nautical Almanac*, vol. IX. Washington, DC: US Government Printing Office.

Russell, J. L. (1964). Kepler's Laws of Planetary Motion: 1609–1666. *British Journal for the History of Science*, **2**, 1–24.

Saliba, G. (2002). Greek Astronomy and the Medieval Arabic Tradition. *American Scientist*, **90**, 360.

Seidelmann, P. K. (1976). Celestial Mechanics. In J. Belzer, A. G. Holzman, & A. Kent, eds., *Encyclopedia of Computer Science and Technology*, vol. 4. New York, NY, and Basel: Marcel Dekker, Inc.

Sharma, M. L. (1977). Āryabhata's Contribution to Indian Astronomy. *Indian J. Hist. Sci.*, **12**, 90.

Standish, E. M. & Williams, J. G. (2009). Orbital Ephemerides of the Sun, Moon, and Planets. In S. E. Urban & P. K. Seidelmann eds., *Explanatory Supplement to the Astronomical Almanac*. Mill Valley, CA: University Science Books, pp. 305–346.

Supplement to the A. E. 1968. (1966). HMNAO, Royal Greenwich Observatory and NAO, US Naval Observatory, Sussex, England and Washington DC, USA.

Streete, T. (1661). *Astronomia Carolina. A New Theorie of the Coelestial Motions*. London: R. Smith and S. Briscoe.

Szebehely, V. G. & Mark, H. (1998). *Adventures in Celestial Mechanics*. New York, NY: John Wiley & Sons.

Tagliaferri, G. & Tucci, P. (2003). *The Dispute between Carlini-Plana and Laplace on the Theory of the Moon*. Springer, 427–441.

Toomer, G. J. (1998). *Ptolemy's Almagest*. Princeton, NJ: Princeton University Press.

Trans. IAU. 2015, XXVIIIB, pp. 35–37.

Turyshev, S. G., Williams, J. G., Nordtvedt, K. Jr., et al. (2004). Years of Testing Relativistic Gravity: Where Do We Go from Here? *Lecture Notes in Physics*, **648**, 311–330.

Will, C. M. (1974). *Experimental Gravitation*, B. Bertotti, ed. New York, NY: Academic Press.

4

Variable Earth Rotation

4.1 Pre 19th Century

The concept of the Earth's rotation was suggested as long ago as the Greek Pythagorean philosophers. The original idea is generally credited to the philosophers of the fourth century BC, Hiketas and Ekphantus of Syracuse, but they may have been influenced by the earlier world system of Philolaus. Aristarchus and his follower, Seleukus, about a century later, also suggested that the Earth rotated on its axis. If the Earth rotated, people and things would be expected to fly off. The general concept was that the Earth was stationary and fixed, and the Sun, Moon, planets, and stars were a limited distance away and moving around the Earth (Dreyer, 1953). That the Earth rotated was not generally accepted until the 15th century.

Although some individuals in the Middle Ages accepted the Earth's motion, it was Copernicus, Galileo, and Newton who promoted the concepts that the Earth moved around the Sun, that the distances to the planets and stars were large, and that the Earth rotated daily on its axis. Newton's gravitation answered the questions of what held people and things onto the Earth, and provided the basis for computing the motions of celestial bodies, but it was not until 1851 that Léon Foucault provided the first observational "proof" of the Earth's rotation with his famous pendulum (Tobin, 2003).

The constancy of the Earth's rotation rate was generally not questioned by the 16th century. Chapter 3 outlines the role of celestial mechanics that led to the suggestion of the possible variation in the Earth's rotation. There we see that John Flamsteed demonstrated its invariability, at least within his observational accuracy. However, Edmond Halley (1695) used Flamsteed's observations to conclude that the Moon's motion seemed to be speeding up, a fact that could not be explained totally by gravitational forces. Newton suggested that the apparent acceleration of the Moon might be due to changes in the Earth's rotation, but this idea was based on a mistaken idea of interplanetary vapors (Newton, 1713). Immanuel Kant (1754) suggested that the action of the Earth's tides should slow the Earth's rotation rate. Pierre-Simon Laplace (1749–1827), however, determined that the observed deceleration of the Moon in its orbit could be explained gravitationally and that it was unnecessary to relate this effect to a change in the Earth's rotation. William Ferrel (1853) showed that Laplace had neglected second-order effects, and that when they were included, a problem remained in explaining the observed lunar motion completely by gravitational forces.

4.2 Secular Variation

The mean longitude of the Moon can be expressed mathematically by the series

$$L = L_0 + n_1 T + n_2 T^2 + \ldots, \qquad (4.1)$$

where T is time reckoned from some arbitrary reference epoch. This expression provides the direction of the Moon as measured along the ecliptic from the vernal equinox. The celestial mechanics explanation for the T^2 term in that expression was provided by Laplace (1786), who tried to explain the entire Moon's motion as due to planetary perturbations. He pointed out that the disturbing force on the Moon depends on the mean distance of the Earth from the Sun and the eccentricity of the Earth's orbit. Knowing that the eccentricity of the Earth's orbit had been decreasing, Laplace was able to provide a numerical value for n_2. John Couch Adams (1853) pointed out that the tangential component of the disturbing force would also affect the value of n_2. Even with Adams's correction, however, it appeared that the value determined from celestial mechanics was only about half of the value derived from ancient eclipse observations.

   Although problems were recognized in explaining the secular acceleration of the Moon's motion, it was not until the mid-19th century that the secular retardation of the rate of rotation of the Earth was suggested independently by Charles-Eugene Delaunay (1859, 1866) and William Ferrel (1864) to explain that the secular acceleration of the mean motion of the Moon, determined from gravitational perturbations, was only about half of the observed acceleration determined independently by Richard Dunthorne (1747, 1749), Tobias Mayer (1753), and Jérôme Lalande (1792) from observations over the previous 2,500 years. The excess observed secular acceleration was ascribed to tidal retardation of the rotation, which is accompanied by variation in the orbital velocity of the Moon, according to the conservation of momentum.

   Tidal deceleration is illustrated in Figure 4.1. The gravitational attraction of the Moon raises an ocean tidal bulge on the Earth. However, the Earth's rotation carries the bulge beyond the line connecting the centers of mass of the two bodies.

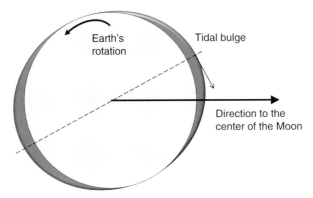

Figure 4.1 Tidal deceleration of the Earth

The gravitational attraction of the Moon on this bulge provides a braking action on the Earth. And we know this as *tidal deceleration*. The conservation of angular momentum in the Earth–Moon system then causes the Moon to be accelerated in its motion around the Earth.

Gerald Clemence (1971) makes the statement:

"The earliest reference that I have found to possible changes in the speed of rotation of the Earth is by Thomson and Tait (1890). They state that Delaunay, who himself made a complete analytical solution of the main problem of the lunar theory (that is, the solution of the three-body problem, in which the bodies are the Earth, the Sun and the Moon), suggested about 1866 that the rotation of the Earth is retarded by tidal friction. They go on to say

'The ultimate standard of accurate chronometry must (if the human race live on the earth for a few million years) be founded on the physical properties of some body or more constant character than the earth: for instance, a carefully arranged metallic spring . . . .' Jacques Lévy, Paris Observatory, based on his personal knowledge of their work, attributes the statement to Thomson" (Lord Kelvin).

Simon Newcomb credits Ferrel with being "the first to publish a correct theory of the retardation produced in the rotation of the earth by the action of the tides and the consequent slow lengthening of the day" (Newcomb, 1903a). Brush (1986) points out that "Ferrel, strictly speaking, deserves priority for this proposal since he made it at a meeting of the American Academy of Arts and Sciences in Boston in December, 1864, a few weeks before Delaunay read his paper to the Académie des Sciences in Paris, but it was Delaunay's reputation that persuaded astronomers to adopt it."

Over the years, the understanding of the forces involved and the numerical estimates of the tidal acceleration have improved significantly. Values of the kinematical tidal acceleration (d^2L/dt^2), which is twice the value of n_2, in arcseconds per century squared in Equations (4.1) compatible with the recent JPL development ephemerides DE 430 and DE 431 of 2014 are $-25''.82$ $\pm 0''.03$/century^2 and $-25''.80$ $\pm 0''.03$/century^2, respectively (Folkner et al., 2014). The most recent value from lunar laser ranging data is $-25.97 +/-0.05''$/century^2 (Williams & Boggs, 2016). A more extensive discussion of tidal perturbations and determinations of the values is given in Chapront, Chapront-Touze, and Francoou (2002). From an analysis of the effects on the orbits of near-Earth satellites and the conservation of angular momentum in the Earth–Moon system, Christodoulidis, Smith, Williamson, and Klosko (1988) provide the following empirical relation between the tidal acceleration of the Moon and the luni-solar tidal deceleration of the Earth's rotation:

$$\dot{\Omega} = (49 \pm 3) \times 0.004869\, \dot{n} \times 10^{-22} \text{rad s}^{-2}. \tag{4.2}$$

4.3 Irregular Variations in the Earth's Rotation

In the latter part of the 19th century, Simon Newcomb and Friedrich Ginzel independently investigated observational evidence for irregular variations in the Moon's mean motion and the possibility that these could be explained by variation of the Earth's rotational speed (Stephenson, 2003). While Ginzel (1899) investigated records of eclipses, Newcomb established a qualitative correlation between the differences between observed and computed positions of the Moon, and Mercury transit observation residuals. Only about one-third of the Moon's residuals could be attributed to variations in the Earth's rotation (Newcomb, 1896). By 1903, Newcomb concluded that the variations in the motion of the moon were really due to causes that had eluded investigation (Newcomb, 1903b). He thought the errors in periodic terms in the theory of the Moon were more likely to explain the differences than variations in the rotation of the Earth. As the Carters point out in their book on Simon Newcomb (2006), in the first chapter of his book *Side-Lights on Astronomy*, with editions from 1882 to 1906, Newcomb discussed the unsolved problems of astronomy. The last problems concern the "deviations in the movements which astronomers cannot always explain, and which may be due to some hidden causes that, when brought to light, shall lead to conclusions of the greatest importance to our race." The first deviation he discussed was the rotation of the Earth. Here he said that: "Sometimes for several years at a time it seems to revolve a little faster, and then again a little slower. The changes are very slight; they can be detected only by the most laborious and refined methods; yet they must have a cause, and we should like to know what that cause is. The moon shows a similar irregularity of motion. For half a century, perhaps through a whole century, she will go around the earth a little ahead of her regular rate, and then for another half century or more she will fall behind. The changes are very small;

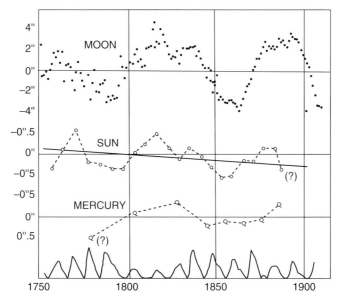

Figure 4.2 Deviations of the longitudes of the Moon, Sun, and Mercury from Brown (1914). The bottom curve is Wolf's sunspot numbers that Brown showed because of the known connection between the sunspot frequency and magnetic disturbances.

they would never have been seen with the unaided eye, yet they exist. What is their cause? Mathematicians have vainly spent years of study in trying to answer the question."

It was not possible to establish the variability of the rotation of the Earth only from observations of the Moon. To do so required comparing the differences between the ephemerides and observations of more than one body. The concept was that irregularities in the Earth's rotation would produce irregularities in the timescales used to make solar system observations that were used to produce the ephemerides. These irregularities would then appear in the ephemerides of all of the solar system objects. However, the variations would be in proportion to their respective mean motions. The accuracies of the observations made it desirable to consider the fastest-moving bodies for this investigation. E. W. Brown, in an address to the British Association in Australia in 1913, showed diagrams of the deviations in the longitude of the Moon from its theoretical orbit, along with similar curves for Mercury and the Sun (Figure 4.2) (Brown, 1914). However, Brown tried to attribute the variations to magnetic forces, and never mentioned the possibility of variations in the rotation of the Earth. Because of the then known connection between sunspot frequency and prevalence of magnetic disturbances, Wolf's sunspot numbers were plotted at the bottom of his diagram. Glauert (1915a, 1915b) further investigated correlations among the Moon, Sun, Mercury, and Venus residuals based on Greenwich observations from 1866 to 1914, but, although his results showed "satisfactory agreement," he felt that further investigation was required.

Carrying out these investigations presented a number of difficulties. Observations were made using different techniques at different observatories, causing systematic errors among series of observations; they were subject to various accidental errors that varied with time; errors in the positions of the Sun led to errors in the positions of the planets; observations of Mercury and Venus taken before and after inferior conjunction differed systematically; there were uncertainties concerning the location of the equinox, the origin of calculated and observed positions, for the observations and the tables. Meridian observations of Mercury were of limited value, so transit observations (Innes, 1925a) had to be used. Observations of Venus were more numerous and more accurate due to Venus's closer proximity to the Earth. In 1917, Frank Ross and Simon Newcomb (1917) re-discussed the observations of Mars, but found systematic errors in the observations, and because the motion of Mars is slower than that of Venus, the observations were less sensitive to variations in the rotation of the Earth. The basis for the theoretical ephemerides had varied over the years. For example, the solar ephemeris was based on Calini's tables from 1836 to 1863, Le Verrier's tables from 1864 to 1900, and Newcomb's tables after 1900. Using these tables led to systematic differences as a function of time.

R. T. A. Innes (1925b) wrote two short notes comparing the time differences indicated by the transits of Mercury, the Moon, eclipses of Jupiter's satellites, and the Sun. He pointed out that, since Einstein had found the explanation for the motion of Mercury's perihelion, all empirical terms ought to be removed and that celestial body motions should be able to be based on gravitational forces only. Empirical terms only caused confusion. He concluded that, while the results were crude, the century 1780 to 1880 was about 30 seconds longer than the century 1800 to 1900. All four sets of data gave qualitatively the same results, indicating variability in the rotation rate of the Earth (Figure 4.3).

Innes (1925a) indicates the need to distinguish between the time observed and the time in a Newtonian sense, which is an absolute, true, and mathematical time as conceived by Newton as flowing at constant rate, unaffected by the speed or slowness of the motion of material things. He suggested calling rotation time, or observed time, *Greenwich Time*, and the Newtonian time *World* or *Universal Time*, a suggestion that was never adopted. He further suggested that there would be a need to publish the differences as corrections to rotation time, which did become necessary.

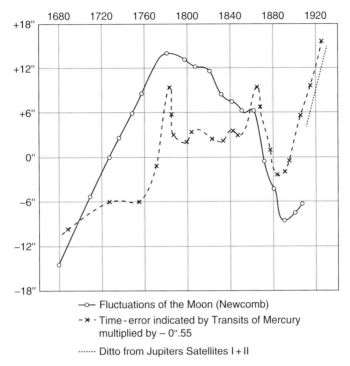

Figure 4.3 Fluctuations in the Moon's motion compared with the time errors determined from transits of Mercury and the motion of Jupiter's satellites from Innes (1925b). With permission John Wiley and Sons, Inc.

Harold Spencer Jones (1926) collected all the data available for a comparison of the residuals of the Sun, Mercury, Venus, and Mars. He then plotted the data, multiplying the planetary residuals by the ratio of the mean motion of the Earth to the mean motion of the planet. If the errors were due to the rotation of the Earth, the four curves should be identical, within the observational errors. Figure 4.4 shows the resulting curves. The curves show a gradual fall from the beginning with a minimum about 1870. Then there is a fairly rapid rise, a slower rise, and a more rapid rise to a maximum at about 1896. Then there is a drop for the rest of the period. The curves are similar qualitatively, and the amplitude changes are about equal. He gave correlation coefficients between the Sun and the planets that range between 0.82 and 0.86, except for the Mercury meridian observations, which are of less accuracy. This indicates that the residuals are from a common cause, namely the variation in the rotation rate of the Earth.

Spencer Jones then prepared a composite curve for the Sun and planets by smoothing the curves and plotting points at two-year intervals. He gave the data weights of 3 for the Sun, 2 for Venus, and 1 for Mercury transits and Mars. That curve is shown in Figure 4.5. The data were compared to longitude residuals of the

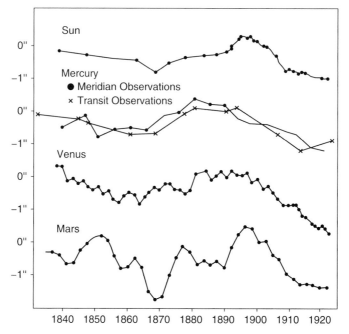

Figure 4.4 Variations in the motion of the Sun and planets from Spencer Jones (1926) © H. Spencer Jones. The Rotation of the Earth. *MNRAS* (1926) 87(1): 4–131, with permission of Oxford University Press, on behalf of the Royal Astronomical Society.

Moon provided by Dyson and Crommelin (1923). These were residuals from observations made at Greenwich of the Moon's longitude compared to Brown's longitude as modified by Fotheringham (1920), who had corrected the theoretical secular acceleration of Brown to fit the ancient eclipse observations. Fotheringham had introduced a term of period 257 years and semi-amplitude of 13.6 arcseconds. In addition, there were minor fluctuations of irregular period remaining for the Moon. These are shown in Figure 4.5 as the Moon residuals. The scale is adjusted to be approximately the same as the composite curve for the Sun and planets. The similarity of the two curves is evident, but it should be emphasized that, if the great empirical term, introduced by Brown to explain an apparent periodic variation of 257 years, were not removed from the residuals, the curves would not be similar. If the changes in the longitude residuals between 1897 and 1924 are considered, the ratios of the changes are in reasonable agreement with the theoretical ratios of the mean motions for the planets and the Sun. Spencer Jones concludes that during the period 1836 to 1924, the longitude fluctuations of the Sun, Mercury, Venus, and Mars, and the minor fluctuations of the Moon's longitude, within limits of observational errors, can be attributed to changes in the rate of rotation of the Earth.

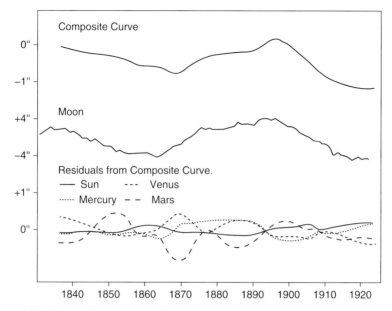

Figure 4.5 Comparison of variations in the Moon's position with the composite
planetary variations from Spencer Jones (1926). © H. Spencer Jones. The Rotation
of the Earth. *MNRAS* (1926) 87(1): 4–131, with permission of Oxford University
Press, on behalf of the Royal Astronomical Society.

De Sitter (1927) addressed two questions concerning the longitude residuals: (1)
Are the fluctuations of the longitudes of the Sun and planets equal to those of the
Moon in the exact ratios of the mean motions, or diminished by some factor? (2)
Does the fluctuation of the Sun and planets agree with the total fluctuations of the
Moon, or only with minor fluctuations after removing the great empirical term?
From this investigation of observational data he found that the secular acceleration
of the Sun and planets could be explained by retardation of the Earth's rotation due
to tidal friction, but would require a tidal deceleration twice that found from
theoretical considerations of the time (Jeffreys, 1924).

Considering the great empirical term, the observations of the Sun and Venus
were limited to 1835–1925, so short a time that the long-period sinusoidal curve
differs little from a straight line. So representations of differences in those cases
could be absorbed in solutions for orbital elements. Thus, the only independent
observations, besides those of the Moon, were the observations of the transits of
Mercury across the Sun's disk. de Sitter concluded that the separation of the great
empirical term from the minor fluctuations of the Moon's longitude was artificial.
The fluctuations could be explained better by a series of straight lines. He found
corrections to Brown's lunar tables, after removing the empirical sine term, and to
Newcomb's tables of the Sun, Mercury, and Venus. The combined observational

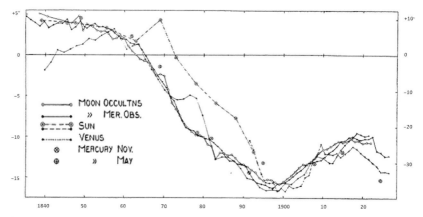

Figure 4.6 de Sitter (1927) plots of observational residuals; the left scale is the Moon's longitude and the right scale is corrections to time.

evidence is shown in Figure 4.6. The scale on the left of the figure refers to the Moon's longitude and the scale on the right gives the corrections to the time, corresponding to corrections to the Moon's longitude multiplied by 1.25.

Spencer Jones (1932) discussed the observations of occultations of stars by the Moon from 1672 through 1908 and concluded that the fluctuations in the Moon's longitude could be interpreted as being due to variations in the rate of rotation of the Earth. In 1939, he (Spencer Jones, 1939) established clearly that the fluctuations in the mean longitudes of the Sun, Mercury, and Venus corresponded to the fluctuations in time required to account for the fluctuations in the Moon's mean longitude (see Figures 4.7 and 4.8). Clemence (1943) adopted Spencer Jones's fluctuations in the mean longitudes of Mercury in his discussion of observations of Mercury, and found that the observations agreed with the adopted fluctuations.

In the 1930s, German scientists used newly developed quartz clocks to demonstrate an apparent annual variation in the Earth's rotational speed (Pavel & Uhink, 1935; Scheibe & Adelsberger, 1936). In France, Stoyko (1937) tabulated differences between time from pendulum clocks and the rotation of the Earth for three years in the period 1934–1937, and also found an annual variation in the rotation rate of the Earth. He determined that the length of day (LOD) in January exceeded that in July by 2 milliseconds (ms)/day. By 1950, quartz crystal clocks were used to determine double amplitudes of annual variations of the length of the day of 2.6, 1.8, and 2.8 ms/day by Scheibe and Adelsberger (1950), Finch (1950), and Stoyko (1950), respectively. However, these values were actually too large by factors of two or three. Systematic errors in the assumed positions of stars used to determine time were cause for concern. Smith and Tucker (1953) determined that the annual variation in length of day was less than ±0.5 ms/day using improved star positions.

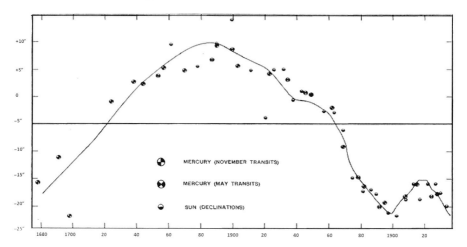

Figure 4.7 Fluctuations in the longitude of the Moon from 1680 from observations of the Moon, Sun, and Mercury from Spencer Jones (1932)

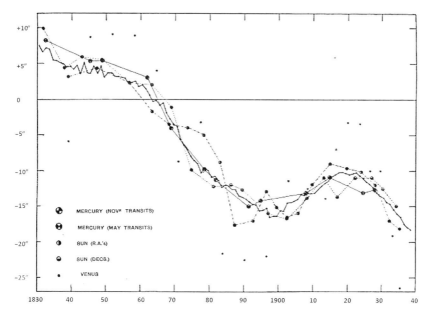

Figure 4.8 Fluctuations in the longitude of the Moon after 1830 from observations of the Moon, Sun, Mercury, and Venus from Spencer Jones (1932)

Markowitz (1955) reported the discovery of fortnightly and monthly variations in the rotational speed, which, together with the theoretical calculations of Mintz and Munk (1953), led him to assert that the causes of these variations were meteorological for the annual term and tidal for the others.

Reviews of the Earth's variable rotation rate can be found in Munk and MacDonald (1960), Munk (1966), Lambeck (1980), and the *Explanatory Supplements* (1961 and 1992).

4.4 Early Explanations for the Variable Rotation

In the face of mounting evidence that the rotational speed of the Earth does change, questions about the physical causes arose. Angular momentum had to be conserved. Computations of the amount of energy, and/or mass displacements, required to cause fluctuations in the Earth's rotation seemed large, and it was hard to explain how such changes could take place without geophysicists noticing other changes. The astronomical observations became major contributors to learning about the geophysics and dynamics of the Earth.

While Kant had suggested that the Earth's rotation might be affected by tidal friction, Julius Robert Mayer (1848) described the consequent reaction on the lunar orbit (Brosche, 1984). At the same time that observational evidence for the variable rotational speed was being pursued, George Howard Darwin, the son of the famous biologist, was beginning his work on tides and their possible relation to the Earth's rotation (Darwin, 1877, 1879, 1880, 1898). His work built on the ideas of Lord Kelvin (Thomson, 1863). Darwin's work was more concerned with the solid Earth tides, but he did deal with ocean tides. Harold Jeffreys (1924) investigated the effect of ocean tides extensively, providing quantitative estimates of the effect on the Earth's rotation.

In his analysis of the motions of the Moon, Sun, Venus, and Mercury, de Sitter (1927) accepted the hypothesis that fluctuations in their longitudes were due to variations in the Earth's rotation and proposed that these variations were due to the combination of two causes. One is a series of abrupt changes in the rate of rotation of the Earth caused by changes in the moment of inertia, perhaps due to expansions and contractions of the Earth. The other cause was proposed to be variability of the coefficient of tidal friction. As Spencer Jones (1939) pointed out in a re-discussion of observations of the Earth's rotation, changes in the moment of inertia of the Earth can increase or decrease its rotation rate and consequently affect the observed longitudes of the Sun, Moon, and planets in proportion to their mean motions. Tidal friction only retards the rotation rate. However, the change in angular momentum of the Earth is compensated by a corresponding change in the angular momentum of the Moon's orbital rotation. The effects on the Moon and planets will be proportional to the mean motions. However, the effect on the longitude of the Moon, being related to the shape of the Moon's orbit, cannot be predicted by theory. Thus, the secular acceleration of the Moon cannot be predicted theoretically.

The apparent irregularities in the Earth's rotation posed a significant problem for early investigations. Brouwer (1952) discussed the hypothesis that the fluctuations in the Earth's rotation were due to cumulative random changes in its angular velocity. In that article, he mentioned that he and T. E. Sterne had discussed this possibility in about 1935, but concluded that the evidence strongly favored large abrupt changes. He also mentioned that Spencer Jones had speculated that the Earth may be like a pendulum in its behavior and that its rate of rotation is liable to frequent and small irregular changes. However, Brouwer does go on to discuss the explanation that the irregularities are due to apparent variations in the moment of inertia about the Earth's rotation axis. Munk and Revelle (1952) advanced the explanation that the irregularities in the rate of rotation of the Earth were due to electromagnetic coupling of the mantle to a turbulent core. At about the same time S. K. Runcorn (1954) came to the same interpretation. Vestine (1953) discussed the possibility that these irregularities might be related to motions in the Earth's magnetic field.

Van Den Dungen, Cox, and van Mieghem (1949) studied the variation in the moment of inertia of the atmosphere about the polar axis for periodic variations. The preliminary theoretical results were able to account for about a fifth of the observed effects. Munk and Miller (1950) did further studies and found about 15% of the observed effects. Mintz and Munk (1953), however, found that the Earth's reaction would be in reasonable agreement.

4.5 Current Understanding of the Earth's Variable Rotation

We now know that the secular deceleration of the Earth results principally from two causes. Lunar tides would be expected to increase the observed length of day by $+2.31 \pm 0.15$ ms day^{-1} cy^{-1}. A second nontidal effect, caused by glacial isostatic adjustment of portions of the Earth's crust related to the last ice age, actually shortens the length of day by -0.69 ± 0.14 ms day^{-1} cy^{-1} (Mitrovica et al., 2015). The combination of these two phenomena results in a change in the length of day of $+1.62 \pm 0.21$ ms day^{-1} cy^{-1}. Figure 4.9 shows the expected change in the LOD corresponding to this combination along with observations (Stephenson, Morrison, & Hohenkerk, 2016).

The combination of the theoretical tidal and nontidal effects leads to an expected variation between a timescale based on the Earth's rotation and one independent of the Earth's rotation of 29.56 ± 3.83 s/century^2. This is displayed in Figure 4.10 along with the observations.

Both of these plots show that there are significant departures from a purely secular deceleration of the Earth's rotation. The decadal irregularities, as shown in Figure 4.11, are attributed to the interactions of the Earth's mantle and the liquid core. This is based on observations of the magnetic field at the Earth's surface along

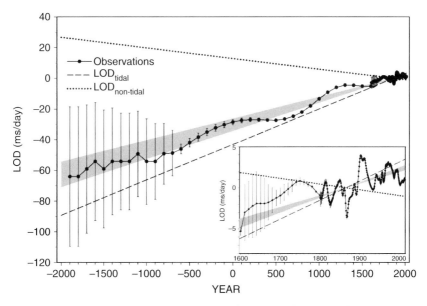

Figure 4.9 Observed and theoretical values of LOD as a function of time drawn by the authors. The observed values are derived from data available at Her Majesty's Nautical Almanac Office's website (see Stephenson et al., 2016). The shaded area represents the theoretical expected variation due to the combination of the tidal and nontidal effects. The more recent portion of the plot is expanded in the inset.

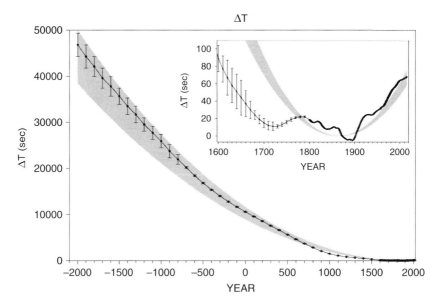

Figure 4.10 Observed and theoretical values of ΔT as a function of time drawn by the authors. The observed values are from ΔT data available at Her Majesty's Nautical Almanac Office's website (see Stephenson et al., 2016). The shaded area represents the theoretical expected variation due to the combination of the tidal and nontidal effects. The more recent portion of the plot is expanded in the inset.

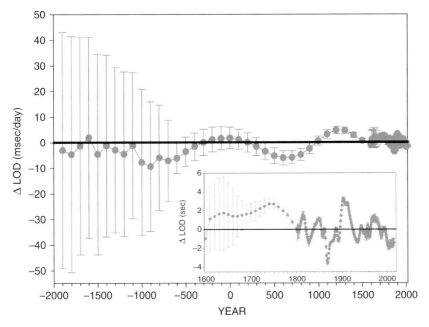

Figure 4.11 Low-frequency variations in LOD

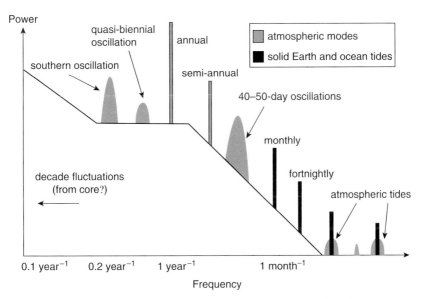

Figure 4.12 Schematic representation of the power spectrum of higher-frequency variations in the Earth's rotation

with theoretical analyses of motions in the Earth's core (Dehant, de Viron, & van Hoolst, 2005; Mound & Buffet, 2003).

We now know that the Earth is subject to variations at many frequencies from decadal to sub-daily, and that these variations have many geophysical and meteorological causes. These were outlined by Eubanks (1993), and Figure 4.12, taken, in part, from that review, shows schematically the various components of the Earth's variable rotation. It shows that the annual and higher frequency variations in the Earth's rotation are closely correlated with the total atmospheric angular momentum (AAM). Other contributors include ocean angular momentum and hydrology. Periodic variations due to tides are also present.

4.6 Consequences

In the early 20th century it became apparent that "astronomical time," based on the Earth's rotation and used for practical astronomical computations, differed from a "uniform" or "Newtonian" time, which is the independent variable for celestial mechanics. Recognizing this fact, de Sitter went on to provide observationally determined corrections to astronomical time to obtain a uniform time. These included both a secular term and irregular corrections given in a table covering the period 1640–1926.5.

With the recognition that the Earth's rotation was not uniform, the method of rating clocks using star observations was limited in accuracy by the variability of the Earth's rotation. So as clocks became more accurate than the Earth, ensembles of clocks were necessary to establish accurate time. Crystal-controlled clocks could smooth out the random observational errors from night to night. The crystal oscillators that were used as time standards varied in frequency from day to day by only a few parts in 10^{10}. To determine time based on the rotation of the Earth, it was necessary to observe selected stars to obtain the hour angle of the equinox based on a catalog of the directions of the stars. So the observations of the positions of stars in their diurnal circuits determined apparent sidereal time referred to observers' instantaneous local meridians. Different instruments were used for these observations, such as meridian or transit circles, photographic zenith tubes, and Danjon astrolabes at optical wavelengths. Optical observations could determine time to a few milliseconds.

The recognition of the variability of the rotation of the Earth also led to some far-reaching consequences. It meant that measuring time based on the Earth's rotation using angle measurements of celestial objects would no longer be suitable for those who required the most accurate timescales. It also meant that an observing program would be necessary to monitor the Earth's rotation (Chapter 5), separate from a uniform time based on mechanical clocks for those who required knowledge of the Earth's orientation in space. So from here on we must separate the concept of time and the determination of timescales into the different types, and trace the developments in the separate cases.

To provide accurate directions to solar system objects, astronomers developed a uniform time for ephemerides and celestial mechanics. This new timescale, defined in terms of the Sun's orbital motion itself, was given the name *Ephemeris Time* (Chapter 6). For everyday practical use, however, time was provided by clocks. The most accurate mechanical clocks were superseded by the introduction of quartz crystal clocks and eventually atomic clocks, providing accuracy improvements by orders of magnitude. Practical timekeeping would no longer be the job of astronomers alone. The familiar connection of time to the Sun continued to require astronomical measurements to "regulate" the more uniform atomic time, but precise time measurements would now be more and more in the hands of the physicists and engineers.

On the other hand, the development of more accurate reference frames, to accommodate the improving solar system ephemerides and the introduction of artificial Earth satellites and interplanetary space travel, made it even more necessary to make observations to determine the rotation of the Earth as one of the components of our knowledge of the Earth's orientation in space. Mathematical models alone could not provide information with the accuracy needed for practical use in these rapidly developing areas. Astronomical observations were required to close the gap between models and observations. This requirement led to the development of new observational techniques to provide improved measurements, which led, in turn, to the development of areas of geophysics to improve our knowledge of Earth systems, which was made possible by the improved observations. The field of Earth orientation sciences bridging astronomy, geodesy, and geophysics came into existence.

References

Adams, J. C. (1853). On the Secular Variation of the Moon's Mean Motion. *Phil. Trans. Royal Soc. London*, **CXLIII**, 397–406.

Brosche, P. (1984). Tidal Friction in the Earth–Moon System. *Phil. Trans. Royal Soc. London A*, **313**, 71–75.

Brouwer, D. (1952). A Study of the Changes in the Rate of Rotation of the Earth. *Astron. J.*, **57**, 125–146.

Brown, E. W. (1914). Address on Cosmical Physics. *British Assoc. Report Australia*, 311–321, *Science, New Series*, **40**, 389–401.

Brush, S. G. (1986). Early History of Selenology. In *Origin of the Moon: Proceedings of the Conference, Kona, HI, October 13–16, 1984*. Houston, TX: Lunar and Planetary Institute, pp. 3–15.

Carter, B. & Carter, M. S. (2006). *Simon Newcomb, America's Unofficial Astronomer Royal*. St. Augustine, FL: Mantanzas Publishing.

Chapront, J., Chapront-Touze, M., & Francoou, G. (2002). A New Determination of Lunar Orbital Parameters, Precession Constant, and Tidal Acceleration from LLR Measurements. *Astron. & Astrophys.*, **387**, 700–709.

Christodoulidis, D. C., Smith, D. E., Williamson, R. G., & Klosko, S. M. (1988). Observed Tidal Braking in the Earth/Moon/Sun System. *J. Geophys. Res.*, **93**, 6216–6236.

Clemence, G. M. (1943).The Motion of Mercury 1765–1937. *Astronomical Papers of the American Ephemeris and Nautical Almanac*, vol. XI, Part I. Washington, DC: US Government Printing Office.

Clemence, G. M. (1971). The Concept of Ephemeris Time: A Case of Inadvertent Plagiarism. *J. History of Astronomy*, **2**, 73–79.

Darwin, G. H. (1877). On the Influence of Geological Changes on the Earth's Axis of Rotation. *Phil. Trans. Royal Soc. London*, **167**, 271.

Darwin, G. H. (1879). On the Precession of a Viscous Spheroid and on the Remote History of the Earth. *Phil. Trans. Royal Soc. London*, **170**, 447–530.

Darwin, G. H. (1880). On the Secular Change of the Orbit of a Satellite Revolving about a Tidally Distorted Planet. *Phil. Trans. Royal Soc. London*, **171**, 713–891.

Darwin, G. H. (1898). *Tides and Kindred Phenomena in the Solar System*. Boston, MA, and New York, NY: Houghton Mifflin and Company.

de Sitter, W. (1927). On the Secular Accelerations and the Fluctuations of the Longitudes of the Moon, the Sun, Mercury and Venus. *Bull. of Astron. Inst. of Netherlands*, **IV**, 21–38.

Dehant, V., de Viron, O., & van Hoolst, T. (2005). Poincaré Flow in the Earth's Core. In N. Capitaine, ed., *Journées 2004 – systèmes de référence spatio-temporels. Fundamental Astronomy: New Concepts and Models for High Accuracy Observations, Paris, 20–22 September 2004*. Paris: Observatoire de Paris.

Delaunay, C.-E. (1859). *Comptes Rendus Acad Sci Paris* (Séance du 25/4/1859) **48**, 817.

Delaunay, C.-E. (1866). *Conférence sur l'astronomie et en particulier sur le ralentissement du mouvement de rotation de la terre*. Paris: G. Bailliere.

Dreyer, J. L. E. (1953). *The History of Astronomy from Thales to Kepler*. Dover Publications.

Dunthorne, R. (1747). A Letter from Mr. Richard Dunthorne, to the Rev. Mr. Cha. Mason, F. R. S. and Woodwardian Professor of Nat. Hist. at Cambridge, Concerning the Moon's Motion. *Philosophical Transactions*, **44**, 412–420.

Dunthorne, R. (1749). A Letter from the Rev. Mr. Richard Dunthorne to the Reverend Mr. Richard Mason F. R. S. and Keeper of the Woodwardian Museum at Cambridge, Concerning the Acceleration of the Moon. *Philosophical Transactions*, **46**, 162–172.

Dyson, F., Sir & Crommelin, A. C. D. (1923). The Greenwich Observations of the Moon. *Monthly Notices Roy. Astron. Soc.*, **83**, 359–370.

Eubanks, T. M. (1993). Variations in the Orientation of the Earth. In David E. Smith & Donald L. Turcotte, eds., *Contributions of Space Geodesy to Geodynamics: Earth Dynamics: Geodynamic Series*, vol. 24. Washington, DC: American Geophysical Union, p. 1.

Explanatory Supplement to The Astronomical Ephemeris and The American Ephemeris and Nautical Almanac (1961). London: Her Majesty's Stationery Office.

Explanatory Supplement to the Astronomical Almanac (1992). P. Kenneth Seidelmann, ed. Mill Valley, CA: University Science Books.

Ferrari, A. J., Sinclair, W. S., Sjogren, W. L., Williams, J. G., & Yoder, C. F. (1980). *J. Geophys. Res.*, **85**, 3939.

Ferrel, W. (1853). On the Effect of the Sun and Moon upon the Rotatory Motion of the Earth. *Astron. J.*, **3**, 138–141.

Ferrel, W. (1864). Note on the Influence of the Tides in Causing an Apparent Secular Acceleration of the Moon's Mean Motion. *Proc. of American Academy of Arts and Sciences*, **VI**, 379–383, 390–393.

Finch, H. (1950). On a Periodic Fluctuation in the Length of the Day. *Monthly Notices Roy. Astron. Soc.*, **110**, 3.

Folkner, W. M., Williams, J. G., Boggs, D. H., Park, R. S., & Kuchynka, P. (2014). The Planetary and Lunar Ephemerides DE430 and DE431. *The Interplanetary Network Progress Report*, **42–196**, 1–81.

Fotheringham, J. K. (1920). The Longitude of the Moon from 1627 to 1918. *Monthly Notices Roy. Astron. Soc.*, **80**, 289.

Ginzel, F. K. (1899). Bemerkungen über den Werth der alten historischen Sonnenfinsterniss für die Mondtheorie. *Astron. Nachr.*, **150**, 1.

Glauert, H. (1915a). The Rotation of the Earth. *Monthly Notices Roy. Astron. Soc.*, **75**, 489–495.

Glauert, H. (1915b). The Rotation of the Earth. *Monthly Notices Roy. Astron. Soc.*, **75**, 685–687.

Halley, E. (1695). Some Account of the Ancient State of the City of Palmyra, with Short Remarks upon the Inscriptions Found There. *Phil. Trans. Roy. Soc.*, **19**, 160.

Innes, R. T. A. (1925a). Transits of Mercury 1677–1924. *Union Observatory Circ. No. 65*.

Innes, R. T. A. (1925b). Variability of the Earth's Rotation. *Astron. Nachr.* **25**, 109.

Jeffreys, H. (1924). *The Earth: Its Origin, History and Physical Constitution*. Cambridge: Cambridge University Press.

Kant, I. (1754). Untersuchung der Frage, ob die Erde in ihrer Umdrehung um die Achse, wodurch sie die Abwechselung des Tages und der Nacht hervorbringt, einige Veränderung seit den ersten Zeiten ihres Ursprungs erlitten habe und woraus man sich ihrer versichern könne, welche von der Königlichen Akademie der Wissenschaften zu Berlin zum Preise für das jetztlaufende Jahr aufgegeben worden, English: Investigation of the Question, Whether the Axial Rotation of the Earth, through Which Day and Night Are Brought About, Has Changed since Its Beginning, and How One Can Be Certain of this, Which the Royal Academy of Sciences in Berlin Has Offered a Prize for the Current Year. In *Wochentliche Königsbergische Frag- und Anzeigungs-Nachrichten* #23 (June 8) and #24 (June 15). [Ak. 1: 185–191]

Lalande, J. *Traité d'astronomie* (2 vols., 1764 enlarged edition, 4 vols., 1771–1781; 3rd edn, 3 vols., 1792).

Lambeck, K. (1980). *The Earth's Variable Rotation: Geophysical Causes and Consequences*. Cambridge: Cambridge University Press.

Laplace P.-S. de (1786). Sur l'équation séculaire de la Lune. *Mém. Acad. Roy. Sci.*, 235.

Markowitz, W. (1955). The Annual Variation in the Rotation of the Earth, 1951–4. *Astron. J.*, **59**, 69.

Mayer, J. R. (1848). *Beiträge zur Dynamik des Himmels*, chap. 8. Heilbronn: Landherr.

Mayer, T. (1753). Novae Tabulae Motuum Solis et Lunae. In *Commentarii Societatis Regiae Scientiarum Gottingensis*, vol. II. Göttingen.

Mintz, Y. & Munk, W. (1953). The Effect of Winds and Bodily Tides on the Annual Variation in the Length of Day. *Monthly Notices Roy. Astron. Soc.*, **113**, 789.

Mitrovica, J. X., Hay, C. C., Morrow, E., Kopp, R. E., Dumberry, M., & Stanley, S. (2015). Reconciling Past Changes in Earth's Rotation with 20th Century Global Sea-Level Rise: Resolving Munk's Enigma. *Science Advances*, 11 Dec. 2015: **1**(11), e1500679. doi:10.1126/sciadv.1500679

Mound, J. E. & Buffett, B. A. (2003). Interannual Oscillations in Length of Day: Implications for the Structure of the Mantle and Core. *Journal of Geophysical Research Solid Earth*, **108**(B7), pp. ETG 2–1, CiteID 2334. Doi:10.1029/2002JB002054.

Munk, W. H. (1966). Variation of the Earth's Rotation in Historical Time. In B. G. Marsden & A. G. W. Cameron, eds., *The Earth–Moon System*. New York, NY: Plenum Press.

Munk, W. H. & MacDonald, G. J. F. (1960). *The Rotation of the Earth*. Cambridge: Cambridge University Press.

Munk, W. H. & Miller, R. L. (1950). Variations in the Earth's Angular Velocity Resulting from Fluctuations in Atmospheric and Ocean Circulation. *Tellus*, **2**, 93–101.

Munk, W. H. & Revelle, R. (1952). On the Geophysical Interpretation of Irregularities in the Rotation of the Earth. *Monthly Notices Roy. Astron. Soc. Geophysical Supplement*, **6**, 331.

Newcomb, S. (1896). *Comptes Rendus Acad Sci Paris*, vol. 1 **cxxii**, 1238.

Newcomb, S. (1903a). *The Reminiscences of an Astronomer*. Boston, MA, and New York, NY: Houghton Mifflin and Company.

Newcomb, S. (1903b). On the Desirableness of Re-Investigation of the Problems Growing Out of the Mean Motion of the Moon, *Monthly Notices Roy. Astron. Soc.*, **63**, 318–324.

Newcomb, S. (1906). *Side-Lights on Astronomy and Kindred Fields of Popular Science. Essays and Addresses*. London and New York, NY: Harper & Brothers.

Newton, I. (1713). *Philosophia Naturalis Principia Mathematica*, 2nd edn. Cambridge, 481.

Pavel, F. & Uhink, W. (1935). Die Quarzuhren des Geodätischen Instituts in Potsdam. *Astron. Nachr.*, **257**, 365–390.

Ross, F. E. & Newcomb, S. (1917). New Elements of Mars and Tables for Correcting the Heliocentric Positions Derived from Astronomical Papers. *Astronomical Papers of the AENA*, **IX**, part II. Washington, DC: US Government Printing Office.

Runcorn, S. K. (1954). The Earth's Core. *Trans. American Geophys. Union*, **35**, 49.

Scheibe, A. & Adelsberger, U. (1936). Nachweis von Schwankungen der astronomischen Tageslange mittels Quarzuhren. *Phys. Zeitschrift*, **37**, 38.

Scheibe, A. & Adelsberger, U. (1950). Die Gangleistungen der PTR-Quarzuhren und die jahrliche Schwankung der astronomischen Tageslange. *Zeitschrift fur Physik*, **127**, 416.

Smith, H. & Tucker, R. (1953). The Annual Fluctuation in the Rate of Rotation of the Earth. *Monthly Notices Roy. Astron. Soc.*, **113**, 251.

Spencer Jones, H. (1926). The Rotation of the Earth. *Monthly Notices Roy. Astron. Soc.*, **87**, 4–31.

Spencer Jones, H. (1932). Discussion of Observations of Occultations of Stars by the Moon, 1672–1908 Being a Revision of Newcomb's "Researches on the Motion of the Moon, Part II." *Annals of the Cape Observatory*, **XIII**.

Spencer Jones, H. (1939). The Rotation of the Earth and the Secular Acceleration of the Sun, Moon, and Planets. *Monthly Notices Roy. Astron. Soc.*, **99**, 541.

Stephenson, F. R. (2003). Historical Eclipses and Earth Rotation. *Astronomy and Geophysics*, **44**, 222–227.

Stephenson, F. R., Morrison, L. V., & Hohenkerk, C. Y. (2016). Measurements of the Earth's Rotation: 720 BC to AD 2015. *Proc. Royal Soc. A*, **472**, 20160404. http://dx.Doi.org/10.1098/rspa.2016.0404.

Stoyko, N. (1937). Sur la périodicité dans l'irrégularité de la rotation de la Terre. *Comptes Rendus Acad. Sciences*, **205**, 79–81.

Stoyko, N. (1950). Sur la variation saisonnière de la rotation de la terre. *Comptes Rendus Acad. Sciences*, **230**, 514.

Thomson, W. (1863). On the Rigidity of the Earth. *Phil. Trans. Royal Soc. London*, **153**, 573–582.

Thomson, W. & Tait, P. G. (1890). *Treatise on Natural Philosophy*. Cambridge: Cambridge University Press, para. 405 (footnote).

Tobin, W. (2003). *The Life and Science of Léon Foucault: The Man Who Proved the Earth Rotates*. Cambridge: Cambridge University Press.

Van den Dungen, F. H., Cox, F. J., & van Mieghem, J. (1949). Sur les fluctuations de periode annuelle de la rotation de la terre. *Bull. Acad. Belg. Cl. Sci.*, **35**, 642–655.

Vestine, E. H. (1953). On Variations of the Geomagnetic Field, Fluid Motions, and the Rate of the Earth's Rotation. *Journ. Geophys. Res.*, **58**, 127.

Williams, J. G. & Boggs, D. H. (2016). Secular Tidal Changes in Lunar Orbit and Earth Rotation. *Celest. Mech. & Dynam. Astron.*, **126**, 89–129.

Williams, J. G., Boggs, D. H., & Folkner, W. M. (2008). DE421 Lunar Orbit, Physical Librations, and Surface Coordinates. IOM 335-JW, DB, WF–20080314–001.

5

Earth Orientation

5.1 Reference Systems

Realization of the complicated motions of the Earth's axis of rotation within the planet and in space along with the discovery of the variability of the Earth's rotational speed eventually made clear that the Earth was becoming less suitable as a practical source for providing a uniform timescale. However, information regarding the Earth's rotation angle and the orientation of its axis is critical in relating reference systems on the Earth to those in space.

A reference system, either terrestrial or celestial, is composed of (1) a specified origin, (2) the directions of fundamental axes, and (3) a set of conventional models, procedures, and constants used to realize the system. A reference frame is the realization of that system through a list of coordinates. *Earth orientation* describes the procedure and models used to relate a terrestrial geodetic reference system to a celestial reference system (CRS). The rigorous details are outlined in the publications of the International Earth Rotation and Reference system Service (formerly the International Earth Rotation Service) (IERS), specifically in the *IERS Conventions (2010)* (Petit & Luzum, 2010) and its updates, which are available electronically (iers.org/IERS/EN/DataProducts/Conventions/conventions.html).

Figure 5.1 shows conceptually a terrestrial reference system (TRS) fixed to the Earth, which is in an orbit in the ecliptic plane defined in a celestial reference system. Celestial reference systems are specified by astronomically defined directions and origins. Most modern systems are considered to have their origins at the barycenter of the solar system, and their polar axes relate in some way to the rotational axis of the Earth. The second axis then lies in the equatorial plane perpendicular to the polar axis and is directed toward a fiducial point in that plane, formerly the vernal equinox. The third axis is chosen to complete a right-handed orthogonal system.

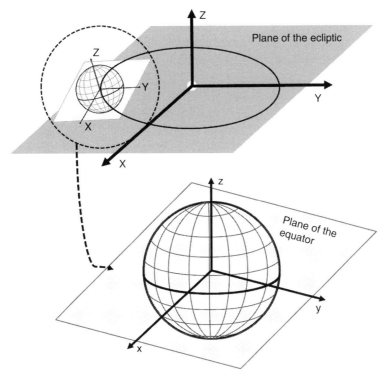

Figure 5.1 Concept of terrestrial and celestial reference systems

Terrestrial reference systems generally have their origins at the center of mass of the Earth with their polar axes related to the direction of an axis fixed with respect to the Earth's crust. The origin of longitudes in the equatorial plane provides the second direction. In 1884, the astronomically defined Greenwich meridian was chosen to provide this origin, but improvements in geodetic accuracy since then have made this definition obsolete. The third axis is chosen to complete a right-handed orthogonal system.

5.1.1 Celestial Reference Frame

Historically, the fixed celestial reference frame was defined by the positions of optically bright, nearby stars at some epoch consistent with a moving reference system based on the solar system dynamics. Thus, the *x*-axis was specified by the vernal equinox, the intersection of the ecliptic and equatorial planes, both of which are in motion. Precession and nutation were described by conventionally adopted models such as the IAU 1976 Astronomical Constants and the IAU 1980 Theory of Nutation. "Mean positions" of stars for specific epochs were based on the adopted precession. "True positions" of date were determined using the adopted theory of nutation in addition.

In 1992, the International Astronomical Union (IAU) designated a standard, fixed, epoch-independent, celestial reference system and frame called the International Celestial Reference System (ICRS) and the International Celestial Reference Frame (ICRF), respectively (Ma & Feissel, 1997). The ICRF is made up of the adopted directions to distant radio sources. Positions of optical stars, consistent with that frame, are provided by the Hipparcos Catalogue (Perryman et al., 1997). See also the recently released GAIA data (gaia.esac.esa.int/documentation/GDR1/).

The ICRF is considered to have its origin at the solar system barycenter. Consequently, it is one component of a Barycentric Celestial Reference System (BCRS), and the ICRS is considered to be a specific realization of a BCRS. Observations are generally made in the terrestrial system whose origin is at the geocenter. In doing this we make use of a Geocentric Celestial Reference System (GCRS) that is kinematically nonrotating with respect to a BCRS, so the orientation of the GCRS is the orientation of the ICRS. The origin of the GCRS moves nonlinearly with respect to the BCRS, but it has fixed directions with respect to the extragalactic sources. Thus, there is a Coriolis-like effect from relativistic theory in the transformations, if referred to the BCRS. This effect is called *geodesic precession and nutation* and is included in the precession/nutation computation.

5.1.2 Terrestrial Reference Frame

Terrestrial frames are comprised of a list of site coordinates and, in more recent times, their possible motions, similar to the use of stellar positions to define the celestial reference system. Historically they were defined using optical astronomical observations made with transit circles, astrolabes, and zenith tubes. These observations made use of the direction of the local vertical defined by the local gravity vector as a reference direction to report observations of astronomical latitude and a local measure of the Earth's rotation angle, UT0-UTC (see Chapter 2). Observations of UT1−UTC derived using these astronomical longitudes were combined by the Bureau International de l'Heure to provide standard values of UT1−UTC for users worldwide. In the 1970s and 1980s, new technologies became available, such as laser ranging to satellites, connected element interferometry, very long baseline interferometry, and lunar laser ranging. Observations made with these techniques were gradually incorporated by the BIH for determining Earth rotation and terrestrial coordinates. In 1984, the BIH established the "BIH Terrestrial System," which listed the longitudes and latitudes of the observing sites whose observations were used to provide the Earth orientation data that they reported. For

Figure 5.2 The Bradley meridian (dotted line), the Airy meridian (dashed line), and the ITRF zero meridian (solid line) at the Royal Greenwich Observatory. Imagery 2014 Google Maps. Infoterra Ltd & Bluesky.

some sites, the longitudes reported to the BIH were adjusted to maintain continuity with previously reported UT1−UTC data, so there would not be an abrupt step in UT1−UTC. The International Terrestrial Reference System (ITRS) and the International Terrestrial Reference Frame (ITRF) (Altamimi, Rebischung, Métivier, & Xavier, 2016), maintained by the IERS, are accepted as the international standards.

The difference between astronomical coordinates determined using astronomical observations with respect to the local vertical as a reference and geodetic coordinates is the deflection of the vertical, abbreviated DoV. It is site dependent due to the local direction of the gravity vector. Astronomical observations are made with respect to the direction of a vertical determined from a mercury basin, equivalent to a plumb line, and it does not necessarily pass through the center of the Earth. Geodetic coordinates are based on a geodetic vertical that is perpendicular to the geoid at the site and that essentially passes through the center of the Earth.

Consequently, a visitor to the Royal Greenwich Observatory in England standing on the metal strip in front of the Airy Transit Instrument marked "The Prime Meridian" with an electronic navigational (e.g., GPS) receiver will not see it display zero degrees longitude. Instead, zero degrees longitude is 102 m, or $5''.3$, east of the marked prime meridian, corresponding to a DoV of $5''.3$ (see Figure 5.2). The ITRF and WGS 84, the basic terrestrial reference frame of the Global Positioning System (GPS), have the same longitude orientation (Malys, Seago, Pavlis, Seidelmann, & Kaplan, 2015). The EGM2008 gravitational model (Pavlis, Holmes, Kenyon, & Factor, 2012) confirms the DoV value at Greenwich. This offset does not cause a rotation of the entire reference frame, rather the differences in the coordinates are unique for each location based on the local values of the deflection of the vertical. Nevertheless, the origin of longitudes is, in practice, very near the site identified by the 1884 decision. The geodetic prime meridian at Greenwich has the same orientation as a function of time as the astronomical meridian. Extended to infinity, these parallel meridian planes sweep past the same stars simultaneously, so that both planes indicate the same UT1.

The terrestrial system is rotating in the celestial system and its orientation in that system is affected by precession, nutation, polar motion and variations in the Earth's rotational speed. Modeling these effects with precision requires considering a number of phenomena. The fact that the Earth is not strictly a rigid body means that nonrigid body effects need to be accounted for in the models. As the Earth's core experiences a free wobble with respect to the mantle, existing geophysical models of nutation may not account for all of the observed motions. Further, motions caused by redistribution of mass in the Earth, its oceans and atmosphere, along with relatively high-frequency variations in global meteorology and hydrology, may need to be taken into account.

5.1.3 Intermediate Reference System

To facilitate the transformation between a barycentric celestial reference frame and a geocentric terrestrial reference frame, a Celestial Intermediate Reference System (CIRS) is used. It is a geocentric reference system related to the GCRS by a time-dependent rotation taking into account precession and nutation. It is defined by the intermediate equator of the Celestial Intermediate Pole (CIP) and the Celestial Intermediate Origin (CIO) for a specific date. The CIP is a geocentric equatorial pole, whose direction results from (i) the part of the precession/nutation model with periods greater than two days, (ii) the retrograde diurnal part of polar motion (including the free core nutation), and (iii) the frame bias. Its ITRS position results from (i) the part of polar motion that is outside the retrograde diurnal band in the ITRS, and (ii) the motion in the ITRS corresponding to nutation components with periods less than two days. The CIP is a conventionally defined pole separating the motion of the pole of the ITRS in the ICRS into the celestial motion of the CIP (precession/nutation), including all the terms with periods greater than two days in the celestial reference system (frequencies between -0.5 cycles per sidereal day [cpsd] and $+0.5$ cpsd), and the terrestrial motion of the CIP (polar motion), including all the terms outside the retrograde diurnal band in the TRS (frequencies lower than -1.5 cpsd or greater than -0.5 cpsd). In the IAU 1980 Theory of Nutation, the pole was called the Celestial Ephemeris Pole (CEP), but the name was changed to be consistent with newer nomenclature. The actual motion of the CIP is realized by the IAU 2000A precession/nutation model, plus time-dependent corrections provided by the IERS. The division of periodic terms with periods more and less than two days is arbitrary.

The CIO is the origin of right ascensions on the intermediate equator in the CIRS. It is the nonrotating origin in the GCRS that was originally set close to the GCRS meridian and throughout 1900–2100 stays within 0.1 arcseconds of this alignment. The CIO was located on the true (CIP) equator of J2000.0 at point 2.012 mas from the ICRS prime meridian at right ascension 00h 00m 0.000 134 16s in the ICRS. As the true equator moves in space, the path of the CIO in space is such that the point has no instantaneous east–west velocity along the true equator. In contrast, the equinox has instantaneous velocity along the equator.

Similarly, a Terrestrial Intermediate Reference System (TIRS) is used in transforming between frames. This is a geocentric reference system defined by the intermediate equator of the CIP and the Terrestrial Intermediate Origin (TIO). It is related to the ITRS by polar motion and a small, slowly varying quantity called the TIO locator. It is related to the CIRS by a rotation called the Earth Rotation Angle (ERA) around the CIP, which defines the common *z*-axis of the two systems. The TIO was originally set at the ITRF origin of longitude and throughout 1900–2100 stays within 0.1 mas of the ITRF zero-meridian.

With the introduction of the new reference system in the 1992–2004 period, the ICRS replaced the reference system based on the positions and motions of the Fifth Fundamental Catalogue (FK5) (Fricke, Schwan, & Lederle, 1988) star catalog. The CIO replaced the moving vernal equinox; the TIO replaced the Greenwich Meridian; the ERA replaced the Greenwich Sidereal Time. The alternative system based on the equinox, mean and true positions, and the Greenwich Mean Sidereal Time is still supported and when properly applied can provide equivalent accuracies (Kaplan, 2005).

5.2 Variations in Earth Orientation

The transformation between celestial and terrestrial frames is specified by five angles called *Earth orientation parameters*. Three would be sufficient, but five angles are used in order to describe the physical processes involved and to make the transformations easier to apply. Two angular coordinates (X and Y) are used to model the changing direction of the CIP due to the precession and nutation of the Earth. These phenomena are driven by the gravitational attraction of the solar system bodies, principally the Sun and the Moon, on the nonspherical Earth. *Precession* refers to the aperiodic portion of the motion and *nutation* refers to the periodic portion. Both motions depend on the positions of the solar system bodies and the internal structure of the Earth, but they can be modeled mathematically with reasonable accuracy.

Figure 5.3 Reference system axes and motions related to the Earth's orientation in space

Two more angles are used to describe the motion of the CIP with respect to the Earth's crust. This phenomenon is called *polar motion* and is driven by geophysical and meteorological variations within the Earth and its atmosphere. Polar motion is difficult to model because the forces driving the motion are difficult to predict. As a result, these angles must be observed astronomically and made available to users operationally.

The last of the five angles characterizes the rotation angle of the Earth and is expressed as the time difference UT1–UTC. Principal variations in the rotation speed of the Earth include a constant deceleration due to tidal deceleration and deglaciation, decadal variations due to changes in the internal distribution of the Earth's mass, largely seasonal meteorologically driven variations, and tidally driven periodic variations. As with polar motion, UT1–UTC is difficult to model and predict, and must be observed astronomically and reported to users routinely. Figure 5.3 displays the motions and the reference system axes related to the changing orientation of the Earth in space.

5.2.1 Precession/Nutation

Gravitational forces exerted by solar system bodies act on the nonspherical Earth to cause its orientation in the celestial reference system to change. In 26,000 years, the

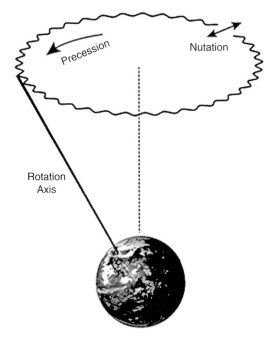

Figure 5.4 Schematic representation of precession and nutation

polar axis appears to describe a cone in space. This motion, called *precession*, is shown in Figure 5.4. Hipparchus is generally credited with discovering this effect in the second century BC, when he found that over time the positions of the equinoxes move westward along the ecliptic compared to the stars. The motion amounts to about 50 seconds of arc per year. *Nutation* refers to the much smaller periodic departures of the motion of the polar axis from the precessional motion. It too is due to gravitational forces acting on the Earth's bulge and was first detected by English astronomer James Bradley (1748). The principal period of the nutational motion is 18.6 years, which is due to the lunar orbit.

In modern terminology, *precession/nutation* is the motion of the CIP in the GCRS, including the free core nutation and other corrections to the standard models. Knowledge of the orbits and masses of solar system objects, principally the Sun and Moon, allows us to determine the torques acting on the Earth. That information along with an understanding of the internal constitution of the Earth permits us to derive mathematical expressions to model the precession/nutation of the Earth's axis in space. The IAU-recommended model for precession, called IAU 2006 Theory of Precession or P03 (Capitaine et al., 2003), provides the position of the CIP in the GCRS using six polynomial terms for both the X and Y components of the representation of this motion. In nutation, however, many more terms are required to represent the motion. The model recommended by the IAU, known as the IAU 2000A precession/nutation model (Petit & Luzum, 2010), contains 1,306 sin-cosine terms for X and 962 for Y, representing periodic motion components with periods ranging from hundreds of years to two days. It provides accuracy at the microarcsecond (μas) level. A shorter version using only 78 terms for each component provides accuracy at the single milliarcsecond (mas) level (McCarthy & Luzum, 2003).

In addition, a small correction obtained by astronomical observations can be applied. The IERS provides these "celestial pole offsets" in the form of the differences, dX and dY, of the CIP coordinates in the GCRS with respect to the IAU 2000A precession/nutation model (i.e., the CIP is realized by the IAU 2000A precession/nutation plus these celestial pole offsets). In parallel, the IERS also provides the offsets, in longitude and obliquity with respect to the older IAU 1976/ 1980 precession/nutation model at www.iers.org/IERS/EN/DataProducts/ EarthOrientationData/eop.html.

5.2.2 Polar Motion

Polar motion refers to the motion of the CIP with respect to the surface of the Earth. It is composed of a set of motions covering a range from secular to sub-daily timescales. Figure 5.5 shows the major periodic components of the power spectrum.

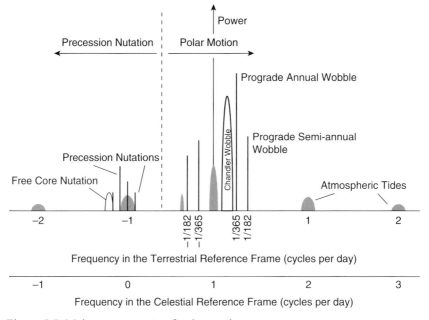

Figure 5.5 Major components of polar motion

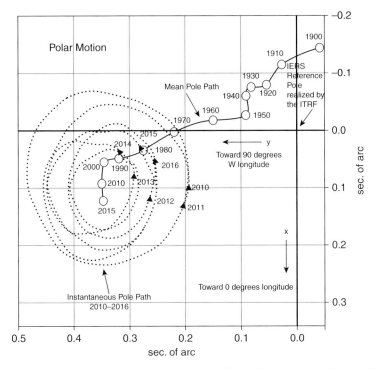

Figure 5.6 The trace of the pole over the Earth's surface from the International Earth Rotation and Reference System Services. The mean and instantaneous positions are shown separately.

Observational evidence for the existence of polar motion, detected as a variation of astronomical latitude, was gathered in the 19th century by astronomers and geodesists, including F. W. Bessel, F. G. W. Struve, C. A. F. Peters, O. W. Struve, J. A. H. Glydén, and M. Nyrén (Dick, 2000; Höpfner, 2000; Verdun & Beutler, 2000). Karl Friedrich Küstner is generally credited with establishing the reality of polar motion in the 1880s (Küstner, 1888, 1890; Brosche, 2000). Seth C. Chandler (1892) was the first to identify the major periodic components (Carter & Carter, 2000). This motion has been an active area of research since the 19th century (Munk & MacDonald, 1960; Lambeck, 1980; McCarthy, 2000). Figure 5.7 shows the major components of the power spectrum.

Figure 5.6 demonstrates the secular drift of the pole since 1900 and the periodic motions from 2010 to 2016. The drift is described by the angular coordinates of a "mean pole." The foremost of the periodic terms is a free motion of the pole, called the *Chandler wobble* after its discoverer, Seth C. Chandler. Its period is approximately 435 days and corresponds to the free motion of a nonrigid Earth that was originally predicted by L. Euler (1765) for a rigid Earth. The second principal component of the polar motion is an annual motion driven by the seasonal redistribution of the Earth's atmospheric and water mass.

5.2.2.1 Drift

Drift is the aperiodic motion roughly in the direction of 75° W longitude thought to be caused by post-glacial rebound and glacial melting. It has been described by angular coordinates of a "mean" pole. The IERS now (2018) provides a linear representation called the "secular" pole derived from a fit to the polar path from 1900 through 2017. The coordinates designated (xs, ys) and are given in milli-arcseconds by xs = 55.0 + 1.677 * (t − 2000), ys = 320.5 + 3.460 * (t − 2000) where t is the date in years of 365.25 days. Drift parameters provide information used to investigate models of the Earth's interior.

5.2.2.2 Chandler and Annual Variations

The power spectrum of polar motion (Figure 5.5) is dominated by the Chandler and annual components. *Chandler motion* refers to the motion of the CIP over the Earth's surface due to the fact that the rotation axis is not aligned with the axis of inertia. This motion was predicted by Euler (1765) for a rigid Earth, but was first described observationally by Seth Chandler (1891). The period for a rigid Earth would be approximately 306 days, but Chandler observed a period of 427 days. This discrepancy was explained later by Simon Newcomb as being due to the fluidity of the oceans and the elasticity of the solid Earth (1891). We now know that the Chandler motion is approximately circular with a period of about 433 days. We also know that the amplitude of the motion is variable, but that it is of the order of about 150 mas. The possible excitation of this motion remains in doubt. Possible causes include variations in pressure at the ocean bottom, variations in atmospheric pressure, groundwater, and seismic excitation, as well as core–mantle torque. Annual or seasonal polar motion is a stable prograde motion with an amplitude of approximately 90 mas. Causes are apparently atmospheric, oceanic, and groundwater excitations. The combination of the somewhat elliptical motion of the seasonal variation and the quasi-circular Chandler motion causes the pattern seen in Figure 5.4.

5.2.2.3 Other Variations

The high-frequency portion of the polar motion spectrum contains a series of quasi-periodic motions with amplitudes of the order of one mas. These motions are poorly understood, but causes are assumed to be atmospheric and oceanic. High-precision observations have also confirmed the existence of diurnal and sub-diurnal polar motion caused by ocean tides. They can be reasonably modeled with

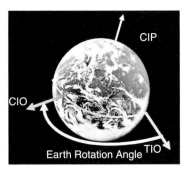

Figure 5.7 Earth Rotation Angle

amplitudes of the order of one mas. Periodic motions driven by gravitational torques with periods less than two days are also included in polar motion.

5.2.3 UT1

UT1 is a measure of the Earth's rotation angle expressed in time units and treated conventionally as an astronomical timescale defined by the rotation of the Earth with respect to the Sun. In practice, UT1 was defined until January 1, 2003 by means of a conventional formula (Aoki et al., 1982). UT1 is now defined as being linearly proportional to the ERA, and the transformation between the ITRS and GCRS is specified using the ERA. The ERA is the angle measured along the intermediate equator of the CIP between the TIO and the CIO, positively in the retrograde direction and increasing linearly for an ideal, uniformly rotating Earth (Figure 5.7). It is related to UT1 by a conventionally adopted expression in which the ERA is a linear function of UT1.

$$ERA = \theta(T_u) = 2\pi(0.779\ 057\ 273\ 264\ 0$$
$$+1.002\ 737\ 811\ 911\ 354\ 48\ T_u) \tag{5.1}$$

where T_u = (Julian UT1 date – 2 451 545.0), and UT1 = UTC + (UT1–UTC), or equivalently

$$\theta(T_u) = 2\pi \left(\begin{array}{l} \text{UT1 Julian Days elapsed since } 2\ 451\ 545.0 + 0.779\ 057\ 273\ 264\ 0 \\ +\ 1.002\ 737\ 811\ 911\ 354\ 48\ T_u \end{array} \right).$$
$$\tag{5.2}$$

Its time derivative is the Earth's angular velocity.

UT1 is determined by astronomical observations (currently from VLBI observations of the diurnal motions of distant radio sources and from integrating estimates of its time rate of change determined from the analysis of the orbits of artificial Earth satellites). It can be regarded as a time determined by the rotation of the

Earth, and is obtained, in practice, by applying the correction UT1–UTC provided by the IERS, to the UTC timescale.

Greenwich Sidereal Time (GST) is an angle related to UT1 composed of the sum of the ERA and the angular distance between the CIO and the equinox along the moving equator called the *equation of the origins*, $E_0(t)$. This angle in right ascension is the sum of the accumulated precession, $E_{prec}(t)$, and accumulated nutation (called the *equation of the equinoxes*, $E_e(t)$) from the epoch of reference to the current date). GST contains both polynomial and periodic terms (Urban & Seidelmann, 2012).

$$E_{prec}(t) = -0''.014\ 506 - 4612''.156\ 653\ 4t - 1''.391\ 581\ 7t^2$$
$$+0''.000\ 000\ 44t^3 + 0''.000\ 029\ 956t^4. \tag{5.3}$$

$$Ee(t) = \Delta\psi \cos \varepsilon_A + 0''.002\ 640\ 96 \sin \Omega + 0''.000\ 063\ 52 \sin 2\Omega$$
$$+0''.000\ 011\ 75 \sin (2F{-}2D{+}3\Omega)$$
$$+0''.000\ 011\ 21 \sin(2F - 2D + \Omega)$$
$$-0''.000\ 004\ 55 \sin(2F - 2D + 2\Omega)$$
$$+0''.000\ 002\ 02 \sin(2F + 3\Omega)$$
$$-0''.000\ 001\ 72 \sin 3\Omega - 0''.000\ 000\ 87t \sin \Omega, \tag{5.4}$$

where $\Delta\psi$ is the nutation in longitude, ε_A is the obliquity of the ecliptic referred to the ecliptic of date, and F, D, Ω are fundamental luni-solar arguments. Greenwich Sidereal Time is then $GST(t_u,t) = \theta(t_u) - E_0(t)$, and Greenwich Mean Sidereal Time is $GMST(t_u,t) = \theta(t_u) - E_{prec}(t)$.

5.3 Variations in Earth Orientation

In a nonrotating system, the torque equation is

$$\dot{\vec{H}} = \vec{L}, \tag{5.5}$$

where $\vec{L}$ is the torque and $\vec{H}$ is the angular momentum. In a reference frame rotating with angular velocity $\vec{\omega}$, this can be written as

$$\dot{\vec{H}} + \vec{\omega} \times \vec{H} = \vec{L}. \tag{5.6}$$

The angular momentum of a system of rotating particles can be expressed as the sum of two parts

$$\vec{H} = I\vec{\omega} + \vec{h}, \tag{5.7}$$

where I is the inertia tensor for matter in the volume V,

$$I = \iiint_V r^2 \rho(\vec{r}) dv, \tag{5.8}$$

dv is the differential volume element with position $\vec{r} = (r_1, r_2, r_3)$ and spatial density $\rho(\vec{r})$. In Equation (5.7), $\vec{h}$ represents the relative angular momentum due to the motions $\vec{u}(\vec{r})$ relative to the system given by

$$\vec{h} = \iiint_V \rho(\vec{r})(\vec{r} \times \vec{u}) dv. \tag{5.9}$$

Equation (5.6) then becomes

$$\frac{d}{dt}\left(I\vec{\omega} + \vec{h}\right) + \vec{\omega} \times \left(I\vec{\omega} + \vec{h}\right) = \vec{L}, \tag{5.10}$$

which is referred to as the *Liouville equation* (Munk & MacDonald, 1960). It shows mathematically that torques, variations in the relative angular momentum, or any phenomenon that acts to modify the inertia tensor will affect the angular velocity vector and consequently the orientation of the Earth.

5.4 Transforming between Reference Frames

Mathematically, the procedure to transform from a TRS to the celestial reference system at the epoch t is written

$$[CRS(t)] = Q(t)\ R(t)\ W(t)\ [TRS(t)], \tag{5.11}$$

where $Q(t)$, $R(t)$, and $W(t)$ are the transformation matrices describing the motion of the precession/nutation, the rotation of the Earth around the axis of the pole, and polar motion, respectively (see *IERS Conventions [2010]*). The parameter t, used in this and all of the following expressions, is defined by

$$t = [TT - 2000\ January\ 1,\ 12h\ TT\ in\ days]\ /36525. \tag{5.12}$$

Note that 2000 January 1.5 TT = Julian Date 2451545.0 TT.

In the following discussion of the rotation matrices to be used in the transformations, we use the notations R_1, R_2, and R_3 to indicate rotations about the x-, y-, and z-axes of the reference system, respectively. That is:

$$R_1(\theta) = \begin{bmatrix} 1 & 0 & 0 \\ 0 & \cos\theta & \sin\theta \\ 0 & -\sin\theta & \cos\theta \end{bmatrix},$$

$$R_2(\theta) = \begin{bmatrix} \cos\theta & 0 & -\sin\theta \\ 0 & 1 & 0 \\ \sin\theta & 0 & \cos\theta \end{bmatrix}, \tag{5.13}$$

$$R_3(\theta) = \begin{bmatrix} \cos\theta & \sin\theta & 0 \\ -\sin\theta & \cos\theta & 0 \\ 0 & 0 & 1 \end{bmatrix}.$$

Referring to Equation (5.11) the precession/nutation matrix $Q(t)$ can be written as

$$Q(t) = \begin{bmatrix} 1-aX^2 & -aXY & X \\ -aXY & 1-aY^2 & Y \\ -X & -Y & 1-a(X^2+Y^2) \end{bmatrix} \cdot R_3(s), \qquad (5.14)$$

with

$$a = 1/2 + 1/8(X^2 + Y^2). \qquad (5.15)$$

Angular "coordinates," X and Y of the CIP in the CRS, are provided by the conventional IAU 2000A or IAU 2000B precession/nutation models, based on geophysical and astronomical theory. It does not include what has been called *planetary precession* in the literature, which is the motion of the ecliptic caused by planetary gravitation. Before the IAU adopted this procedure in 2000 to describe the Earth's precession and nutation, the ecliptic was used as a fundamental reference plane and its motion was described by a conventional expression also based on astronomical theory. The combined precession /nutation of the Earth's pole and the ecliptic was called *general precession*.

The quantity s specifies the position of the CIO on the equator of the CIP. It is given mathematically by the expression

$$s = -XY/2 + \sum_i S_i t^j \sin(n_{1i}\ell + n_{1'}\ell' + n_{F_i}F + n_{D_i}D + n_{\Omega_i}\Omega)$$
$$+ \sum_i C_i t^j \cos(n_{1i}\ell + n_{1'}\ell' + n_{F_i}F + n_{D_i}D + n_{\Omega_i}\Omega), \qquad (5.16)$$

where the coefficients S_i, C_i, $n\ell$, $n\ell'$, n_F, n_D, and n_Ω are given in Table 5.1 for those terms with coefficients greater than 0.5 µas, and

ℓ = mean anomaly of the Moon
$= 134°.963\ 402\ 51 + 1\ 717\ 915\ 923''.217\ 800t + 31''.879200\ t_2 + 0''.05163500\ t^3$
$- 0.0002447000\ t^4,$

ℓ' = mean anomaly of the Sun
$= 357°.529\ 109\ 18 + 129\ 596\ 581''.048\ 100\ t - 0''.553\ 200\ t^2 + 0''.000\ 1360\ 0\ t^3$
$- 0''.000\ 011\ 490\ 0\ t^4,$

$F = L - \Omega$
$= 93°.272\ 090\ 62 + 1\ 739\ 527\ 262''.847\ 800t - 12''.751\ 200\ t^2 - 0''.001\ 037\ 00\ t^3$
$+ 0''.000\ 004\ 170\ 0\ t^4,$

Table 5.1 *Polynomial coefficients used to determine* s *with coefficients greater than 0.5 μas. Units are given in microarcseconds.*

i	j	Si	Ci	n_ℓ	$n_{\ell'}$	n_F	n_D	n_Ω
1	0	−2,640.73	0.39	0	0	0	0	1
2	0	0.00	94.00	0	0	0	0	0
3	0	−63.53	0.02	0	0	0	0	2
4	0	−11.75	−0.01	0	0	2	−2	3
5	0	−11.21	−0.01	0	0	2	−2	1
6	0	4.57	0.00	0	0	2	−2	2
7	0	−2.02	0.00	0	0	2	0	3
8	0	−1.98	0.00	0	0	2	0	1
9	0	1.72	0.00	0	0	0	0	3
10	0	1.41	0.01	0	1	0	0	1
11	0	1.26	0.01	0	1	0	0	−1
12	0	0.63	0.00	1	0	0	0	−1
13	0	0.63	0.00	1	0	0	0	1
14	1	0.00	3,808.65	0	0	0	0	0
15	1	−0.07	3.57	0	0	0	0	2
16	1	1.73	−0.03	0	0	0	0	1
17	2	743.52	−0.17	0	0	0	0	1
18	2	0.00	−122.68	0	0	0	0	0
19	2	56.91	0.06	0	0	2	−2	2
20	2	9.84	−0.01	0	0	2	0	2
21	2	−8.85	0.01	0	0	0	0	2
22	2	−6.38	−0.05	0	1	0	0	0
23	2	−3.07	0.00	1	0	0	0	0
24	2	2.23	0.00	0	1	2	−2	2
25	2	1.67	0.00	0	0	2	0	1
26	2	1.30	0.00	1	0	2	0	2
27	2	0.93	0.00	0	1	−2	2	−2
28	2	0.68	0.00	1	0	0	−2	0
29	2	−0.55	0.00	0	0	2	−2	1
30	2	0.53	0.00	1	0	−2	0	−2
31	3	0.00	−72,574.11	0	0	0	0	0
32	3	0.30	−23.42	0	0	0	0	1
33	3	−0.03	−1.46	0	0	2	−2	2
34	4	0.00	27.98	0	0	0	0	0
35	5	0.00	15.62	0	0	0	0	0

D = mean elongation of the Moon from the Sun

$\quad$ = $297°.850\ 195\ 47 + 1\ 602\ 961\ 601''.209\ 000\ t - 6''.370\ 600\ t^2 + 0''.006\ 593\ 00\ t^3$
$\quad - 0.000\ 031\ 690\ 0\ t^4,$

Ω = mean longitude of the ascending node of the lunar orbit

$\quad$ = $125°.044\ 555\ 01 - 6\ 962\ 890''.543100\ t + 7''.472\ 200\ t^2 + 0''.007\ 702\ 00\ t^3$
$\quad - 0''.000\ 0593\ 900\ t^4,$ and

where L is the mean longitude of the Moon.

The IAU has recommended that the precession/nutation model IAU 2000A, or the shorter version IAU 2000B for those who need a model accurate only to the level of one mas, be used to describe this motion. Software to implement these models can be found at www.iausofa.org/.

Referring again to Equation (5.11), the rotation of the Earth is given by

$$R(t) = R_3(-\theta), \tag{5.17}$$

θ being the Earth Rotation Angle at date t on the equator of the CIP. It is obtained from the conventional relationship to UT1 (Equation 5.1). The polar motion rotation in Equation (5.11) is given by

$$W(t) = R_3(-s')R_1(y)\ R_2(x), \tag{5.18}$$

where x and y are the angular polar motion coordinates.

The motion of the CIP within the TRF cannot be modeled. Instead, this motion must be observed and accounted for appropriately in the transformation between coordinate systems. The size of these motions is small, but very significant for precise transformation between reference frames. The polar coordinate information must be observed and reported. The data are available from the IERS. The IERS provides a series of files containing the latest data and predictions for the future (see www.iers.org/iers/products). s' can be approximated for the 21st century as a function of time by

$$s' = -47\ \mu\text{as}\ t. \tag{5.19}$$

5.5 Determination of Earth Orientation

Astronomical observations of polar motion have been carried out for more than 100 years. A variety of instruments and techniques have been employed during that time. Optical observations of stars were first used to establish the existence of polar motion and to continue its monitoring. Transit circles, astrolabes, and zenith telescopes were the primary optical instruments, but both visual and photographic zenith telescopes were employed until the late 1980s as a significant source of polar motion information. Doppler observations of navigational satellites were used to improve determinations of polar motion during the 1970s and 1980s. Modern techniques came into existence in the 1970s when laser ranging to artificial Earth satellites, connected element, and very long baseline interferometry (VLBI) began to make important contributions to our observational knowledge of polar motion. Currently, polar motion information is obtained principally from observations of the satellites of the GPS and VLBI.

Observations of the Earth's rotation have been made for centuries using a host of astronomical instrumentations. Ancient civilizations used the direction of the Sun's shadow. Following the invention of the telescope, this instrument was used to make observations of celestial bodies to measure time. As with polar motion, transit circles, astrolabes, and photographic zenith telescopes were used operationally for this purpose. Astronomical observations continue to be made to determine the difference between UT1 and UTC. VLBI currently provides the only true source of UT1−UTC information. Other space techniques, such as lunar laser ranging, laser ranging to satellites, and the determination of GPS satellite orbits, provide either an estimate of UT1 that depends on independent observations of polar motion using other techniques or the length of day, but not the actual Earth Rotation Angle.

Precession and nutation information historically was determined using transit circle observations. The most recent models, though, are based on VLBI observations of distant radio sources.

5.6 Earth Orientation Data

The International Latitude Service (ILS) was the first source of operational estimates of polar motion, using instrumentation consisting only of visual zenith telescopes. As more optical instrumentation became available, the ILS was transformed into the International Polar Motion Service (IPMS), which included not only the visual zenith telescopes of the ILS but also the visual and photographic telescope data of other international institutions (Yokoyama, Manabe, & Sakai, 2000). The BIH began the routine collection of polar motion data to facilitate its mission to determine UT1. Determination of UT1 relied on accurate determination of polar motion, and the BIH collected and published polar motion information to provide a consistent set of UT1 and polar motion data. The optical data from 1899 to 1992 were analyzed for Earth Rotation Parameters by Vondrak (1999). The IERS began operation in 1988 as the International Earth Rotation Service following the recommendation of the MERIT Working Group. Essentially this service combined the operation of the IPMS and the BIH into one organization that made optimal use of the most accurate astronomical observations available to provide the ICRS, the ITRS, and the means to relate them to each other. Current polar motion and UT1 data are available routinely from the IERS at www.iers.org.

References

Altamimi, Z., Rebischung, P., Métivier, L., & Xavier, C. (2016). ITRF2014: A New Release of the International Terrestrial Reference Frame Modeling Nonlinear Station Motions. *J. Geophys. Res. Solid Earth*, **121**, 6109–6131.

Aoki, S., Guinot, B., Kaplan, G. H., Kinoshita, H., McCarthy, D. D., & Seidelmann, P. K. (1982). The New Definition of Universal Time. *Astron. Astrophys.*, **105**, 359–361.

Bradley, J. (1748). A Letter to the Right Honourable George Earl of Macclesfield Concerning an Apparent Motion Observed in Some of the Fixed Stars. *Philosophical Transactions*, **45**, 1–43.

Brosche, P. (2000). Küstner's Observations of 1884–85: The Turning Point in the Empirical Establishment of Polar Motion. In S. Dick, D. McCarthy, & B. Luzum, eds., *Polar Motion: Historical and Scientific Problems*. San Francisco, CA: Astronomical Society of the Pacific Conference Series, 101–108.

Carter, M., & Carter, W. (2000). Seth Carlo Chandler Jr.: The Discovery of Variation of Latitude. In S. Dick, D. McCarthy, & B. Luzum, eds., *Polar Motion: Historical and Scientific Problems*. San Francisco, CA: Astronomical Society of the Pacific Conference Series, 109–122.

Chandler, S. C. (1891). On the Variation of Latitude, I. *Astron. J*, **11**, 59–61.

Dick, S. J. (2000). Polar Motion: A Historical Overview on the Occasion of the Centennial of the International Latitude Service. In S. Dick, D. McCarthy, & B. Luzum, eds., *Polar Motion: Historical and Scientific Problems*. San Francisco, CA: Astronomical Society of the Pacific Conference Series, 3.

Euler, L. (1765). Du mouvement de rotation des corps solides autour d'un axe variable. *Mémoires de l'académie des sciences de Berlin* **14**, 154–193.

Fricke, W., Schwan, H., & Lederle, T. (1988). *Fifth Fundamental Catalogue, Part I*. Heidelberg: Veröff. Astron. Rechen Inst.

Höpfner, J. (2000). On the Contributions of the Geodetic Institute Potsdam to the ILS. In S. Dick, D. McCarthy, & B. Luzum., eds., *Polar Motion: Historical and Scientific Problems*. San Francisco, CA: Astronomical Society of the Pacific Conference Series, 139–146.

Kaplan, G. H. (2005). *The IAU Resolutions on Astronomical Reference Systems, Time Scales, and Earth Rotation Models: Explanation and Implementation, U. S. Naval Observatory Circular 179*. Washington, DC: US Naval Observatory.

Küstner, F. (1888). Neue Methode zur Bestimmung der Aberrations-Constante nebst Untersuchungen über die Veränderlichkeit der Polhöhe. *Beobachtungs-Ergebnisse der Koniglichen Sternwarte zu Berlin*, **3**, 1–59.

Küstner, F. (1890). Über Polhöhen – Aenderungen beobachtet 1884 bis 1885 zu Berlin und Pulkowa. *Astron. Nachr.*, **125**, 273.

Lambeck, K. (1980). *The Earth's Variable Rotation: Geophysical Causes and Consequences*. Cambridge and New York, NY: Cambridge University Press.

Ma, C. & Feissel, M. (1997). *Definition and Realization of the International Celestial Reference System by VLBI Astrometry of Extragalactic Objects, International Earth Rotation Service Tech. Note 23*. Paris: Observatoire de Paris.

Malys, S., Seago, J. H., Pavlis, N. K., Seidelmann, P. K., & Kaplan, G. H. (2015). Why the Greenwich Meridian Moved. *J. of Geodesy*, **89**, 1263–1272.

McCarthy, D. (2000). Polar Motion – An Overview. In S. Dick, D. McCarthy, B. Luzum, eds., *Polar Motion: Historical and Scientific Problems*. San Francisco, CA: Astronomical Society of the Pacific Conference Series, 223.

McCarthy, D. D. & Luzum, B. J. (2003). An Abridged Model of the Precession–Nutation of the Celestial Pole. *Celest. Mech. & Dynam. Astron.*, **85**, 37–49.

Munk, W. H. & MacDonald, G. J. F. (1960). *The Rotation of the Earth: A Geophysical Discussion*. Cambridge: Cambridge University Press.

Newcomb, S. (1891). On the Periodic Variation of Latitude and the Observations with the Washington Prime Vertical Transit. *Astron. J.*, **11**, 81–82.

Pavlis, N. K., Holmes, S. A., Kenyon, S. C., & Factor, J. K. (2012). The Development and Evaluation of the Earth Gravitational Model 2008 (EGM2008). *J. Geophys. Res. Solid Earth*, **117**, B04406.

Perryman, M. A. C., Lindegren, L., Kovalevsky, J., et al. (1997). The Hipparcos Catalogue. *Astron. Astrophys.*, **323**, L49–L52.

Petit, G. & Luzum, B., eds. (2010). *IERS Conventions (2010)*. Frankfurt am Main: Verlag des Bundesamts für Kartographie und Geodäsie.

Urban, S. E. & Seidelmann, P. K., eds. (2012). *Explanatory Supplement to the Astronomical Almanac*. Mill Valley, CA: University Science Books.

Verdun, A. & Beutler, G. (2000). Early Observational Evidence of Polar Motion. In S. Dick, D. McCarthy, & B. Luzum, eds., *Polar Motion: Historical and Scientific Problems*. San Francisco, CA: Astronomical Society of the Pacific Conference Series, 67.

Vondrak, J. (1999). Earth Rotation Parameters 1899.7:1992.0 After Reanalysis within the Hipparcos Frame. *Surveys in Geophysics*, **20**, 169–195.

Yokoyama, K., Manabe, S., & Sakai, S. (2000). History of the International Polar Motion Service/International Latitude Service. In S. Dick, D. McCarthy, & B. Luzum, eds., *Polar Motion: Historical and Scientific Problems*. San Francisco, CA: Astronomical Society of the Pacific Conference Series, 147–160.

6

Ephemeris Time

6.1 Need for a Uniform Timescale

From the time of Newton the independent variable of time for all astronomical ephemerides and planetary theories had been assumed to be uniform. When it was established that the Earth's rotation, and the mean solar time derived from the Earth's rotation, were not uniform, the need arose for a new timescale to serve as the time argument for ephemerides and almanacs. Astronomers needed to provide a concept for an ideal timescale, a definition that approached the ideal, and a practical realization that made the timescale available.

In the 1930s and 1940s, the clocks providing standard time were pendulum clocks. Quartz crystal clocks were coming into use, but there were concerns about the accuracy and reliability of the clocks as the source of uniform time. There was a desire to introduce an astronomical timescale that was thought to be accurate, reliable, and uniform. Instead of using the rotation of the Earth, astronomers turned for a uniform timescale to the concept of a time-like argument based on the orbital motion of solar system bodies, and particularly to the apparent motion of the Sun that reflects the orbital motion of the Earth. This new timescale, eventually to be called *Ephemeris Time*, came into existence through a series of events. The definition of the new timescale would be given in terms of Newcomb's theory of the Sun. The practical realization would be from observations of solar system bodies.

However, this took more than 20 years to come to completion. During that time, there was significant progress in the development of new clocks based on advances in physics. The introduction and application of the theory of relativity for practical astronomical applications was slow, because generally the accuracies did not require the complexities of the theory. This would change rapidly with the advent of artificial satellites and high-speed digital computers.

6.2 Danjon Proposal

In 1948, the Comite International des Poids et Mesures (CIPM) referred a proposal
to establish a uniform fundamental time standard to the International Astronomical
Union (IAU). This proposal was considered at the International Colloquium on the
Fundamental Constants of Astronomy in Paris in 1950 (Colloque international sur
les constantes fondamentales de l'astronomie, 1950). There, suggestions were
made that led to the establishment of Ephemeris Time (ET). However, in 1929,
André Danjon, an astronomer at the Observatory of Strasbourg, had already written
a paper, basically proposing the creation of a dynamical timescale. This paper and
its contents apparently were either unknown or forgotten by the attendees of the
Paris conference, with one possible exception. Danjon, who in 1950 was director of
l'Observatoire de Paris and chairman of the meeting, made no reference to his
earlier paper. Why he did this is unknown. It could have been because he felt the
meeting was to make recommendations, and not review history. It could have been
from modesty, or a sense of *noblesse oblige.*

   In 1967, at an international colloquium on the gravitational problems of n-bodies
(Colloque sur le gravitationnel problème des n corps, 1968), M. Jacques Lévy of
l'Observatoire de Paris read a paper on the French contributions to the development
of celestial mechanics in the past three centuries. He included a statement that can
be translated as:

"It is A. Danjon who seems to have been the creator of the present ephemeris time; about
1929 he wrote 'we should not delay longer in adopting a new practical definition of time . . .
in order to fix the time by the position of the planets in their orbits it is necessary to make the
calculations in celestial mechanics according to the law of Newton. A certain future date
will be introduced in the calculations; the positions of the planets and the Moon will be
obtained for this date; when observations show that these bodies pass through the calculated
positions, the instant in question will be attained. It is thus that the bodies of the solar system
mark the time in their orbits, graduated like a sun-dial'."

Further on in the paper, Danjon continues as translated by Clemence (1971):

"Let us restrict ourselves to wishing that some day we may discover a good terrestrial
standard of time, and leave these difficulties of pure logic, since also we are reassured about
their practical repercussions. The dreaded objection falls before results of observation; no
contradiction is to be feared, since the Moon and the planets give the same time. As the time
so defined is furnished by the law of Newton, it is called Newtonian time; until experiments
prove the contrary this time will be called absolute. How do we read the time on the dial of
the planetary orbits?"

   "There everything is yet to be done, or nearly everything. Of course we shall continue to
employ terrestrial time derived from meridian passages of stars, perhaps as a first approximation, or perhaps as a means of interpolation, or rigorously for the needs of geodesy and
navigation. But for the other applications, celestial mechanics, the study of spectroscopic
binaries and of short period variables, *etc.* we shall correct it by the differences between

terrestrial time and Newtonian time, the latter being derived from the observations of the Moon, the Sun, and the interior planets."

Clemence did not discover this information until 1970, and he wrote a paper on "The Concept of Ephemeris Time: A Case of Inadvertent Plagiarism" (Clemence, 1971), from which the preceding information is taken.

6.3 Clemence Proposal

In 1948, Clemence, in a paper on astronomical constants that included some pages on time measurement, wrote:

"The time used in the national ephemerides is the kind of time that satisfies the equations of celestial mechanics, often called Newtonian time, while the time used by astronomers in practice depends on the variable rotation of the earth. The use of one kind of time for the theory and another for the observations produces discrepancies between the observation and theory which are proportional to the mean motions of the objects observed . . . the solar and planetary theories give the coordinates of the Sun and planets as functions of Newtonian time, whereas the observed positions are functions of variable time; hence the theoretical coordinates differ from the observed . . . It therefore seems logical to continue the use of mean solar time . . . for civil purposes, and to introduce Newtonian time for the convenience of astronomers and other scientists only . . . Astronomers will continue to set their clocks to the mean solar time of the Greenwich meridian . . . and to publish their observations on this standard of reference. Whenever an observation is to be compared with a theoretical position, the time of the observation will first be corrected to Newtonian time. The corrections to Universal time needed to reduce it to Newtonian time can be published in the national ephemerides."

In that paper, Clemence (1948) notes the earlier work of Spencer Jones (1939) in which Spencer Jones determined corrections to Newcomb's tabular longitude of the Sun (1895) that are required to match actual observations. These corrections are due to the fact that the Earth's variable rotational speed produces a nonuniform timescale that is not suitable for the precise dynamical description of the Sun's motion. He then went on to note that using these corrections has the effect of making the day defined by his "Newtonian" time equivalent to a mean solar day of 1900. This is not correct since Newcomb's Tables, while providing the solar longitude in terms for the year 1900, are actually based on much earlier observations of the Sun, and so the timescale in Newcomb's work more closely resembles the mean solar second of the mid-19th century (Nelson et al., 2001).

   At the International Colloquium on the Fundamental Constants of Astronomy in Paris in 1950, Clemence presented his paper and pointed out that an inconsistent mean solar second presented a problem in providing a consistent set of astronomical constants. He proposed that the national ephemerides routinely provide tabular values to apply to the astronomical Universal Time as observed (based on the mean solar time) in order to obtain what he called "Universal Time (1900)" which he considered to be compatible with the mean solar second of 1900. He, therefore, proposed for discussion the proposition that the mean solar second of 1900 be used in all cases where the mean solar second of date was not appropriate. In the formal write-up of his paper in *Colloque international sur les constantes fondamentales de l'astronomie* (1950), Clemence acknowledges Spencer Jones for pointing out that what Clemence had called Universal Time (1900) is not the time at any one epoch but is more like the mean time during the 19th century. Spencer Jones in his paper offered the opinion that it would be better to "specify the correction required to astronomical time, which is what is needed in investigations in dynamical astronomy, rather than to base everything on the length of the mean solar second at one particular epoch, because the data are not sufficiently accurate to permit of a precise specification of the length at a particular epoch."

In discussions at that meeting Clemence (1971) recalled that Spencer Jones made what Clemence considered to be the first suggestion that an atomic frequency along with a suitable means of integrating the measures might be used to provide time. Consistent with his 1929 paper, Danjon suggested that the motion of the Earth around the Sun might provide a more suitable definition of time than the solar day and proposed the term "dynamical time." D. Brouwer proposed the name "Ephemeris Time." Recommendation 6 of the conference read:

"The Conference recommends that, in all cases where the mean solar second is unsatisfactory as a unit of time by reason of its variability, the unit adopted should be the sidereal year at 1900.0; that the time reckoned in these unit [*sic*] be designated *Ephemeris Time*; that the change of mean solar time to Ephemeris Time be accomplished by the following correction

$$\Delta t = 24.349s + 72.3165sT + 29.949sT^2 + 1.821B,$$

where *T* is reckoned in Julian centuries from 1900.0 January 0 Greenwich Mean Noon and *B* has the meaning given by Spencer Jones in *Monthly Notices R.A.S.* (vol 99, 1939, p.541) and that the above formula define also the second.

No change is contemplated or recommended in the measure of Universal Time, nor in its definition."

6.4 Adoption and Definition

This 1950 Paris recommendation was adopted by the IAU Commission on Time (*Trans. Int. Astron. Union*, 1954) at the IAU General Assembly in Rome in 1952 with a slight modification of the numerical expression to read:

$$\Delta T = 24.349s + 72.318sT + 29.950sT^2 + 1.82144B. \tag{6.1}$$

However, Danjon later pointed out that the tropical year is a better choice than the sidereal year, because, unlike the sidereal year, the tropical year's length is independent of the adopted value of precession. Therefore, the 10th meeting of the CGPM in 1954, following an earlier recommendation of the CIPM, proposed the following definition of the second: "The second is the fraction 1/31 556 925.975 of the length of the tropical year for 1900.0" (*Trans. Int. Astron. Union*, 1957). The numerical value of the fraction was based on Newcomb's formula for the geometric mean longitude of the Sun for the epoch of January 0, 1900, 12h UT (Newcomb, 1895) given by

$$L = 279° \ 41' \ 46''.04 + 129 \ 602 \ 768''.13T + 1''.089T^2, \qquad (6.2)$$

where T is the time reckoned in Julian centuries of 36,525 days. From the value of the linear coefficient in Newcomb's formula, the tropical year of 1900 would then contain $[(360 \times 60 \times 60)/129 \ 602 \ 768.13] \times 36 \ 525 \times 86 \ 400 = 31 \ 556 \ 925.975$ s. This was subsequently adopted by the IAU at the General Assembly in 1955. Once again, Danjon offered a later comment, noting that the fraction ought to have a slightly more precise value in order to provide exact numerical agreement with Newcomb's formula. Consequently, in 1956, the CIPM adopted the slightly more precise value with the words: "La second est la fraction 1/31 556 925.9747 de l'année tropique pour 1900 janvier 0 a 12 heures de temps des ephemerides." The Comite also established the Comite Consultatif pour la Définition de la Seconde (CCDS) to coordinate the work of physicists on atomic standards and of astronomers on the astronomical standard of Ephemeris Time (Procès Verbaux des Séances, 1957). The ephemeris second, as a fraction of the tropical year, was formally adopted by the 11th CGPM in 1960. Finally, Ephemeris Time was adopted by the 10th General Assembly of the International Astronomical Union in Moscow in 1958 with the following statement: "Ephemeris Time is reckoned from the instant, near the beginning of the calendar year A.D. 1900, when the geometric mean longitude of the Sun was 279° 41' 48".04, at which instant the measure of Ephemeris Time was 1900 January $0^d \ 12^h$ precisely."

With these definitions, Ephemeris Time was equivalent to the system of time in Newcomb's Tables of the Sun. Thus, ephemerides determined from Newcomb's Tables of the Sun and planets could be considered to have Ephemeris Time as the independent argument. The mean longitude of any planet, or the Moon, could have been used to define the epoch and rate of a uniform time system, and Ephemeris Time could be used as the independent variable in their tables and theories. The definition of the ephemeris second was the first official definition of the second as a unit of measure.

6.5 Observational Determination

The recommendations provided a basis for a new, more uniform timescale, but its usefulness depended on the means to access this timescale. In principle, it would be possible to determine the position of the Sun in the sky with respect to a star catalog, determine the time in the adopted solar ephemeris when it should have been in that position, and then call that time the Ephemeris Time of the observation. This time could then be compared with the time determined from astronomical observations of the Earth's rotation angle, Universal Time (UT), to determine the corrections to be applied to the Earth's rotational time in order to provide the more uniform timescale. Although Ephemeris Time was based on the ephemeris of the Sun, observing the Sun directly is difficult, particularly with the precision needed to define a timescale. The faster-moving and more precisely observed Moon was much more useful in determining Ephemeris Time from real observations.

Observations of the Moon with respect to star catalogs could be made using transit circles in the traditional manner. Timed observations of occultations of stars by the Moon also provided valuable ΔT ($= ET - UT$) data. A new instrument was developed in the 1950s by William Markowitz that was designed to determine positions of observing sites on the Earth, but found useful application in the determination of Ephemeris Time (Markowitz, 1954, 1961). It was designed to photograph the Moon against a stellar background using a rotating filter to keep the position of the Moon on the photographic plate constant during the exposure. Figure 6.1 shows the camera along with Markowitz. The observations provided a position of the Moon in the reference frame of the star catalog, whose star positions were used in the analysis of the lunar position.

Figure 6.1 Markowitz moon camera

Many difficulties remained with determining Ephemeris Time from the Moon. There were problems with the ephemeris of the Moon, specifically the secular acceleration value, the completeness of the gravitational theory of its motion, the irregularity of the limb profile, and the implied position of the equinox in the lunar theory.

For observations of the Moon, the accuracy depends on the empirical value of the tidal secular acceleration in the mean longitude of the Moon included in the lunar ephemeris. Prior to lunar laser ranging data, Spencer Jones's value of $-22''.4$ /cy^2 was used; modern investigations use a value of $-25''.97$ /cy^2 (Williams & Boggs, 2016). Distinctions are also necessary concerning the lunar ephemerides used in the analyses. As we have seen, in the period 1960 to 1984, improvements were introduced into the lunar theory. The resulting values of ΔT could differ slightly depending on the Ephemeris Time used.

Observations of the Moon also required small corrections for the irregular surface features called "limb corrections." These features introduced errors in observations of the position of the Moon with transit circles and the timings of occultations of stars. These irregularities are responsible for the diamond ring and Bailey's beads effects during solar eclipses. C. B. Watts (1963) published his lunar limb charts so that occultation observations could be corrected for the irregular limb of the Moon. These charts could provide corrections of as large as 6 s in the determinations of Ephemeris Time.

In addition, Brown's (1897–1908) lunar theory, the conventional basis for the lunar ephemeris of the 1950s, could not be used directly to determine Ephemeris Time, because the theory was not strictly gravitational or completely in accord with Newcomb's tables. This was corrected by removing the great empirical term from the lunar mean longitude, and correcting the tabular mean longitude by

$$\Delta L = 8''.72 - 26''.74T - 11''.22T^2, \tag{6.3}$$

where T is measured in centuries of 36,525 ephemeris days from 1900.0. Additional corrections were required for some of the periodic terms in longitude, latitude, and parallax. Beginning in 1960, the lunar ephemeris was calculated directly from an amended version of Brown's theory, not from the tables. The *Improved Lunar Ephemeris* (1954) provided the information for dates from 1952 through 1959 using the same procedure. From 1960 to 1984, various improvements were introduced into the lunar theory. Different versions were designated as $j = 0$, 1, or 2, and the corresponding Ephemeris Times were designated ET0, ET1, or ET2.

Systematic differences in the positions of the equinox as realized in the theory of the Sun, the lunar theory, and the star catalogs became a problem, leading to a number of investigations, corrections, and papers concerning the equinox. For example, it was recognized that the origin in right ascension of the FK4 star catalog (Fricke & Kopff, 1963) was not the same at all declinations, and different from the equinox of the almanac ephemerides of the Sun, the Moon, or numerically integrated ephemerides. The terms *catalog equinox* and *dynamical equinox* were introduced to distinguish between the two types of equinoxes. Corrections were introduced for the equinox in the lunar theory to determine Ephemeris Time. Equinox position and motion corrections were introduced between the FK4 and FK5 (Fricke et al., 1988) star catalogs.

When ET was first instituted, observations of stars using visual and photographic zenith tubes, transit circles, and astrolabes were used to provide Universal Time based on the variable rotation of the Earth. In more recent times, observations using lunar laser ranging and very long baseline interferometry (VLBI) observations that began in 1969 have been employed to measure Universal Time.

For dates before telescopic observations were available, observations of eclipses have been able to provide the long-term behavior of ΔT. For observations of total or near-total solar eclipses without accurate timing information, it is still possible to determine ΔT. We can calculate the path of totality on the Earth's surface for any total solar eclipse in the past, assuming a constant rotational speed for the Earth. If an observer at that time is able to report that the eclipse actually occurred in a different location, we can interpret that geographical difference as being due to the fact that our assumption of the Earth's rotation angle, based on a constant rotational speed, was in error. The difference in the longitude between the predicted and actual locations of the eclipse observation can then be used to determine the value of ΔT that would be required to see the eclipse where it actually occurred. Figure 6.2 shows the method graphically.

6.6 The Ephemeris Second and Atomic Time

Until 1960, the second was defined as 1/86400 of the mean solar day, ignoring the variability in the Earth's rotation and assuming that the Earth rotation was uniform. In 1960, the ephemeris second was introduced as the replacement for the second defined in terms of mean solar time. The ephemeris second was defined as 1/31556925.9747 of the tropical year for 1900 January 0^d 12^h ET, as determined from Newcomb's theory of the Sun. This definition was more precise, but it was difficult to realize from observations and to implement using operational clocks.

By 1960, Markowitz, Hall, Essen, and Perry (1958) had already determined the value of the atomic second in terms of the ephemeris second. So in 1967, the Système International (SI) second was introduced as the duration of 9,192,631.770 periods of the radiation corresponding to the transition between

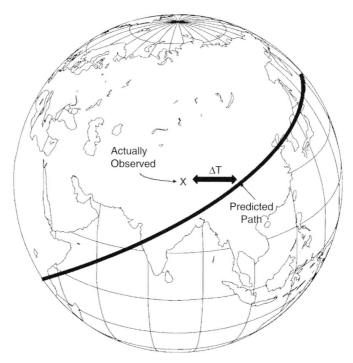

Figure 6.2 Determining ΔT from ancient eclipses

two hyperfine levels of the ground state of the caesium 133 atom. This numerical value is such that the SI second is equal to the value of the ephemeris second, as determined from the lunar observations from 1956 to 1965. This corresponds approximately to the mean solar second of the mid-19th century. The ephemeris second had only lasted as the definition of the second for seven years.

The availability of operational atomic timescales beginning in July 1955 meant that Ephemeris Time could be determined from atomic timescales much more accurately than from solar system observations. Ephemeris Time could thus be made readily available. The atomic time second had been defined to be equal to the Ephemeris Time second through the work of Markowitz and his colleagues (1958). The actual realization of atomic timescales, however, was accomplished independently at a number of laboratories around the world. In 1967, the CCDS standardized atomic timescales by establishing the origin of International Atomic Time (TAI) to be in approximate agreement with 0^h UT2 on January 1, 1958 (*BIPM Com. Cons. Déf. Seconde*, 1970). In doing this, the CCDS followed the recommendation of IAU Commissions 4 (Ephemerides) and 31 (Time) in 1967. The result was that, since the atomic time second differed from the variable Universal Time second, atomic time immediately began to deviate from Universal Time. The offset between TAI and Ephemeris Time, and its dynamical time successors, is now

established as 32.184 seconds. Consequently we can effectively extend the definition of Ephemeris Time into the era of atomic time using the relationship:

$$ET = TAI + 32.184s. \tag{6.4}$$

The assumption is generally made that a timescale based on physical phenomena
on Earth does not differ nonlinearly from a timescale based on solar system
dynamics. We will see that atomic clocks will become a source of accurate and
reliable time and frequency, which will be the basis for many timescales and the
most accurate physical measurements.

6.7 Historical ΔT

Historical records of the time-varying ΔT provide valuable information about long-
term variations in the Earth's rotation. Brouwer (1952) analyzed lunar occultation
observations to determine values of ΔT from 1820 to 1950. He detected a change in
the length of day of about two milliseconds per day per century during that interval
and decadal fluctuations of several milliseconds per day. McCarthy and Babcock
(1986) used lunar occultations for the early period, and lunar laser ranging and
VLBI observations, when available, to determine values of ΔT from 1657 to 1984.

F. R. Stephenson and L. V. Morrison have investigated extensively the historical
values of ΔT back to ancient times (Morrison & Stephenson, 2001, 2004, 2005;
Stephenson, 1997, 2003, 2007a, 2007b; Stephenson & Morrison, 1984, 1994;
Stephenson, Morrison, & Hohenkerk, 2016). In the latest publication,
Stephenson, Morrison, and Hohenkerk analyzed measurements of the Earth's
rotation from 720 BC to AD 2015, including timed and untimed eclipse and
lunar occultation observations. They find a change in the length of day (LOD) of
an average rate of increase of 1.8 ms per century, which is less than the predicted
tidal friction value of 2.3 ms per century. This is thought to be due to an acceleration
of $1.5 \pm 0.4 \times 10^{-22}$ rad s^{-2}, possibly due to a combination of post-glacial rebound
and core–mantle coupling. They find fluctuations on time scales of decades to
centuries, with an indication of an oscillation with a period of about 1,500 years.
Figures 6.3 and 6.4 show the results of their analysis. The impact of these
observations on studies of the Earth's rotation is discussed in Chapter 5.

6.8 Problems with Ephemeris Time

Ephemeris Time was introduced as the time argument for ephemerides and astronomical almanacs in 1960. It was quickly apparent that atomic time provided
a more accurate and available source of Ephemeris Time than solar system observations. It was also soon recognized that there were difficulties concerning the

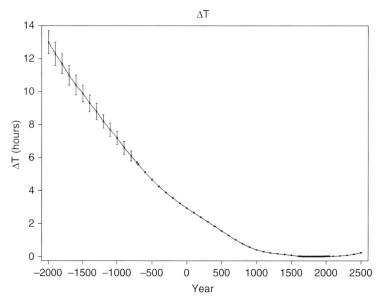

Figure 6.3 ΔT from 2000 BC to AD 2500 from data in Stephenson, Morrison, and Hohenkerk (2016).

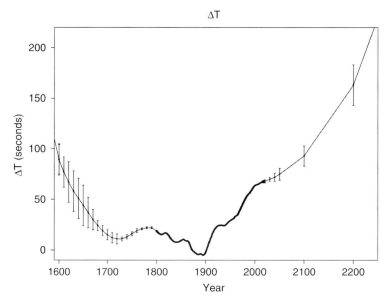

Figure 6.4 ΔT from AD 1600 to AD 2300 from data in Stephenson, Morrison, and Hohenkerk (2016)

definition based on the theory of the Sun and the actual determinations based on observations of the Moon. The ephemerides also had problems accurately modeling the effects of tidal friction.

The definition of the second of Ephemeris Time was specified for 1900, based on Newcomb's theory, but the observational data to which the theory had been fit dated from the 1700s and 1800s. So the ephemeris second corresponds roughly to the solar second for about 1820. In addition, the definition depends on values that Newcomb adopted for the aberration and precession constants. Hence, any improvements in the astronomical constants and reference systems would theoretically change the definition of Ephemeris Time. In practice, however, this was a conventionally adopted definition and any such modifications to try to model a true mean solar second of 1900 were never considered.

No specified relationships between the versions of Ephemeris Time could be determined from the observations of different solar system bodies, i.e. the Sun, Moon, Mercury, or Venus. Each determination depended on the constants used in the ephemerides and the analyses of the observations. Inevitably further, more accurate observations of the motions of solar system bodies would lead to improvements of the ephemerides and a resulting change in the realizations of Ephemeris Time.

Beside these technical problems with the definition and realization of Ephemeris Time, there was a fundamental difficulty in the availability of the ephemeris second and Ephemeris Time for practical applications. There was, of necessity, a process of observation, analysis, and the comparison with ephemerides in order to determine either Ephemeris Time or the ephemeris second. This also meant that both could only be determined to a low level of precision. Atomic time would provide a much more available and precise timescale and second. At the same time that Ephemeris Time was introduced, atomic timescales were becoming more available, and the coordination of radio time signals using Universal Time was beginning to be more widespread. So the standard timescales were being steered, or stepped, to agree with Earth rotation times without consideration of the less available Ephemeris Time.

The timing of the introduction of Ephemeris Time in 1960 was additionally unfortunate, as this was the beginning of the space age, the introduction of digital computers and a number of scientific developments requiring improvements in accuracies. Ephemerides based on numerical integrations, rather than analytical theories, were being developed, made possible by expanding computer capabilities. The astronomical constants and solar system theories of Newcomb were now more than 60 years old and in need of improvement, which could, in theory, change Ephemeris Time.

Even more important, the theory of relativity had not really been applied to celestial mechanics and timekeeping so far, but the accuracies being sought and achieved required the introduction of relativistic corrections. The equations of motion being used for numerical integrations of all the bodies of the solar system were beginning to include terms for relativity. It was recognized that the Lorentz transformations of general relativity distinguished between timescales at the solar system barycenter, the geocenter, and the Earth's geoid. Ephemeris Time was not specific as being barycentric or geocentric, and was used in both cases, without distinction. Ephemeris Time was neither a proper nor a coordinate time according to relativistic theory. At IAU Colloquium 9 in Heidelberg, Germany in August 1970, Irwin I. Shapiro, then of Massachusetts Institute of Technology, said: "The current definition of Ephemeris Time is philosophically repugnant, aesthetically horrifying, and completely inadequate" (Mulholland, 1972). However, it would take until 1976 for the International Astronomical Union to reach agreement to introduce improvements and changes. A number of changes became effective in 1984, including new dynamical timescales to replace Ephemeris Time.

6.9 Relativity

Prior to 1960, in much of the astronomical community the accuracies of ephemerides, star catalogs, timescales, and observations did not require the application of the theory of relativity. So, although Einstein had published his paper on relativity in 1905, it had not been applied in most cases by the 1960s. In addition, there were some competing theories, and questions concerning the application of the theory. The result was the development of the parameterized post-Newtonian (PPN) formulizations of the theory of relativity (Will, 1974). PPN introduced variable parameters that could be solved for or set to specific values according to the adopted assumptions. PPN was introduced by ephemerides developers in their numerical integrations in the 1960s and thereafter. Thus, they also had to introduce relativistic transformations for the reference frames and timescales of the observations.

The 1970s saw the formation of a number of IAU Working Groups to develop resolutions for changes in the reference system, fundamental star catalogs, planetary masses, astronomical constants, theory of nutation, and timescales. The decision was reached to introduce all the changes at the same time so there would be a clean break between the old and the new. The resolutions were adopted at the 1976 IAU General Assembly in Grenoble, France, and were to be introduced in astronomical publications in 1984 (Kaplan, 1982). Relativity was introduced, although with some reluctance to commit completely to the Einstein theory of relativity.

6.10 Dynamical Timescales

The improving observational accuracy pointed out the need to replace Ephemeris Time with other times that were conceptually based on solar system dynamics and the theory of relativity, properly defined, and realizable as the time arguments for ephemerides and almanacs. There was also the requirement to somehow extend timescales for times before 1955, when atomic time first became available, to be used in investigations of observations prior to 1955. Therefore, two new time-like arguments were introduced in 1984. These were Barycentric Dynamical Time (TDB) and Terrestrial Dynamical Time (TDT). The acronyms are given according to their French names. These time-like arguments were to be continuations of Ephemeris Time, but defined to correspond with their respective reference frames, either the geocenter or the barycenter. To achieve the continuity, TDT was specified with respect to TAI with an offset and rate to match ET. It was specified that TDB would only differ from TDT by periodic terms, so the epoch and rate of the two would agree. This specification would prove impossible to satisfy accurately, and naming difficulties would arise, as seen in Chapter 7.

References

BIPM Com. Cons. Déf. Seconde (1970). **5**, 21–23. Reprinted in B. E. Blair, ed., *Time and Frequency: Theory and Fundamentals*, Natl. Bur. Stand. (U.S.) Monograph 140, Washington, DC: US Government Printing Office.

Brouwer, D. (1952). A Study of the Changes in the Rate of Rotation of the Earth. *Astron. J.*, **57**, 125–146.

Brown, E. W. (1897–1908). Theory of the Motion of the Moon. Containing a New Calculation of the Expressions for the Coordinates of the Moon in Terms of the Time. *Mem. R. Astron. Soc.*, **53**, 39–116, 163–202; **54**, 1–63; **57**, 51–145; **59**, 1–103.

Clemence, G. M. (1948). On the System of Astronomical Constants. *Astron. J*, **53**, 169–179.

Clemence, G. M. (1971). The Concept of Ephemeris Time: A Case of Inadvertent Plagiarism. *J. History of Astronomy*, **2**, 73–79.

Colloque sur le gravitationnel problème des n corps. Paris, France, 16–18 Août, 1967. (Conference on the Gravitational N-Body Problem, Paris, France, August 16–18, 1967) (1968). Paris, Publication du Centre National de la Recherche Scientifique.

Colloque international sur les constantes fondamentales de l'astronomie (1950). In *Bull. Astron.*, **15**, 163–292.

Danjon, A. (1929). Le temps, sa définition pratique, sa mesure. *L'astronomie*, **43**, 13–22.

Essen, L. & Perry, J. V. L. (1957). The Cesium Resonator as a Standard of Frequency and Time. *Phil. Trans R. Soc. London Series A*, **250**, 45–69.

Fricke, W., Kopff, A., in collaboration with Gliese, W., Gondolatsch, F., Lederle, T., Nowacki, H., Strobel, W., & Stumpff, P. (1963). Fourth Fundamental Catalogue (FK4). *Veröff. Astron. Rechen-Inst. Heidelberg*, **10**.

Fricke, W., Schwan, H., Lederle, T. in collaboration with Bastian, U., Bien, R., Burkhardt, G., du Mont, B., Hering, R., Jährling, R., Jahreiß, H., Röser, S., Schwerdtfeger, H. M., & Walter, H. G. (1988). Fifth Fundamental Catalogue (FK5). Part I. The Basic Fundamental Stars. *Veröff. Astron. Rechen-Inst. Heidelberg*, **32**.

Improved Lunar Ephemeris 1952–1959 (1954). Washington, DC: US Government Printing Office.

Kaplan, G. H. (1982). *The IAU Resolutions on Astronomical Constants, Time Scales, and the Fundamental Reference Frame*. Washington, DC: US Naval Observatory Circular 163.

Markowitz, W. (1954). Photographic Determination of Moon's Position, and Applications to the Measure of Time, Rotation of the Earth, and Geodesy. *Astron. J.*, **59**, 69–73.

Markowitz, W. (1961). The Photographic Zenith Tube and the Dual-Rate Moon-Position Camera. In G. P. Kuiper & B. M. Middlehurst, eds., *Telescopes, Stars and Stellar Systems*. Chicago, IL: University of Chicago Press, 88.

Markowitz, W., Hall, R. G., Essen, L., & Perry, J. V. L. (1958). Frequency of Cesium in Terms of Ephemeris Time. *Phys. Rev. Lett.* **1**, 105.

McCarthy, D. D. & Babcock, A. (1986). The Length of Day since 1656. *Physics of the Earth and Planetary Interiors*, **44**, 281–292.

Morrison, L. V. & Stephenson, F. R. (2001). Historical Eclipses and the Variability of the Earth's Rotation. *J. Geodynamics*, **32**, 247–265.

Morrison, L. V. & Stephenson, F. R. (2004). Historical Values of the Earth's Clock Error ΔT and the Calculation of Eclipses. *J. History of Astronomy*, **35**, 327–336.

Morrison, L. V. & Stephenson, F. R. (2005). ADDENDUM: Historical Values of the Earth's Clock Error. *J. History of Astronomy*, **36**, 339.

Mulholland, J. D. (1972). Measures of Time in Astronomy. *Pub. Astron. Soc. Pacific*, **94**, 357–364.

Nelson, R. A., McCarthy, D. D., Malys, S., et al. (2001). The Leap Second: Its History and Possible Future. *Metrologia*, **38**, 509–529.

Newcomb, S. (1895). *Astronomical Papers of the American Ephemeris and Nautical Almanac*, **VI**, Part I: Tables of the Sun. Washington, DC: US Government Printing Office, 9.

Proces Verbaux des Seances, deuxieme serie (1957). **25**, 77.

Spencer Jones, H. (1939). The Rotation of the Earth and the Secular Acceleration of the Sun, Moon, and Planets. *Monthly Notices Roy. Astron. Soc.*, **99**, 541.

Stephenson, F. R. (1997). *Historical Eclipses and Earth's Rotation*. Cambridge, UK: Cambridge University Press.

Stephenson, F. R. (2003). Harold Jeffreys Lecture 2002: Historical Eclipses and Earth's Rotation. *Astronomy and Geophysics*, **44**, 2.22–2.27.

Stephenson, F. R. (2007a). Babylonian Timings of Eclipse Contacts and the Study of the Earth's Past Rotation. *J. History of Astronomy*, **9**, 145–150.

Stephenson, F. R. (2007b). Variations in the Earth's Clock Error ΔT between AD 300 and 800 as Deduced from Observations of Solar and Lunar Eclipses. *J. History of Astronomy*, **10**, 211–220.

Stephenson, F. R. & Morrison, L. V. (1984). Long-Term Changes in the Rotation of the Earth: 700 B.C. to A.D. 1980. In R. Hide, ed., *Rotation in the Solar System* (London, UK, 1984); reprinted in *Phil. Trans. Royal Soc London, A*, **313**, 47–70.

Stephenson, F. R. & Morrison, L. V. (1994). Long-Term Changes in the Rotation of the Earth: 700 B.C. to A.D. 1990. *Phil. Trans. Royal Soc London, A*, **351**, 165–202.

Stephenson, F. R., Morrison, L. V., & Hohenkerk, C. Y. (2016). Measurements of Earth Rotation 720 BC to AD 2015, *Proceedings of Royal Society A*, **472**, 20160404.

Trans. Int. Astron. Union, Vol. VIII, Proc. 8th General Assembly, Rome, 1952. (1954). P. T. Oosterhoff, ed., New York, NY: Cambridge University Press, 66.

Trans. Int. Astron. Union, Vol. IX, Proc. 9th General Assembly, Dublin, 1955. (1957).
P. T. Oosterhoff, ed., New York, NY: Cambridge University Press, 451.

Trans. Int. Astron. Union, Vol. X B, Proc. 10th General Assembly, Moscow, 1958. (1960).
D. H. Sadler, ed. Cambridge, UK: Cambridge University Press.

Watts, C. B. (1963). The Marginal Zone of the Moon. *Astronomical Papers for the American Ephemeris and Nautical Almanac*, vol. **17**. Washington, DC: US Government Printing Office.

Will, C. M. (1974). *Experimental Gravitation*, B. Bertotti ed. New York, NY: Academic Press.

Williams, J. G. & Boggs, D. H. (2016). Secular Tidal Changes in Lunar Orbit and Earth Rotation. *Celest. Mech. & Dynam. Astron.*, **126**, 88–129.

7

Relativity and Time

7.1 Newtonian Reference Systems

Although Einstein's theory of relativity was published in 1905 and a number of consequences of the theory were predicted, 50 years later it was not being applied for timekeeping or reference systems. The accuracies at that time just did not require the complications of the theory. Special relativity might have been applied in restricted circumstances, but Newtonian gravitational theory was adequate, and it was easily understood.

However, in the 1960s, the accuracies were improving due to technology developments and the requirements of the space age. In 1976, the International Astronomical Union introduced relativistic concepts for time and the transformations between timescales. Reference systems based on the theory of relativity were not introduced until 1991.

In Newtonian systems, we accept the Newtonian universal law of gravity and mechanics. Time is an independent absolute progression in terms of some unit of time. The coordinate system is in the framework of Euclidean geometry. All systems of coordinates are equivalent, or interchangeable, as long as they move with a constant velocity. Transformations between timescales and coordinate systems are linear functions or angular rotations. The concepts are generally easily visualized.

7.2 Special Relativity

In 1905, Einstein proposed the theory of special relativity to explain several effects that seemed to contradict Newtonian physics and mechanics. It deals with motions in inertial reference frames. A reference frame is inertial if a free particle that is not subject to any external force moves without acceleration with respect to the frame. In Newtonian mechanics, such free particles move

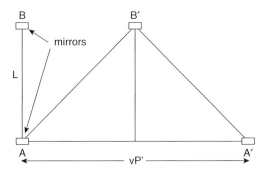

Figure 7.1 Time dilation

with constant speed along straight lines. In the Galilean principle of relativity, if a reference frame is inertial, all reference frames that move with a constant speed with respect to that reference frame are also inertial. Einstein generalized this principle by stating that all inertial frames are totally equivalent for the performance of all physical experiments.

In 1887, Michelson and Morley (1887) tested the law of addition of velocities. It was thought that light was propagated through ether, so a test was devised to determine the orbital velocity of the Earth with respect to the ether, by measuring the velocity of light along the Earth's motion, opposite to it, and perpendicular to it. The speed was found to be the same in all directions. The conclusion was that there was no ether and no velocity could be larger than the speed of light. Thus, light is seen to be traveling with a constant speed, whether the source and/or the observer are fixed or moving. The independence of light velocity on the motion of the source can also be demonstrated using binary star observations.

This development has implications for precise timing. Referring to Figure 7.1, if we have mirrors at A and B separated by length L, in a reference system (S) and a light pulse repeatedly reflected between them, an observer at A with a clock would expect to measure the periodic return of the pulse with a period, P, equal to the distance the light pulse has to travel divided by the speed of light or $P = 2L/c$, where c is the speed of light in vacuum.

Another observer in a system (S′), moving with constant speed v from A to A′ perpendicular to the line L connecting the mirrors, will also observe a periodic response with a period equal to the distance the light travels divided by the speed of light. However, in this case, we would expect the light to return after being reflected at B′ to A′. Applying the Pythagorean Theorem, the distance traveled will be:

$$d = 2\sqrt{\left(\frac{vP'}{2}\right)^2 + L^2}.$$ (7.1)

We would then expect to observe a response with period:

$$P' = \frac{d}{c} = \frac{2}{c}\sqrt{\left(\frac{vP'}{2}\right)^2 + L^2}.$$ (7.2)

From this we find:

$$P' = \frac{2L}{c}\frac{1}{\sqrt{1 - v^2/c^2}}.$$ (7.3)

Then the apparent period for the moving observer is $P' = P\,\gamma$, with:

$$\gamma = \frac{1}{\sqrt{1 - v^2/c^2}}$$ (7.4)

Thus, the moving clock ticks slower than the fixed clock by $1/\gamma$. Due to the fixed speed of light, a time interval measured in a moving frame is longer than the same time interval measured in a frame fixed with respect to the events. This phenomenon is known as *time dilation*. A similar fundamental result is the contraction of moving objects. A rod in motion relative to an observer will have its length contracted along the direction of motion as compared to an identical rod that is stationary. If L is the length of the rod at rest, the apparent length of the rod moving with velocity v is $L' = L/\gamma$.

7.3 Lorentz Transformations

The relationships just discussed were formalized by Lorentz into transformations between spatial coordinates $\mathbf{r}$ (x,y,z) and time t in a reference frame S and $\mathbf{r}'$ (x',y',z') and time t' in a reference frame S′ moving with a constant velocity $\mathbf{V}$ with respect to S.

The vector form of the Lorentz transformation is:

$$t' = \gamma\left(t - \mathbf{V}\cdot\mathbf{r}/c^2\right),$$

$$\mathbf{r}' = \mathbf{r} - \left[\gamma t - (\gamma - 1)\frac{\mathbf{V}\cdot\mathbf{r}}{|\mathbf{V}|^2}\right]\mathbf{V},$$ (7.5)

where γ is given by Equation (7.4). The inverse formulas are:

$$t = \gamma(t' + \mathbf{V} \cdot \mathbf{r}'/c^2),$$

$$\mathbf{r} = \mathbf{r}' + \left[\gamma t' + (\gamma - 1)\frac{\mathbf{V} \cdot \mathbf{r}'}{|\mathbf{V}|^2}\right]\mathbf{V}. \qquad (7.6)$$

If we assume the velocity $\mathbf{V}$ is along the x axis, $\mathbf{V} = (v_x, 0, 0)$, then the y and z components are identities. The terms independent of γ vanish and Equation (7.5) becomes:

$$
\begin{aligned}
t' &= \gamma\left(t - \frac{v_x x}{c^2}\right), \\
x' &= \gamma(x - v_x t), \\
y' &= y, \\
z' &= z,
\end{aligned}
\qquad (7.7)
$$

the form in which the Lorentz transformation is usually given.

The space coordinates are x, y, and z. ct has the dimension of length, and t is then called *coordinate time*. So in special and general relativity, an event can be specified in four coordinates, three spatial and one time coordinate.

If Δt and Δx, Δy, Δz are the coordinate differences between two events, P and Q, the quantity Δs^2 defined by

$$\Delta s^2 = -c^2\Delta t^2 + \Delta x^2 + \Delta y^2 + \Delta z^2 \qquad (7.8)$$

is invariant under the change of coordinates. Δs is the interval between the two events, and Equation (7.8) defines the Minkowski metric. This shows that the interval is a quantity that can be associated with two events without referring to any particular coordinate system. The quantity Δs^2 is a generalized distance called the *invariant interval*. It is usually given as an infinitesimal displacement and written as:

$$ds^2 = -c^2 dt^2 + dx^2 + dy^2 + dz^2. \qquad (7.9)$$

7.4 Coordinate and Proper Time

In special and general relativity, there are two types of time, proper and coordinate. Proper time is measured along the trajectory of an observer in space-time ("world line"). In practice, it is measured by a physical clock accompanying the observer. The clock must be insensitive to environmental conditions, gravity, and accelerations. Proper time is invariant in any coordinate change. Since the second of the Système International (SI) is defined only in terms of the periods of radiation for the caesium atom, it contains no indication of a specific, gravitational potential, or state of motion. Thus, any observer can realize the SI second as the unit for proper times. Proper time cannot be used to describe phenomena in extended domains, in which cases coordinate time must be used.

In special and general relativity, a four-dimensional space-time reference system uses three spatial coordinates (x_1, x_2, x_3) and a fourth, $x_o = ct$, where t is the coordinate time in this reference system. Coordinate time is an unambiguous way of dating in a specific reference system and is to be used as the time basis in the theory of motion in the system. In metrology, it can be argued that coordinate time cannot be measured, but only computed. The relation of proper time of an observer to coordinate time is provided by the metric, which takes into account the surrounding masses and energy.

If two events occur separated by a time dt at the same place in a reference frame, the interval (Equation [7.9]) reduces to:

$$ds^2 = -c^2 dt^2. \tag{7.10}$$

The time t is linked with the place and the new time is called the proper time, τ, which describes the local physics of the point. Then the interval (Equation [7.9]) can be written for the general case as:

$$ds^2 = -c^2 d\tau^2 = -c^2 dt^2 + dx^2 + dy^2 + dz^2. \tag{7.11}$$

This means that for the interval s^2 between two events, the quantity $\sqrt{s^2}/c$ equals the difference in readings of a clock moving at a constant velocity between two events. To derive the relationship between proper and coordinate time, consider the components of the constant velocity as measured with respect to coordinate time.

$$c^2 \left(\frac{d\tau}{dt}\right)^2 = c^2 - \left[\left(\frac{dx}{dt}\right)^2 + \left(\frac{dy}{dt}\right)^2 + \left(\frac{dz}{dt}\right)^2\right], \tag{7.12}$$

from which

$$\frac{d\tau}{dt} = \sqrt{1 - v^2/c^2} = \frac{1}{\gamma}. \tag{7.13}$$

This is the same expression as that shown in Section 7.2 for the time dilation relationship between clocks in reference frames moving with a velocity v with respect to each other. So for the moving observer the period of the clock is in coordinate time, while the fixed observer measures in proper time. For an observer at rest in the reference frame, the proper time coincides with the coordinate time in

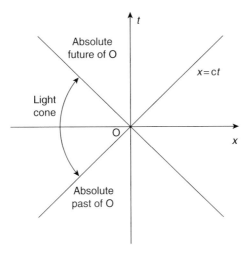

Figure 7.2 Minkowski diagram

that reference frame in special relativity. This is not the case for observations of moving objects. This concept is visualized with the Minkowski diagram.

7.5 Minkowski Diagrams

The Minkowski diagram, developed in 1908 by Herman Minkowski (1908), provides an illustration of the properties of space and time in the special theory of relativity. It allows a quantitative understanding of the corresponding phenomena like time dilation and length contraction without mathematical equations. It is a space-time diagram with usually only one spatial dimension. Referring to Figure 7.2, distance is displayed on the x-axis and time on the y-axis. Events happening on a horizontal path in space can then be represented on a horizontal line in the diagram. Each point in the diagram represents an event in space and time, and the curve it follows is called its *world line*. It is called an event whether or not anything actually happens at that position. Objects plotted on the diagram can be thought of as moving from bottom to top as time passes, so a vertical line corresponds to a fixed point in space.

The world lines of light pulses are represented by straight lines defined by $x = ct$. The world line of a photon traveling through the origin to the right is a straight line with a slope of 45° if the scales on both axes are chosen appropriately. If light cones are drawn in the positive and negative time directions from a certain event (O), space-time is separated into distinct "future" and "past" regions. The future is the locus of all events that have not yet happened and can be affected. An event cannot affect anything outside of its future light cone, because, in order to do so, it would

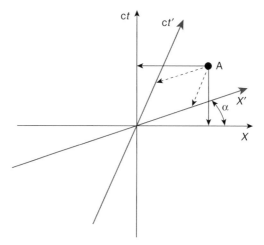

Figure 7.3 Minkowski diagram showing the superposition of two coordinate systems moving relative to each other with a constant velocity

have to send some sort of message to the desired location faster than the speed of light. The past is the locus of events that contributed to the current state. Anything that happened before and is not in the past light cone could not have affected the present, because its future light cone does not encompass the present.

We can superpose the coordinate systems for two observers moving relative to each other with constant velocity v in a Minkowski diagram to allow the space and time coordinates, x and t, used by one observer to be read off immediately with respect to the corresponding x' and t' used by the other, and vice versa (Figure 7.3). Both observers would assign an event at A to different times and locations. Events, which are estimated to happen simultaneously from the viewpoint of one observer, happen at different times for the other. In the Minkowski diagram, this simultaneity corresponds with the introduction of a separate path axis for the moving observer. Each observer interprets all events on a line parallel to his path axis as simultaneous. The sequence of events from the viewpoint of an observer can be illustrated graphically by shifting this line in the diagram from bottom to top. If ct is assigned on the time axes, the angle α between both path axes will be identical with that between both time axes. This follows from the speed of light being the same for all observers, regardless of their relative motion. α is given by:

$$\tan \alpha = \frac{v}{c}, \tag{7.14}$$

where v is the relative velocity between the reference frames. The corresponding transformation from x and t to x' and t' and vice versa is described mathematically by the Lorentz transformation.

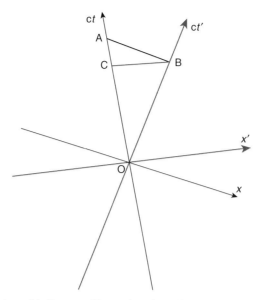

Figure 7.4 Minkowski diagram illustrating time dilation

Time dilation is illustrated in Figure 7.4. In this case, both observers consider the clock of the other as running slower. Relativistic time dilation means that a clock moving relative to an observer is running slower, as well as the time in this system. This can be read immediately from Figure 7.4. The observer and clock at A is assumed to move from the origin O toward A, and the clock from O to B. For the observer at A, all events happening simultaneously in this moment are located on a straight line parallel to its path axis passing A and B. Since OB < OA, the observer concludes that the time passed on the clock moving relative to him is less than that passed on his own clock, since they were together at O. A second observer having moved together with the clock from O to B will argue that the other clock has reached only C, and, therefore, this clock runs slower. The reason for the apparently paradoxical statements is the different determination of the events happening synchronously at different locations. Due to the principle of relativity, the question of who is right has no answer and does not make sense.

7.6 Time in Special Relativity

The coordinate time in a reference frame must be given by a clock that is fixed relative to the reference frame. The second viewed as the unit of proper time is independent of the frame. So the synchronization of clocks in the frame must involve clocks fixed with respect to that frame. Clocks fixed with respect to the frame are synchronized according to Einstein's synchronization rule with respect to clocks in a moving frame due to time dilation. Hence, corrections are necessary to compare clocks with different velocities.

Likewise an observer in a fixed frame observing a frequency from a source in a frame moving with velocity **v** will detect a frequency difference due to the Doppler Effect. The difference between the emitted frequency, f, measured in the source's frame and the received frequency, f', measured in the observer's frame is:

$$f' = f \frac{\left(1 + {}^{\mathbf{v} \cdot \mathbf{n}}/_c\right)}{\sqrt{1 - {}^{v^2}/_{c^2}}} \tag{7.15}$$

where **n** is the unit vector along the straight path of the signal in the direction of the propagation.

7.7 General Relativity

While special relativity applies to an empty and, hence, unreal universe, general relativity is a model of the real universe, where mass and energy find their proper place. General relativity is a theory that accounts for accelerations such as gravity. Unlike the fixed framework of Newtonian space-time, here gravitational effects are not viewed as action at a distance; instead, they appear as local geometrical properties of a Riemannian space-time, curved by the presence of energy and mass.

In Einstein's theory of relativity, the inertial frame is generalized by the concept of the free-falling isolated local frame, which is electrically and magnetically shielded, sufficiently small that inhomogenities in the external fields can be ignored throughout the volume, and in which self-gravitating effects are negligible. In this free-falling frame, any local nongravitational test experiment is independent of where and when in the universe it is performed and of the velocity of the frame.

In general relativity, there are no privileged coordinate systems, although some may be more convenient than others. The geometrization of the gravitational field is a formulation of a general principle of general relativity, the equivalence principle.

7.7.1 Metrics in General Relativity

In Newtonian mechanics, an object located at a point **P** with coordinates (x, y, z) in a three-dimensional Cartesian reference frame, in the presence of a gravitational potential U (x,y,z), would experience a force with components of:

$$\frac{\partial U}{\partial x}, \quad \frac{\partial U}{\partial y}, \quad \frac{\partial U}{\partial z}. \tag{7.16}$$

So the force, **f**, at point, **P**, can be related to U by:

$$\mathbf{f} = \nabla U. \tag{7.17}$$

If there are N masses, m_i, located at $\mathbf{Q}_i$ acting on the point, **P**, the Newtonian potential is:

$$U = G \sum_{i=1}^{N} \frac{m_i}{|\mathbf{P} - \mathbf{Q}_i|}, \tag{7.18}$$

where G is the universal constant of gravity. The potentials for finite gravitating bodies, the spherical harmonic development for point outside the sphere, the additional terms for relativity, and gravito-magnetic fields can be found in Kovalevsky and Seidelmann (2004).

For general relativity, the general four-dimensional metric is:

$$ds^2 = \sum_{i=0}^{3} \sum_{j=0}^{3} g_{ij} dx_i dx_j, \tag{7.19}$$

where $g_{ij} = g_{ji}$ and the index 0 corresponds to the time coordinate. In general relativity, if the potential vanishes and there is no spatial rotation and no acceleration of the origin, the expression reduces to the space-time special relativity metric.

$$ds^2 = -c^2 dx_0^2 + dx_1^2 + dx_2^2 + dx_3^2. \tag{7.20}$$

The coefficients g_{ij} can be expanded as functions of small parameters. To the accuracies of microarcseconds in direction and 10^{-17} s in time for a weak field like the solar system, the International Astronomical Union in 2000 recommended expansions as given in Section 7.8.

7.7.2 The Equivalence Principle

In classical mechanics, there are two distinct types of matter. Inertial mass m' is the mass appearing in Newton's second law: $F = m'a$; the inertial mass is the measure of inertia of a body, a measure of how difficult it is to change its velocity. The gravitational mass m is the mass appearing in the expression for Newtonian attraction: $F = G\, m_1\, m_2/r^2$. Gravitational mass is the measure of the magnitude of the gravitational force F generated by a body. To the accuracy of observations, it has long been recognized that $m = m'$, and this weak equivalence principle is used in classical mechanics.

Einstein generalized this weak equivalence principle to a free-falling laboratory near a gravitating mass being equivalent to a laboratory in free motion, as on an artificial satellite or a space probe outside the solar system. Generally, the equivalence principle states that there is no experiment that allows one to distinguish whether a laboratory is in a uniformly accelerating elevator, or if it is a fixed laboratory in a gravitational field with a uniform strength over the space of the laboratory.

Thus, there is no preferred local inertial reference frame. All are equivalent and all nongravitational laws of nature take the same form as in special relativity. Thus, for a locally inertial reference frame, there is a metric that has the form and properties of the special relativity metric describing a flat four-dimensional space-time. In general relativity, there is an extended metric, which depends on the distribution and motion of masses in the universe, and at each point there is a flat special relativity space-time, tangent to the general relativity space-time such that the four coordinates and their derivatives with respect to the two space-times can be set to correspond. In general relativity space-time, a test particle moves on a geodesic of the metric. So light bends in a gravitational field, because the geodesic is not a straight line. Also, light traveling through a gravitational field experiences a blue or red shift depending on whether the gravitational potential is larger at the point of emission or at the point of reception. This corresponds to a gravitational time "dilation," and adds to the time dilation due to velocity.

The metric of general relativity depends on the distribution and motion of masses and is expressed through the gravitational potential, U, and the gravito-magnetic vector potential **W**. Then:

$$ds^2 = g_{\alpha\beta}dx^\alpha dx^\beta, \tag{7.21}$$

with $dx^\alpha = (cdt, dx, dy, dz)$ and $g_{\alpha\beta} = g_{\alpha\beta}(t, x, y, z, U, \mathbf{W})$.

7.8 IAU Resolutions

In 2000, the IAU adopted three resolutions specifying the metrics for general relativity for fundamental astronomy. These are reproduced in what follows from https://syrte.obspm.fr/IAU_resolutions/Resol-UAI.htm:

Resolution B1.3 Definition of Barycentric Celestial Reference System and Geocentric Celestial Reference System

The XXIVth International Astronomical Union General Assembly,

Considering
1. that the Resolution A4 of the XXIst General Assembly (1991) has defined a system of space-time coordinates for (a) the solar system (now called the Barycentric Celestial Reference System, (BCRS)) and (b) the Earth (now called

the Geocentric Celestial Reference System (GCRS)), within the framework of General Relativity,

2. the desire to write the metric tensors both in the BCRS and in the GCRS in a compact and self-consistent form,

3. the fact that considerable work in General Relativity has been done using the harmonic gauge that was found to be a useful and simplifying gauge for many kinds of applications,

Recommends

1. the choice of harmonic coordinates both for the barycentric and for the geocentric reference systems,

2. writing the time-time component and the space-space component of the barycentric metric $g_{\mu\nu}$ with barycentric coordinates $(t, \mathbf{x})$ $(t = $ Barycentric Coordinate Time (TCB)) with a single scalar potential $w(t, \mathbf{x})$ that generalises the Newtonian potential, and the space-time component with a vector potential w^i $(t, \mathbf{x})$; as a boundary condition it is assumed that these two potentials vanish far from the solar system,

explicitly,

$$g_{00} = -1 + \frac{2w}{c^2} - \frac{2w^2}{c^4},$$

$$g_{0i} = -\frac{4}{c^3} w^i,$$

$$g_{ij} = \delta_{ij} \left(1 + \frac{2}{c^2} w \right),$$

with

$$w(t, \mathbf{x}) = G \int d^3 x' \frac{\sigma(t, \mathbf{x}')}{|\mathbf{x} - \mathbf{x}'|} + \frac{1}{2c^2} G \frac{\partial^2}{\partial t^2} \int d^3 x' \sigma(t, \mathbf{x}')|\mathbf{x} - \mathbf{x}'|,$$

$$w^i(t, \mathbf{x}) = G \int d^3 x' \frac{\sigma^i(t, \mathbf{x}')}{|\mathbf{x} - \mathbf{x}'|}.$$

here, σ and σ^i are the gravitational mass and current densities, respectively,

3. writing the geocentric metric tensor G_{ab} with geocentric coordinates $(T, \mathbf{X})$ $(T = $ Geocentric Coordinate Time (TCG)) in the same form as the barycentric one but with potentials $W(T, \mathbf{X})$ and $W^a(T, \mathbf{X})$; these geocentric potentials should be split into two parts – potentials W_E and W_E^a arising from the gravitational action of

the Earth and external parts W_{ext} and W^a_{ext} due to tidal and inertial effects; the external parts of the metric potentials are assumed to vanish at the geocenter and admit an expansion into positive powers of $\mathbf{X}$, explicitly,

$$G_{00} = -1 + \frac{2W}{c^2} - \frac{2W^2}{c^4},$$

$$G_{0a} = -\frac{4}{c^3} W^a,$$

$$G_{ab} = \delta_{ab}\left(1 + \frac{2}{c^2} W\right),$$

the potentials W and W^a should be split according to

$$W(T, \mathbf{X}) = W_E(T, \mathbf{X}) + W_{ext}(T, \mathbf{X}),$$

$$W^a(T, \mathbf{X}) = W^a_E(T, \mathbf{X}) + W^a_{ext}(T, \mathbf{X}),$$

the Earth's potentials W_E and W^a_E are defined in the same way as w and w^i but with quantities calculated in the GCRS with integrals taken over the whole Earth,
4. using, if accuracy requires, the full post-Newtonian coordinate transformation between the BCRS and the GCRS as induced by the form of the corresponding metric tensors,
explicitly, for the kinematically non-rotating GCRS (T = TCG, t = TCB, $r^i_E \equiv x^i - x^i_E(t)$, and a summation from 1 to 3 over equal indices is implied),

$$T = t - \frac{1}{c^2}\left[A(t) + v^i_E r^i_E\right] + \frac{1}{c^4}\left[B(t) + B^i(t)r^i_E + B^{ij}(t)r^i_E r^j_E + C(t, x)\right] + O(c^{-5}),$$

$$X^a = \delta_{ai}\left[r^i_E + \frac{1}{c^2}\left(\frac{1}{2}v^i_E v^j_E r^j_E + W_{ext}(x_E)r^i_E + r^i_E a^j_E r^j_E - \frac{1}{2}a^i_E r^2_E\right)\right] + O(c^{-4}),$$

where

$$\frac{d}{dt}A(t) = \frac{1}{2}v^2_E + W_{ext}(x_E),$$

$$\frac{d}{dt}B(t) = -\frac{1}{8}v^4_E - \frac{3}{2}v^2_E W_{ext}(x_E) + 4v^i_E w^i_{ext}(x_E) + \frac{1}{2}w^2_{ext}(x_E),$$

$$B^i(t) = -\frac{1}{2}v_E^2 v_E^i + 4w_{ext}^i(\mathbf{x}_E) - 3v_E^i w_{ext}(\mathbf{x}_E),$$

$$B^{ij}(t) = -v_E^i \delta_{aj} Q^a + 2\frac{\partial}{\partial x^j}w_{ext}^i(\mathbf{x}_E) - v_E^i \frac{\partial}{\partial x^j}w_{ext}(\mathbf{x}_E) + \frac{1}{2}\delta^{ij}\dot{w}_{ext}(\mathbf{x}_E),$$

$$C(t,\mathbf{x}) = -\frac{1}{10}r_E^2\left(a^{\cdot i}{}_E r_E^i\right),$$

here x_E^i, v_E^i, and a_E^i are the components of the barycentric position, velocity and acceleration vectors of the Earth, the dot stands for the total derivative with respect to t, and

$$Q^a = \delta_{ai}\left[\frac{\partial}{\partial x_i}w_{ext}(\mathbf{x}_E) - a_E^i\right].$$

The external potentials, w_{ext} and w_{ext}^i, are given by

$$w_{ext} = \sum_{A\neq E} w_A, \quad w_{ext}^i = \sum_{A\neq E} w_A^i,$$

where E stands for the Earth and w_A and w_A^i are determined by the expressions for w and w^i with integrals taken over body A only.

Notes

It is to be understood that these expressions for w and w^i give g_{00} correct up to $O(c^{-5})$, g_{0i} up to $O(c^{-5})$, and g_{ij} up to $O(c^{-4})$. The densities σ and σ^i are determined by the components of the energy momentum tensor of the matter composing the solar system bodies as given in the references. Accuracies for G_{ab} in terms of c^{-n} correspond to those of $g_{\mu\nu}$.

The external potentials W_{ext} and W_{ext}^a can be written in the form

$$W_{ext} = W_{tidal} + W_{iner},$$

$$W_{ext}^a = W_{tidal}^a + W_{iner}^a.$$

W_{tidal} *generalises the Newtonian expression for the tidal potential. Post-Newtonian expressions for W_{tidal} and W_{tidal}^a can be found in the references. The potentials W_{iner}, W_{iner}^a are inertial contributions that are linear in X^a. The former is determined mainly by the coupling of the Earth's nonsphericity to the external potential. In the kinematically non-rotating Geocentric Celestial Reference System, W_{iner}^a describes the Coriolis force induced mainly by geodetic precession.*

Finally, the local gravitational potentials W_E *and* W_E^a *of the Earth are related to the barycentric gravitational potentials* w_E *and* w_E^i *by*

$$W_E(T, \mathbf{X}) = w_E(t, \mathbf{x})\left(1 + \frac{2}{c^2}\, v_E^2\right) - \frac{4}{c^2}\, v_E^i w_E^i(t, \mathbf{x}) + O(c^{-4}),$$

$$W_E^a(T, \mathbf{X}) = \delta_{ai}\left(w_E^i(t, \mathbf{x}) - v_E^i w_E(t, \mathbf{x})\right) + O(c^{-2}).$$

Resolution B1.4 Post-Newtonian Potential Coefficients
The XXIVth International Astronomical Union General Assembly,

Considering
1. that for many applications in the fields of celestial mechanics and astrometry a suitable parametrization of the metric potentials (or multipole moments) outside the massive solar system bodies in the form of expansions in terms of potential coefficients are extremely useful, and
2. that physically meaningful post-Newtonian potential coefficients can be derived from the literature,

Recommends
1. expansion of the post-Newtonian potential of the Earth in the Geocentric Celestial Reference System (GCRS) outside the Earth in the form

$$W_E(T, \mathbf{X}) = \frac{GM_E}{R}\left[1 + \sum_{l=2}^{\infty}\sum_{m=0}^{+l}\left(\frac{R_E}{R}\right)^l P_{lm}(\cos\theta)\left(C_{lm}^E(T)\cos m\varphi + S_{lm}^E(T)\sin m\varphi\right)\right].$$

here C_{lm}^E and S_{lm}^E are, to sufficient accuracy, equivalent to the post-Newtonian multipole moments introduced by Damour *et al.*(1991). θ and ϕ are the polar angles corresponding to the spatial coordinates X^a of the GCRS and $R = |\mathbf{X}|$, and
2. expression of the vector potential outside the Earth, leading to the well-known Lense-Thirring effect, in terms of the Earth's total angular momentum vector $\mathbf{S}_E$ in the form

$$W_E^a(T, \mathbf{X}) = -\frac{G}{2}\frac{(\mathbf{X} \times \mathbf{S}_E)^a}{R^3}.$$

Resolution B1.5 Extended Relativistic Framework for Time Transformations and Realisation of Coordinate Times in the Solar System
The XXIVth International Astronomical Union General Assembly,

Considering

1. that the Resolution A4 of the XXIst General Assembly (1991) has defined systems of space-time coordinates for the solar system (Barycentric Reference System) and for the Earth (Geocentric Reference System), within the framework of General Relativity,

2. that Resolution B1.3 entitled "Definition of Barycentric Celestial Reference System and Geocentric Celestial Reference System" has renamed these systems the Barycentric Celestial Reference System (BCRS) and the Geocentric Celestial Reference System (GCRS), respectively, and has specified a general framework for expressing their metric tensor and defining coordinate transformations at the first post-Newtonian level,

3. that, based on the anticipated performance of atomic clocks, future time and frequency measurements will require practical application of this framework in the BCRS,

4. that theoretical work requiring such expansions has already been performed,

Recommends

that for applications that concern time transformations and realisation of coordinate times within the solar system, Resolution B1.3 be applied as follows:

1. the metric tensor be expressed as

$$g_{00} = -\left(1 - \frac{2}{c^2}\left(w_0(t, \mathbf{x}) + w_L(t, \mathbf{x})\right) + \frac{2}{c^4}\left(w_0^2(t, \mathbf{x}) + \Delta(t, \mathbf{x})\right)\right),$$

$$g_{0i} = -\frac{4}{c^3} w^i(t, \mathbf{x}),$$

$$g_{ij} = \left(1 + \frac{2w_0(t, \mathbf{x})}{c^2}\right)\delta_{ij},$$

where ($t \equiv$ Barycentric Coordinate Time (TCB), $\mathbf{x}$) are the barycentric coordinates, $w_0 = G\sum_A {}^{M_A}\!/_{r_A}$, with the summation carried out over all solar system bodies A, $\mathbf{r}_A = \mathbf{x} - \mathbf{x}_A$, $\mathbf{x}_A$ are the coordinates of the center of mass of body A, $r_A = |\mathbf{r}_A|$, and where w_L contains the expansion in terms of multipole moments [see their definition in the Resolution B1.4 entitled "Post-Newtonian Potential Coefficients"] required for each body. The vector potential $w^i(t, \mathbf{x}) = \sum_A w^i_A(t, \mathbf{x})$, and the function $\Delta(t, \mathbf{x}) = \sum_A \Delta_A(t, \mathbf{x})$ are given in note 2.

1. the relation between TCB and Geocentric Coordinate Time (TCG) can be expressed to sufficient accuracy by

$$\text{TCB} - \text{TCG} = c^{-2} \left[\int_{t_0}^{t} \left(\frac{v_E^2}{2} + w_{0\text{ext}}(\mathbf{x}_E) \right) \, dt + v_E^i r_E^i \right] \tag{1}$$

$$-c^{-4} \left[\int_{t_0}^{t} \left(-\frac{1}{8} v_E^4 - \frac{3}{2} v_E^2 w_{0\text{ext}}(\mathbf{x}_E) + 4 v_E^i w_{\text{ext}}^i(\mathbf{x}_E) + \frac{1}{2} w_{0\text{ext}}^2(\mathbf{x}_E) \right) dt \right.$$

$$\left. - \left(3 w_{0\text{ext}}(\mathbf{x}_E) + \frac{v_E^2}{2} \right) v_E^i r_E^i \right],$$

where v_E is the barycentric velocity of the Earth and where the index ext refers to summation over all bodies except the Earth.

Notes
1. *This formulation will provide an uncertainty not larger than 5×10^{-18} in rate and, for quasi-periodic terms, not larger than 5×10^{-18} in rate amplitude and 0.2 ps in phase amplitude, for locations farther than a few solar radii from the Sun. The same uncertainty also applies to the transformation between TCB and TCG for locations within 50000 km of the Earth. Uncertainties in the values of astronomical quantities may induce larger errors in the formulas.*
2. *Within the above mentioned uncertainties, it is sufficient to express the vector potential $w_A^i(t, \mathbf{x})$ of body A as*

$$w_A^i(t, \mathbf{x}) = G \left[\frac{-(\mathbf{r}_A \times \mathbf{S}_A)^i}{2 r_A^3} + \frac{M_A v_A^i}{r_A} \right],$$

where $\mathbf{S}_A$ is the total angular momentum of body A and v_A^i are the components of the barycentric coordinate velocity of body A. As for the function $\Delta_A(t, \mathbf{x})$ it is sufficient to express it as

$$\Delta_A(t, \mathbf{x}) = \frac{GM_A}{r_A} \left[-2v_a^2 + \sum_{B \neq A} \frac{GM_B}{r_{BA}} + \frac{1}{2} \left(\frac{(r_A^k v_A^k)^2}{r_A^2} + r_A^k a_A^k \right) \right] + \frac{2G v_A^k (\mathbf{r}_A \times \mathbf{S}_A)^k}{r_A^3},$$

where $\mathbf{r}_{BA} = |\mathbf{x}_B - \mathbf{x}_A|$ and a_A^k is the barycentric coordinate acceleration of body A. In these formulas, the terms in $\mathbf{S}_A$ are needed only for Jupiter ($S \approx 6.9 \times 10^{38}$ $m^2 s^{-1}$ kg) and Saturn ($S \approx 1.4 \times 10^{38}$ $m^2 s^{-1}$ kg), in the immediate vicinity of these planets.
3. *Because the present Recommendation provides an extension of the IAU 1991 recommendations valid at the full first post-Newtonian level, the constants L_C*

and L_B that were introduced in the IAU 1991 recommendations should be defined as $<TCG/TCB> = 1 - L_C$ and $<TT/TCB> = 1 - L_B$, where TT refers to Terrestrial Time and $<>$ refers to a sufficiently long average taken at the geocenter. The most recent estimate of L_C is (Irwin and Fukushima, 1999)

$$L_C = 1.48082686741 \times 10^{-8} \pm 2 \times 10^{-17},$$

From the Resolution B1.9 on "Redefinition of Terrestrial Time TT," one infers $L_B = 1.55051976772 \times 10^{-8} \pm 2 \times 10^{-17}$ by using the relation $1 - L_B = (1 - L_C)(1 - L_G)$. L_G is defined in Resolution B1.9.

Because no unambiguous definition may be provided for L_B and L_C, these constants should not be used in formulating time transformations when it would require knowing their value with an uncertainty of order 1×10^{-16} or less.

4. If TCB-TCG is computed using planetary ephemerides which are expressed in terms of a time argument (noted T_{eph}) which is close to Barycentric Dynamical Time (TDB), rather than in terms of TCB, the first integral in Recommendation 2 above may be computed as

$$\int_{t_0}^{t} \left(\frac{v_E^2}{2} + w_{0ext}(\mathbf{x}_E) \right) dt = \left[\int_{T_{eph_0}}^{T_{eph}} \left(\frac{v_E^2}{2} + w_{0ext}(\mathbf{x}_E) \right) dt \right] \bigg/ (1 - L_B).$$

7.9 Timescales

7.9.1 International Atomic Time

International Atomic Time (TAI) was established originally using caesium atomic clocks operating in laboratories on the surface of the Earth. There was no consideration of relativistic effects in the first applications of the new atomic standards in establishing an atomic timescale. However, as accuracies improved and atomic time standards were located at many different sites on Earth and in orbit around the Earth, corrections based on differences in potentials had to be introduced to compare atomic standards (see Chapter 13).

7.9.2 Dynamical Timescales

With the adoption of the resolutions defining the Barycentric and Geocentric Celestial Reference Systems, two timescales, Barycentric Coordinate Time (TCB) and Geocentric Coordinate Time (TCG), were also introduced and the relationships between these timescales and Terrestrial Time (TT) were clarified. Barycentric Dynamical Time (TDB) was defined in 2006 as a linear scaling of TCB to have the approximate rate of TT. These timescales are discussed in Chapter 9.

7.10 Relativistic Effects in Time Transfer

Time transfer can be considered in two different reference frames, either a geocentric, Earth-fixed, rotating frame or a geocentric, nonrotating, local inertial frame. Clearly, clock comparisons and synchronization must be made in a common coordinate system. For Earth-based purposes and for orbiting satellites, the rotating or nonrotating geocentric coordinate system should be used. Since they have the same coordinate time, either may be used. In practice, for ground-based clocks, the rotating system is preferable, while for orbiting clocks, the nonrotating system is preferred. (See Chapter 15 for details on time transfer.)

References

Damour, T., Soffel, M., & Xu, C. (1991). General-Relativistic Celestial Mechanics. I. Method and Definition of Reference Systems. *Phys.Rev. D.*, **43**, 3273–3307.

Irwin, A. & Fukushima, T. (1999). A Numerical Time Ephemeris of the Earth. *Astron. Astrophy.*, **348**, 642–652.

Kovalevsky, J. & Seidelmann, P. K. (2004). *Fundamentals of Astrometry*. Cambridge: Cambridge University Press.

Michelson, A. A. & Morley, E. W. (1887). On the Relative Motion of the Earth and the Luminiferous Ether. *American Journal of Science*, **34**, 333–345.

Minkowski, H. (1908). Space and Time. In H. A. Lorentz, A. Einstein, H. Minkowski, & H. Weyl, eds., *The Principle of Relativity*, trans. W. Perrett and G. B. Jeffery, 1923. London: Methuen & Company, Ltd., pp. 73–91.

Trans. Int. Astron. Union, Vol. XXI B Proc. 21st General Assembly Buenos Aires (1991). J. Bergeron, ed. Alphen aan den Rijn, Netherlands: Kluwer Academic Publishers, 1992.

Trans. Int. Astron. Union, Vol. XXIV B Proc. 24th General Assembly Manchester (2000). H. Rickman, ed. San Francisco, CA: Astronomical Society of the Pacific, 2001.

8

Time and Cosmology

8.1 Introduction

Cosmology is the study of the scientific aspects of the origin, structure, evolution, and space–time relationships of the Universe. Historically, some of these issues have been addressed as long ago as ancient Greece, and later by Copernicus and Newton. The general theory of relativity created a new beginning of cosmology in 1917 (Einstein, 1917), and in 1922, Alexander Friedman first suggested an expanding Universe. It was subsequently corroborated by Edwin Hubble's discovery of the red shift in 1929. The Big Bang model was first proposed by Georges Lemaître in 1927. Recent astronomical observations now tell us that the Universe is 13.8 billion years old and composed of 4.9% atomic matter, 26.6% dark matter, and 68.5% dark energy (Planck Collaboration, 2014). Both historically and today the concept of time has always been an important part of cosmology.

8.2 Space-Time Metric

The simplest assumptions to make about the Universe on scales larger than local inhomogeneities, based on the Symmetry Principle that the time coordinate, t, can be synchronized everywhere, are:

(1) it is homogeneous (meaning that properties of matter and of the geometry of space-time are the same at every point in space); and
(2) it is isotropic (meaning that to a hypothetical observer, it looks exactly the same in whatever direction he or she might be looking).

Given those assumptions, the geometry of space-time can be described by the Friedman-Lemaître-Robertson-Walker metric (see Chapter 7 regarding space-time metrics), sometimes called the FLRW metric or the Standard Model (Friedman, 1924; Lemaître, 1933, Robertson, 1935, 1936a, 1936b; Walker, 1937):

$$ds^2 = c^2 dt^2 - [a(t)]^2 \left[\frac{dr^2}{1 - kr^2/R_{curve}^2} + r^2 \left(d\theta^2 + \sin^2\theta d\phi^2 \right) \right], \quad (8.1)$$

in co-moving polar coordinates (r, θ, ϕ) where ds is the line element, k is the curvature constant, and $a(t)$ is the dimensionless time dependent cosmic scale factor describing the expansion (or contraction) of the Universe. R_{curve} defines the Gaussian radius of curvature of the space, and k defines the sign of the curvature, and s taken as -1, 0, $+1$ for negative, flat, or positive curvature, respectively. If space sections were flat, they would have zero curvature. Positive curvature would have finite, closed geometry, and negative curvature would describe an open, infinite, Universe (in both cases, providing it extends everywhere beyond our horizon). For a positive curvature, a good deal of dark matter, which we do not observe in optical images of galaxies, would be required. $a(t)$ is normalized so that, at the present time, we have by definition $a(t) = 1$. The metric is coordinate frame dependent and describes a homogeneous, isotropic Universe only for a set of co-moving observers. They have constant values of (r, θ, ϕ) with the physical distance between them increasing proportional to $a(t)$. The time, t, considered to be a proper time called "cosmic" time, can be measured by a co-moving clock.

8.3 The Expanding Universe

The first observational evidence of the expanding universe was provided by Edwin Hubble in 1929. Hubble's Law is based on the observation that objects observed at great distances have a red shift due to a relative velocity away from the Earth. That velocity is approximately proportional to the galaxy's distance from Earth, up to a few hundred megaparsecs. Hubble's Law is expressed as $v = H_0 D$, where v is the galaxy velocity, D is the distance to the galaxy, and H_0 is the Hubble Constant. In 1929, Hubble confirmed Hubble's Law and determined a value of the Hubble Constant using red shifts that were mostly measured by Vesto Slipher in 1917 (Slipher, 1917; Longair, 2006). In the expression of Hubble's Law, v and D are current values, and not when the light was emitted. H_0 is usually quoted in (km/s)/Mpc, which is the speed in km/s of a galaxy at 1 megaparsec (3.09×10^{19} km) away. The reciprocal of H_0 is the Hubble Time, t_H. Values have been determined over time with the direct measurements calibrated using the Hubble Space Telescope to be $H_0 = 74$ km/s/Mpc, and $t_H = 14$ G (billion) years (Freedman, Kennicutt, & Mould, 2009).

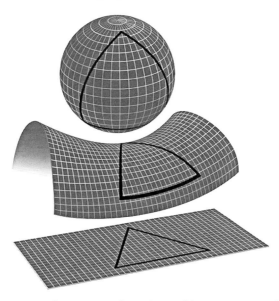

Figure 8.1 Curvature of space. Top: k = +1 – positive curvature; middle: k = −1 – negative curvature; bottom: k = 0 – flat.

8.4 Age of the Universe

Determining the age of the Universe is done in the context of the Lambda-Cold Dark Matter (ΛCDM) model. This model assumes that the Universe is made up of normal (baryonic) matter, cold dark matter, dark energy, and radiation (including both photons and neutrinos). The age is computed knowing H_0 and the density parameters for each of the components. The density parameter refers to the ratio of the actual density to the critical density given by:

$$\rho_c = \frac{3H_0^2}{8\pi G},$$

(8.2)

where G is Newton's Gravitational constant. The full ΛCDM model is described by a number of other parameters, but for the purpose of computing the age of the Universe, only these are required. The determined age is 13.799±0.021 billion years (Planck Consortium, 2014). It is reassuring that independent estimates of the cosmic age, though less accurate, give similar results. For example, the age of our galaxy can be estimated from the decay of heavy elements, leading to an age of 12.5 ±3 Gyrs. The mean age of the oldest globular clusters is 11.5±1.3 Gyrs (Chaboyer, Demarque, Kernan, Krauss, 1998), based on the luminosities of the most massive main sequence stars.

8.5 Evolution of the Universe

Based on the age of the Universe, its measured properties, and our knowledge of the basic laws of physics, our current understanding of the evolution of the Universe is shown in Table 8.1. Since there is not a viable theory to unify quantum phenomena with gravity, the Planck scale is the limit to knowledge. The Planck scale corresponds to an energy of 10^{19} GeV, a mass of 10^{-5} kilograms, a size of 10^{-35} meters, and a time of 10^{-43} seconds. Following a very early, though still speculative, period of exponentially accelerating expansion called *inflation*, the cosmic expansion entered a phase of deceleration that lasted roughly half the current cosmic age. The deceleration was caused by radiation until matter/radiation equality near 47 kyr, and by matter after that. Matter–vacuum equality occurs at three-fourths of the current age, while the change from deceleration to acceleration occurs earlier, at about 56% of the current age. The current cosmic constituents are about 30% matter, with 5% baryons and 27% dark matter (Spergel, 2015). Dark energy, which constitutes 68% of today's Universe, causes an accelerating expansion in the past 6 billion years and an exponential increase in the scale factor, $R = R_0 e^{Ht}$.

8.6 Cosmic Time

In view of the current understanding of the Universe, as outlined in Sections 8.3, 8.4, and 8.5, we can ask what is the meaning of "t" in the FLRW metric (Equation 8.1). This quantity is considered to be the proper time of co-moving observers in the Universe, such that (1) every co-moving observer records the same cosmic history, and (2) at every time, the Universe appears the same in every direction to every co-moving observer. This "cosmic time" is used in order to unambiguously measure time, and/or synchronize observations, by recording the average density of matter in the Universe or the temperature of the cosmic micro-wave background (CMB) radiation. It corresponds to time as measured by clocks at rest relative to expanding space. Recall that this metric presumes an isotropic homogeneous universe and that, following the Copernican principle, we are not located in a privileged place in that Universe. So, assuming symmetry, observers co-moving with the expansion of the Universe record the same cosmic history.

It seems the notion of proper cosmic time is legitimate back to at least 10^{-6} seconds, where the laws of physics are well known and well behaved. In the duration of time that separates this epoch and a postulated Planck time, some have argued that the notion of time becomes problematic, especially near phase transitions and/or inflation (e.g., Rugh & Zinkernagel, 2008, 2016). In addition to these concerns, the occurrence of inflation may, in practice, mark the true barrier to our following time backward, since the pre-inflation realm is, at present, almost entirely unknown.

Table 8.1 *The evolution of the Universe*

Epoch	Time	Temperature	Description
Planck epoch	$< 10^{-43}$ s	$>10^{32}$ K	Current physics is not predictive.
Grand unification epoch	$< 10^{-36}$ s	$>10^{29}$ K	Electromagnetic, weak, and strong interactions are unified.
Inflationary epoch	$< 10^{-32}$ s		Era of accelerating expansion
Electroweak epoch	$> 10^{-32}$ s	10^{28} K $\rightarrow 10^{22}$ K	The Strong Nuclear Force becomes distinct from the Electroweak Force.
Quark epoch	$> 10^{-12}$ s	$> 10^{12}$ K	Quark-gluon plasma. Forces of the Standard Model separated.
Quark confinement	10^{-5} s	10^{12} K	Formation of protons and neutrons.
Hadron epoch	10^{-5} s $\rightarrow 1$ s	10^{12} K $\rightarrow 10^{10}$ K	Quarks are bound into hadrons.
Neutrino decoupling	1 s	10^{10} K	Neutrinos cease interacting with baryonic matter.
Lepton epoch	1 s $\rightarrow$ 10 s	10^{10} K $\rightarrow 10^{9}$ K	Leptons and anti-leptons remain in thermal equilibrium. Electron/positron annihilation marks end of lepton era.
Big Bang nucleosynthesis	10 s $\rightarrow 10^{3}$ s	3×10^{9} K $\rightarrow 5\times10^{8}$ K	Protons and neutrons form primordial atomic nuclei.
Photon epoch	10 s $\rightarrow 1.2 \times 10^{13}$ s	10^{9} K $\rightarrow 10^{4}$ K	Plasma of nuclei, electrons, and photons.
Matter-dominated era	47 kyr $\rightarrow$ 9 Gyr	10^{4} K $\rightarrow$ 3.6 K	Energy density of matter dominates radiation density and dark energy
Recombination	380 kyr	3,000 K	Electrons and atomic nuclei first begin to form neutral atoms.
Dark Ages	380 kyr $\rightarrow$ 150 Myr	3,000 K $\rightarrow$ 60 K	Time between recombination and formation of the first stars. Only radiation emitted was the hydrogen line.
Stelliferous Era	150 Myr $\rightarrow$ 100 Gyr	60 K $\rightarrow$ 0.03 K	Star formation
Reionization	150 Myr $\rightarrow$ 1 Gyr	60 K $\rightarrow$ 19 K	The most distant observable object
Galaxy formation and evolution	1 Gyr $\rightarrow$ 10 Gyr	19 K $\rightarrow$ 4 K	Galaxies coalesce into "proto-clusters."
Dark-energy-dominated era	> 10 Gyr	< 4 K	Matter density falls below dark energy density.
Present time	13.8 Gyr	2.7 K	Observable universe.

8.7 Time's Arrow

The concept of the "arrow of time" refers to the apparent one-way direction of the flow of time. The fundamental microscopic laws of physics appear to work equally well in both directions. However, the second law of thermodynamics says that in a closed system, entropy, which is a measure of the number of microscopic configurations in a thermodynamic system specified by certain macroscopic variables, tends to increase with time. If each microscopic configuration is equally probable, the entropy of the system is the natural logarithm of that number of configurations multiplied by the Boltzmann constant. It is a measure of the disorder of a system having the dimension of energy divided by temperature, and is not invariant under any time reversal. In normal thermodynamics, with no gravity, equilibrium smoothness is the high-entropy state, whereas in the cosmic situation, in which gravity is important, it is collapsed objects and disequilibrium that constitute the high-entropy state (black holes being the maximum). It was inflation that generated the large-scale smoothness on which gravity could ultimately act, and it was inflation that created a relatively low-entropy universe, one with the potential for future complexity and life. The universe evolves by going from simplicity and uniformity to complexity and lumpiness, which seems to violate the second law of thermodynamics. However, as the Universe evolves, processes go out of equilibrium, and with gravitational degrees of freedom, total entropy increases.

8.8 Future of the Universe

The future of the Universe depends on the cosmological parameters. We now expect expansion will continue. However, just as in the earliest phases of the Universe, the far future of physics is also uncertain. The means of measuring time periods that we use now looking into the past of the Universe are main sequence stars and radioactive decay of heavy atoms. Looking forward, we might expect to be using pulsars, cooling dwarfs, and evaporating black holes. Millisecond pulsars are accurate to 10^{-15} seconds, so although atomic clocks can be more accurate, they may not last as long, and timescales constructed from atomic clocks could be subject to small drifts over long time periods caused by inadequate clock modeling. However, pulsars lose accuracy as they lose energy through gravitational radiation, so they might be useful for 10^{16} years, while radioactive isotopes may last for timescales of 10^{25} years.

The expansion of the Universe produces an event horizon at 16 billion light-years, where distant galaxies disappear from view. Star births decrease as less gas is available to form new stars. In 100 Gyears from now, more galaxies leave the horizon. At 500 Gyears, star formation occurs from binary systems; the Milky Way dims, and stellar remnants go into intergalactic space or feed central black holes. At 10^{25} years, matter in the Milky Way is likely to be in isolated stellar remnants and black holes. The timescale for evaporation of a stellar mass black hole is 10^{68} years, and 10^{100} years for a black hole in the center of a large galaxy. The end of the Universe is heat death, not due to temperature, but due to the lack of thermodynamic free energy to sustain any processes that can increase entropy (Dyson, 1979).

Entropy remains a central issue. As the Universe evolves into black holes, the entropy will grow. Why the Universe began with entropy as low as $10^{10} - 10^{15}$ is unknown. The special initial conditions lead to the consideration of eternal cosmologies and multiverse concepts. The Universe still has secrets to be studied (Impey, 2017).

References

Chaboyer, B., Demarque, P., Kernan, P. J., & Krauss, L. M. (1998). The Age of the Globular Clusters in the Light of Hipparcos: Resolving the Age Problem? *Astrophys. J.* **494**, 96–110.

Dyson, F. J. (1979). Time without End: Physics and Biology in an Open Universe. *Reviews of Modern Physics*, **51**, 447–460.

Einstein, A. (1917). Kosmologische betrachtungen zur allgemeinen relativitätstheorie. Sitzungsber. K. Preuss. Akad. Wiss. 142–152. English translation in H. A. Lorentz et al., eds. 1952. *The Principle of Relativity*. Mineola, NY: Dover Publications, 175–188.

Freedman, W. L., Kennicutt, R. C., & Mould, J. R. (2009). Measuring the Hubble Constant with the Hubble Space Telescope. *Proceedings of the International Astronomical Union*, **5**, Issue H15, Highlights.

Friedman, A. (1922). Über die Krümmung des Raumes. *Zeitschrift für Physik*, **10**, 377–386, English translation in Friedman, A. (1999) On the curvature of space. *General Relativity and Gravitation*, **31**, 1991–2000.

Friedman, A. (1924). Über die Möglichkeit einer Welt mit konstanter negativer Krümmung des Raumes. *Zeitschrift für Physik A*, **21**, 326–332, English translation in *General Relativity and Gravitation*, **31**, 31.

Hubble, Edwin (1929). A Relation between Distance and Radial Velocity among Extra-Galactic Nebulae. *Proceedings of the National Academy of Sciences of the United States of America*, **15**, 168–173.

Impey, C. (2017). Cosmic Time: From the Big Bang to the Eternal Future. In E. F. Arias, L. Combrinck, P. Gabor, C. Hohenkerk, & P. K. Seidelmann, eds., *The Science of Time 2016*. Springer.

Lemaître, G. (1927). Un univers homogène de masse constante et de rayon croissant rendant compte de la vitesse radiale des nébuleuses extragalactiques. *Annals of the Scientific Society of Brussels* (in French). **47A**, 41. Translated in: (1931) A Homogeneous Universe of Constant Mass and Growing Radius Accounting for the Radial Velocity of Extragalactic Nebulae. *M. N. Roy. Astron. Soc.*, 91, 483–490.

Lemaître, G. (1933), l'Univers en expansion, *Annales de la Société Scientifique de Bruxelles*, **A53**, 51–85.

Longair, M. S. (2006). *The Cosmic Century.* Cambridge: Cambridge University Press, 109.

Misner, C. W., Thorne, K. S., & Wheeler, J. A. (1973). *Gravitation.* New York, NY: W. H. Freeman.

Planck Collaboration. (2014). Planck 2013 results. XVI. Cosmological parameters, arXiv:1303.5076 [astro-ph.CO].

Robertson, H. P. (1935). Kinematics and World Structure. *Astrophysical J.*, **82**, 284–301.

Robertson, H. P. (1936a). Kinematics and World Structure II. *Astrophysical J.*, **83**, 187–201.

Robertson, H. P. (1936b). Kinematics and World Structure III. *Astrophysical J.*, **83**, 257–271.

Rugh, S. E. & Zinkernagel, H. (2008). On the physical basis for cosmic time. arXiv:0805.1947v1 [gr-qc].

Rugh, S. E. & Zinkernagel, H. (2016). Limits of time in cosmology. arXiv:1603.05449v1 [gr-qc].

Slipher, V. M. (1917). Radial Velocity Observations of Spiral Nebulae. *The Observatory*, **40**, 304–306.

Spergel, D. N. (2015). The Dark Side of Cosmology: Dark Matter and Dark Energy. *Science*, **347**, 1100–1102.

Walker, A. G. (1937). On Milne's Theory of World-Structure. *Proceedings of the London Mathematical Society 2*, **42**, 90–127.

9

Dynamical and Coordinate Timescales

9.1 Replacing Ephemeris Time

With the recognition of the problems with Ephemeris Time (ET) and the acknowledged need for its replacement, new timescales needed to be defined. It was becoming clear in the 1960s and 1970s that atomic time, originally considered a uniform timescale for use in celestial dynamics and possibly in other practical applications, provided a much more accessible timescale for practical use. For dynamical applications, although ET was the independent argument of the equations of motion for the ephemerides used in its definition, it could not be considered the independent argument in the fundamental equations of motion, due to its definition relying on outdated astronomical constants and Newtonian mechanics. So, when changes in the celestial reference system were to be considered at the International Astronomical Union (IAU) General Assembly in Grenoble, France, in 1976, the time was appropriate to discuss the possibility of improved definitions of timescales. It was clearly desirable to make new dynamical time scales continuous with Ephemeris Time within the accuracy of the previous timescale. Also, any new timescales needed to be consistent with the theory of relativity.

In the context of this chapter, dynamical time is understood as the time-like argument of dynamical theories. Dynamical timescales represent the independent variable of the equations of motion of solar system bodies and depend on the reference system being used. The equations of motion can be referred to the barycenter of the solar system, or to the geocenter. Each system would require its appropriate coordinate timescale. The process of arriving at the correct definitions involved some confusion, mistakes, misunderstandings, and corrections.

Some of the choices for possible new dynamical timescales were (1) to improve the definition of ET, (2) to adopt International Atomic Time (TAI), or (3) to seek new definitions for timescales, or time-like arguments. The IAU considered a number of factors in approaching the problem. Coordinated Universal Time (UTC), the adopted timescale, was based on atomic time but subject to one-second discontinuities in order to maintain its relationship to the Earth's rotation. However, a uniform timescale was needed to compute ephemerides in the past as well as the future, and atomic time was not available before 1955. International Atomic Time (the abbreviation TAI stands for the French "Temps Atomique International"), the conventional realization of atomic time, was (and remains) a timescale based on contributed clock data from around the world, and subject to the uncertainties inherent in combining many different time standards (see Chapter 14). The known systematic discrepancies in the ephemerides of solar system objects made the realization of a dynamical timescale using solar system ephemerides problematic. The defining relationships between timescales needed to account for the effects of relativity, and there was a question of the equivalence of time based on quantum physics and time based on dynamics.

   With these considerations, it was decided to recommend new time-like arguments for the independent variable for solar system body theories. The definitions would distinguish between coordinate and proper timescales, be related to TAI at some epoch, and be continuous with ET within its accuracy. Consequently, several timescales were required to accommodate applications in the geocentric and barycentric reference systems.

9.2 Terrestrial Dynamical Time (TDT) and Barycentric Dynamical Time (TDB)

The IAU began to address the issue in its General Assembly in 1976 at Grenoble, France. There, it adopted Recommendation 5:

"Recommendation 5, Time Scale for Dynamical Theories and Ephemerides
   It is recommended that

(a) at the instant 1977 January $01^d\,00^h\,00^m\,00^s$ TAI, the value of the new time scale for the apparent geocentric ephemerides will be 1977 January $1.^d0003725$ exactly ($1^d\,00^h\,00^m\,32.^s184$);
(b) the unit of this time scale will be a day of 86,400 SI seconds at mean sea level;
(c) the time scales for equations of motion referred to the barycenter of the solar system will be such that there will be only periodic variations between these time scales and that for the apparent geocentric ephemerides; and
(d) no time-step be introduced in International Atomic Time."

The recommendations also included notes of explanation that were not part of the recommendation itself. Items of interest included in the notes are the following:

"(1) The time-like arguments of dynamical theories and ephemerides are referred to as dynamical time scales. While it is possible and desirable to base the unit of dynamical time scales on the SI second, it is necessary to recognize that in relativistic theories there will be periodic variations between the unit of time for an apparent geocentric ephemeris and the unit of the corresponding time scale of the equation of motion, which may, for example, be referred to the center of mass of the solar system. (In the terminology of the theory of general relativity such time scales may be considered to be proper time and coordinate time, respectively.) The time scales for an apparent geocentric ephemeris and for the equations of motion will be related by a transformation that depends on the system being modeled and on the theory being used. The arbitrary constants in the transformation may be chosen so that the timescales have only periodic variations with respect to each other. Thus, it is sufficient to specify the basis of a unique time scale to be used for new, precise, apparent geocentric ephemerides.

The dynamical time scale for apparent geocentric ephemerides of recommendation 5 (a) and (b) is a unique time scale independent of theories, while the dynamical time scales referred to the barycenter of the solar system are a family of time scales resulting from the transformation of various theories and metrics of relativistic theories.

"(2) This recommendation specifies a particular dynamical time scale for apparent geocentric ephemerides that is effectively equal to TAI +32.184s. (There are formal differences arising from random, and possibly systematic, errors in the length of the TAI second and the method of forming TAI, but the accumulated effect of such errors is likely to be insignificant for astronomical purposes over long periods of time.) The scale is specified with respect to TAI in order to take advantage of the direct availability of UTC (which is based on the SI second and is simply related to TAI) and to provide continuity with the current values and practice in the use of Ephemeris Time. Continuity is achieved since the chosen offset between the new scale and TAI is the current estimate of the difference between ET and TAI, and since the SI second was defined so as to make it equal to the ephemeris second within the error of measurement. It will be possible to use most available ephemerides as if the arguments were on the new scale. Before 1955, when atomic time is not available, the determinations of ET can be considered to refer to the new time scale. The offset has been expressed in the recommendation as an exact decimal fraction of a day since the arguments of theories and ephemerides are normally expressed in days.

"(3) In view of the desirability of maintaining continuity of TAI and of avoiding the confusion that would arise, if it were to be redefined retrospectively, no step in TAI is proposed. Although the recommendation is in terms of TAI, in practice astronomers will use UTC and convert directly to the dynamical time scale.

"(4) The terminology and notation for dynamical time scales require further consideration in due course.

"(5) Recognizing that the TAI second differed from the SI second between 1969 and the present by $(10 \pm 2) \times 10^{-13}$, a step will be introduced in the scale interval of TAI. Therefore, the epoch of the dynamical time scale for apparent geocentric ephemerides was adjusted to 1977 at Grenoble." (*Trans. IAU*, 1977)."

In 1979 at the IAU General Assembly in Montreal, although there was discussion about the word "dynamical" and possible ambiguities or misunderstandings, the names of Terrestrial Dynamical Time (TDT) and Barycentric Dynamical Time (TDB) were adopted as the names of the two timescales (*Trans. IAU*, 1980). The abbreviations correspond to the French names "Temps Dynamique Terrestre" and "Temps Dynamique Barycentrique," respectively.

As a result of the IAU recommendations of 1976 and 1979, TDT was defined to provide continuity with ET, with the chosen offset thought to be an accurate value of the difference in TAI and ET at the time of the introduction of TDT. The extrapolation of TDT backward, prior to the availability of TAI, must be made by the use of ephemerides of the solar system as was done for ET. TDB was defined to be the independent variable of the equations of motion for solar system bodies with respect to the solar system barycenter. In practice, TDB was to be determined from TDT by a mathematical expression, which depends upon the gravitational theory being used, the astronomical constants, and the positions and motions of the solar system bodies (see Figure 9.1). Mathematically, these timescales could be realized for any TDB epoch, t_{TDB}, with the expressions (9.1) (Kaplan, 1981).

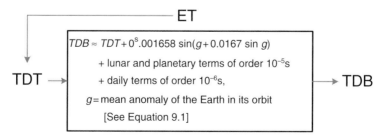

Figure 9.1 Dynamical timescales

$$TDT = TAI + 32.184 \ s,$$
$$TDB \approx TDT + 0.001658 \ s \ \sin(g + 0.0167 \ \sin g)$$
$$+ \text{ lunar and planetary terms of order } 10^{-5}s$$
$$+ \text{ daily terms of order } 10^{-6}s,$$
$$g = \text{mean anomaly of the Earth in its orbit}$$
$$= (357°.528 + 35999°.050° \ \text{T}) \times 2\pi/360°,$$
$$T = (t_{TDB} - 2451545.0)/36525. \tag{9.1}$$

More accurate expressions are available for TDB (Moyer, 1981; Hirayama et al., 1987; Fairhead et al., 1988). These two time-like arguments were to be introduced in use for practical applications in astronomy in the beginning of 1984.

9.3 Problems with TDT and TDB

Although the 1976 and 1979 resolutions were not ambiguous, a number of problems with their definitions arose. Because of the requirement that only periodic differences exist between TDB and TDT, spatial coordinates in the barycentric frame would have to be rescaled in order to maintain the same speed of light in the barycentric and geocentric frames (Standish, 1998; Soffel et al., 2003; Petit & Luzum, 2010). The definitions also did not specify over what period of time there should only be periodic differences and the nature of periodic terms that would be acceptable.

There was also confusion about whether the unit of time of 86400 SI seconds was specified only for the epoch in 1977, or as being constant for all times. The realization of a unit of time in a dynamical system could be expected to vary due to incomplete modeling of the dynamics and possible errors in the model used in preparing the ephemerides.

The use of the word "dynamical" to designate the timescales led to different interpretations and confusion. This issue was discussed in a paper by Guinot and Seidelmann (1988), where they pointed out that the word had many meanings. They noted that TDT is defined by its origin with respect to TAI and its unit, which at any instant is equal to the day of 86400 Système International (SI) seconds, at sea level. Thus, TDT is an ideal form of atomic time (of which TAI with an offset is a practical realization), but it has no relationship with the motion of solar system bodies. Thus, the use of the word "dynamical" in the name was inappropriate, if we understand that the use of the word "dynamical" implies a relationship to dynamical theory, as "atomic" time would imply a relationship with atomic theory. The notes to the Grenoble recommendation used the word "dynamical" both as a time-like argument for dynamical theories and ephemerides, and as a unique timescale independent of theories (i.e., as an ideal atomic time).

Guinot and Seidelmann noted the widely different requirements for timescales, and that it was necessary to distinguish between (1) the need for timescales for the apparent motions of celestial bodies in a space and time reference system for which realizations are available and (2) the guidance to theoreticians concerning the time-like arguments for dynamical theories. They suggested that for future recommendations regarding timescales the IAU should consider the following characteristics for an ideal time:

"(a) This ideal time must have realizations (time scales) available to all terrestrial observers, with a good precision, and such that the deviation between the ideal time and the corresponding realized time scales remains as small as possible and negligible for most of the applications.
(b) The definition must be unambiguous and must contain all the information which is needed so that the ideal time can be considered to be the argument of the ephemerides."

They then went on to propose a recommendation designed to improve the IAU recommendations of 1976 and 1979.

9.4 New Reference System

In the 1980s, it became apparent that there was a need to reconsider the celestial reference system. Radio frequency observations using interferometers were achieving accuracies much better than the optical positional observations, and there were plans for astrometric space missions that would also provide more accurate angular positions and motions than the historical fundamental optical catalogs. Accuracies were improving from tenths of an arcsecond to milliarcseconds. An IAU Working Group on Reference Systems was established with four subgroups: one on reference frames and their origin, another on timescales, a third on astronomical constants, and a fourth on the theory of nutation. IAU Colloquium 127 on Reference Systems (Hughes et al., 1991) produced nine recommendations that defined space-time coordinates within the framework of general relativity for the barycenter of any ensemble of masses. The recommendations were adopted by the IAU in Buenos Aires in 1991 (*Trans. IAU*, 1992).

These recommendations explicitly introduced the general theory of relativity as the theoretical background for the definition of space-time reference frames. It was recommended that the celestial reference frames with origins at the solar system barycenter and the center of mass of the Earth should show no global rotation with respect to a set of distant extragalactic objects, and work was requested to be initiated for the observational basis for the reference frame. The IAU recommended that the equator of this new conventional celestial barycentric reference frame should be as near as possible to the mean equator of J2000.0, and that the origin on this plane be as near as possible to the dynamical equinox of J2000.0, so that there would be no discontinuity of the new reference frame with the FK5 (Fricke et al., 1988). After the initial alignment of the new frame, no further rotation was to be applied to match the frame with a better determination of the mean equator and dynamical equinox of J2000.0. The resolutions began a series of activities that led to the International Celestial Reference System (ICRS), International Celestial Reference Frame (ICRF), Hipparcos reference star catalog, fainter reference star catalogs consistent with the ICRF, the IAU2000A precession–nutation model, IAU 2006 precession theory, and continuing reference system improvements.

9.5 New Timescales

Considering general relativity, concepts were required for celestial reference systems having origins at either the geocenter or barycenter of the solar system, called the Geocentric Celestial Reference System (GCRS) and the Barycentric Celestial Reference System (BCRS), respectively. Coordinate timescales appropriate for both coordinate systems were also required.

In the framework of general relativity, coordinate time is one of the coordinates in a four-dimensional space-time system. In contrast, proper time is time interval measured by a clock between events that occur at the same place as the clock. Coordinate times can be specified for any space-time event. They cannot be measured; they can only be computed from readings of real clocks. The computation must be based on the theoretical relation between proper and coordinate time from the principles of general relativity. The relation involves the model of the solar system, the motions of the observer and the massive bodies, and the mass parameters. The coordinate times corresponding to the coordinate systems, which have their spatial origins at the center of mass of the Earth and at the solar system barycenter, were designated as Geocentric Coordinate Time (TCG) and Barycentric Coordinate Time (TCB), respectively.

The IAU recommended that the coordinate times should be derived from a timescale realized by atomic clocks operating on the Earth. They were to be chosen so they would be consistent with the proper unit of time, the SI second, and for proper length of the SI meter, related to the SI second by the speed of light in vacuum. The SI second is specified as the duration of a specified number of radiation periods of caesium, but it contains no specification of location, gravitational potential, or state of motion of the observer. It is the unit of proper time for any location and defined by:

"The second is the duration of 9 192 631 770 periods of the radiation corresponding to the transition between the two hyperfine levels of the ground state of the cesium 133 atom."

(Bureau International des Poids et Mesures, 2014)

The readings of the coordinate times were set to be 1977 January 1, $0^h\ 0^m\ 32^s.184$ TAI exactly at the geocenter.

Since we are on the Earth, geocentric reference frames and timescales have a special importance, and so it is particularly relevant to establish the relationships among timescales within that context. The center of mass of the Earth can be considered as freely falling in space. Despite its irregular motion within the Earth, amounting to a centimeter or so, we can consider it as a well-defined point and neglect the internal forces and motions that affect it. If we have a clock at rest on the surface of the Earth experiencing a potential U and measuring a proper time interval, $d\tau$, then we can relate that time interval to a geocentric coordinate time interval, dt, neglecting terms in c^{-4}, by using the expression

$$\frac{d\tau}{dt} = 1 - \frac{U}{c^2},\qquad(9.2)$$

where c is the speed of light and U is the total potential composed of the gravitational potential, U_G, and the centrifugal potential, U_R, due to the Earth's rotation as experienced at the clock's location (Audoin & Guinot, 2001). The latter component is given by:

$$U_R = \tfrac{1}{2}\omega^2\left(x^2 + y^2\right),\qquad(9.3)$$

for a clock with geocentric coordinates x, y, z, (z being the distance above the plane of the equator) and the Earth's rotational speed being given by ω.

9.5.1 Terrestrial Time (TT)

The coordinate time to be used with a reference system on the rotating geoid is called Terrestrial Time (TT). It is a coordinate time, and as such, cannot be measured directly but only computed using the readings of real clocks. The proper time of an observer on the rotating geoid is, indeed, close to TT up to terms of order 10^{-17} and up to periodic terms with an amplitude of 1 picosecond (ps) (Klioner, 2008). Thus, TT has only a small potential drift with respect to the statistical timescale TAI derived from physical clock readings. It is not a dynamical timescale, but, as it is related to TCG by a constant rate, it serves as the bridge from TAI to the coordinate timescales.

TAI, which is, in practice, derived by the Bureau International des Poids et Mesures (BIPM) by comparing the readings of many atomic clocks reduced to the geoid, is a realization of the coordinate time "TT-32.184s" (Klioner et al., 2009). The constant 32.184s is the result of the fact that, when TAI was initiated in 1958, it was set equal to the epoch of Universal Time at that time instead of Ephemeris Time, because Ephemeris Time was so poorly determined. However, it was later decided that it would be preferable to have TAI be continuous with ET. Consequently, the constant 32.184s was adopted as the best estimate of the accumulated difference between TAI and ET as of January 1, 1977. At the instant 1977 January 1, $0^h \, 0^m \, 0^s$ TAI exactly, TT had the reading 1977 January 1, $0^h \, 0^m \, 32^s$. 184 exactly. Considering current clock accuracies, the mean rate of TT is not significantly different from the proper time of an observer on the rotating geoid (Klioner, 2008), and so for practical purposes:

$$TT = TAI + 32.184\text{s.} \tag{9.4}$$

Since TAI is almost computed in real time, with operational constraints, it is not an optimal realization of TT. Hence, BIPM computes TT(BIPM) based on a weighted average of the TAI frequency from primary frequency standards. TT(BIPM) is computed each January and an extrapolation is published each month and available at ftp://tai.bipm.org/TFG/TT(BIPM)/TTBIPM.12.ext (Arias, 2013; Petit & Arias, 2015).

9.5.2 Geocentric Coordinate Time (TCG)

TCG is the time coordinate for the four-dimensional geocentric coordinate system, and in terms of the theory of relativity, differs slightly in rate from TT. It was set to coincide with TT on 1977 January 1, $0^d \, 0^h \, 32^s.184$ (JD 2442144.5003725). TT and TCG differ only by a scaling factor so that:

$$TCG = TT + L_G \times (JD - 2443144.5) \times 86400s. \tag{9.5}$$

$L_G = 6.969290134 \times 10^{-10}$ exactly, and is specified as a defining constant that will not change with future improved Earth models. This overcomes difficulties due to temporal changes of the geoid and the intricacy of its definition.

9.5.3 Barycentric Coordinate Time (TCB)

A particularly important reference system in astronomy is that centered at the barycenter of the solar system. This point is chosen in such a way that its motion in the local galaxy is linear, if one neglects the gravitational potential of the galaxy. One of the problems of astrometry is to refer observations to this barycentric system whose associated coordinate timescale is TCB. TCB is the time coordinate for the four-dimensional barycentric coordinate system of the solar system, differing both in secular and periodic effects from TCG, according to the relativistic metric being used. A complication in computing TCB for any event in the solar system is that the time-varying gravitational potentials, which must be accounted for, depend on knowing the positions of the planets.

If TCB and x_0 are the barycentric coordinates of an event in a barycentric reference system, the potential is described by the sum of potentials produced by N significant solar system bodies with assumed point-like masses M_i and positions x_i:

$$U_0 = G \sum_{i=1}^{N} \frac{M_i}{|x_i - x_0|}. \qquad (9.6)$$

If the event refers to the center of mass of the Earth, the potential is given by the summation presented earlier, but would not include the gravitational potential of the Earth itself. This potential can be designated U_0^{ext}. The relationship between TCB and TCG can be given by the full four-dimensional transformation:

$$TCB - TCG = c^{-2} \left[\int_{t_0}^{t} \left(v_e^2/2 + U_0^{ext}(x_e) \right) dt + v_e \bullet (x - x_e) \right] + O(c^{-4}), \qquad (9.7)$$

where c is the speed of light, x_e and v_e denote the barycentric position and velocity vectors of the Earth's center of mass, and x is the barycentric position of the observer. In the integral, $t = $ TCB and t_0 is chosen to agree with the epoch of Terrestrial Time. At the geocenter $x - x_e$ would be zero. The additional terms would be of order c^{-4}. This expression is ephemeris dependent, since it depends on x_e and v_e. Consequently, there is a "time ephemeris" associated with every spatial ephemeris of solar system bodies expressed in TCB. It would be expected that spatial ephemerides will be accompanied by "time ephemerides" for the user.

As an approximation to TCB − TCG in seconds, one might use (McCarthy & Petit, 2004):

$$TCB - TCG = \frac{L_C \times (TT - TT_0) + P(TT) - P(TT_0)}{1 - L_B} + c^{-2} v_e \bullet (x - x_e), \qquad (9.8)$$

where P(TT) denotes nonlinear terms having a maximum amplitude of around 1.6 ms. They may be provided by numerical time ephemerides (Fairhead & Bretagnon, 1990; Irwin & Fukushima, 1999; Moisson & Bretagnon, 2001). Additional terms in c^{-4} can be found in recommendation B1-4 of the IAU (*Trans. IAU*, 2001). This expression has a linear term that represents the difference of their mean rates.

$$(TCB - TCG)_{secular} = L_C (JD - 2\ 443\ 144.5) \times 86400s, \qquad (9.9)$$

where the best value at present of $L_C = 1.48082686741 \times 10^{-8}$ with an uncertainty of 10^{-17}. In addition, there is a nonlinear variation described by a number of periodic terms depending on the various periods present in the motion of planets.

They are discussed in Fukushima (1995), who finds that there are 515 terms that are greater in amplitude than 0.1 ns. The most important terms in TCB − TCG are, in seconds,

$$TCB - TCG = 0.001658 \sin g + 0.000\,014 \sin 2\,g, \qquad (9.10)$$

where $g = 357.53° + 0.985003°$ (JD − 2451545.0) represents essentially the mean anomaly of the Earth's orbit. The epoch of TCB was set to be equal to TT and TCG at 1977 January 1, $0^d\ 0^h\ 32^s.184$ (JD 2443144.5003725).

9.5.4 TDB Redefined

In 1991, the continued use of TDB was authorized, in cases where discontinuity with previous work is deemed undesirable. The convenience of T_{eph} (Section 9.5.5) for use in operational ephemerides and problems with the original definition of TDB prompted the IAU in 2006 to redefine TDB as well. Hence, the IAU at the 2006 General Assembly in Prague redefined TDB for appropriate applications with the following recommendation.

"In situations calling for the use of a coordinate time scale that is linearly related to Barycentric Coordinate Time (TCB) and remains close to Terrestrial Time (TT) at the geocenter for an extended time span, TDB is defined as the following linear transformation of TCB

$$TDB = TCB - L_B \times (JD_{TCB} - T_0) \times 86400 + TDB_0, \qquad (9.11)$$

where $T_0 = 2443144.5003725$, and $L_B = 1.550519768 \times 10^{-8}$ and $TDB_0 = -6.55 \times 10^{-5}$ s are defining constants."

There were the following notes to the resolution (*Trans. IAU*, 2007):

"1. JD_{TCB} is the TCB Julian date. Its value is $T_0 = 2443144.5003725$ for the event 1977 January 1 00h 00 m 00s TAI at the geocenter, and it increases by one for each 86400s of TCB.

2. The fixed value that this definition assigns to L_B is a current estimate of $L_C + L_G - L_C \times L_G$, where L_G is given in IAU Resolution B1.9 (2000) and L_C has been determined (Irwin & Fukushima, 1999) using the JPL ephemeris DE405. When using the JPL Planetary Ephemeris DE405, the defining L_B value effectively eliminates a linear drift between TDB and TT, evaluated at the geocenter. When realizing TCB using other ephemerides, the difference between TDB and TT, evaluated at the geocenter, may include some linear drift, not expected to exceed 1 ns per year.

3. The difference between TDB and TT, evaluated at the surface of the Earth, remains under 2 ms for several millennia around the present epoch.

4. The independent time argument of the JPL ephemeris DE405, which is called T_{eph} (Standish, 1998) [see section 9.5.5], is for practical purposes the same as TDB defined in this Resolution.

5. The constant term TDB_0 is chosen to provide reasonable consistency with the widely-used TDB − TT formula of Fairhead & Bretagnon (1990). *n.b.* The presence of TDB_0 means that TDB is not synchronized with TT, TCG, and TCB at 1977 Jan 1.0 TAI at the geocenter.
6. For solar system ephemerides development the use of TCB is encouraged."

The difference between TDB and TT does not exceed 0.002 seconds and, for many practical applications, can be neglected (Klioner, 2008).

9.5.5 *Barycentric Ephemeris Time (T_{eph})*

Through the series of revisions of timescales, changes in names of the timescales, and further clarifications and improvements in the applications of relativity to the timescales, the Jet Propulsion Laboratory followed an independent, but somewhat parallel path. The basic calculations of the barycentric ephemerides were per-formed using equations of motion, including the needed relativistic terms and using an independent variable that was a relativistic timescale for the solar system barycenter. Barycentric Ephemeris Time is a coordinate time related to TCB by an offset and a scale factor. Ephemerides based upon the coordinate time T_{eph} are automatically adjusted in the ephemeris creation process so that the rate of T_{eph} has no overall difference from the rate of TT (Standish, 1998), and also no overall significant difference from the rate of the redefined TDB. For this reason, space coordinates obtained from the ephemerides are consistent with TDB. The differ-ence between T_{eph} and TDB is that T_{eph} depends on the ephemeris used, while TDB has been defined by a particular relationship with respect to TCB. Specifically, this mean that the constants L_B and TBD_0 may be different for different ephemerides from those values that define TDB as shown in Equation (9.11). The numerical values will depend on the definition of the transformation between TT and T_{eph} used during the construction of the particular ephemeris (Klioner, 2008).

9.6 ΔT and Ephemeris Time Revised

The development of coordinate timescales further refined the original concept of Ephemeris Time (Chapter 6) to meet the growing needs for improved accuracy. In the recommendations that implemented the dynamical timescales no mention was made of officially changing the definition of ΔT. It was originally defined as ΔT = ET − UT, and, strictly speaking, that remains the definition for times prior to 1983. From 1984 to 2000, after TDT was introduced as a replacement for ET in 1984, it was assumed that ΔT = TDT − UT. Similarly, following the introduction of TT in 2001, ΔT became TT − UT. The existence of atomic time since 1956 together with expression ΔT = TDT − UT and equation (9.1) also allowed us to estimate

ΔT for the years following the introduction of atomic time by $\Delta T = TAI + 32.184s$ $-$ UT. Although TAI did not formally exist before 1971, it was considered an extension of the Bureau International de l'Heure atomic timescale that had been continuous back to 1955 (Nelson et al., 2001).

Having defined the new set of dynamical and coordinate timescales and resolved the various problems that had existed, there remained the problem that there was no atomic timescale prior to 1955. New numerically integrated ephemerides replaced the theories of Newcomb, and they did not have a numerical expression for the mean longitude of the Sun. Hence, while ET is to be used prior to 1955, there has not been a formal revised definition of Ephemeris Time, as it is currently being used for the reduction of observations and the determinations of ΔT prior to 1955. A definition for Ephemeris Time Revised (ET_R), which matches what is being used in practice, was proposed by Guinot and Seidelmann (1988):

a). Ephemeris Time Revised (ET_R) is reckoned from the instant 1958 January 0^h TAI, at which time ET_R has the value 1958 January 0^h 0^m $32^s.184$, and

b). The unit of time of ET_R is the SI second, in its present definition (atomic second).

This would make Ephemeris Time continuous with Terrestrial Time prior to 1955 to the accuracies involved. It would also be continuous with TDT, but it would not be continuous with TCB or TCG, which differ secularly from TT. No formal action has been taken defining ET_R, but the practice for determining a dynamical time and values of ΔT prior to 1955 has basically followed the foregoing definition.

Values of ΔT are given annually in *The Astronomical Almanac* from 1620 to the present. Since UTC has steps due to leap seconds, the values of TAI $-$ UTC are also given from 1972 to the present, and extrapolated values are given for a few years into the future. Values are also available from the International Earth Rotation and Reference system Service (IERS) and US Naval Observatory (USNO) websites.

9.7 Relationships among Coordinate Timescales

Dynamical timescales can be related to TT by mathematical expressions. Barycentric Dynamical Time has periodic differences from TT, but has a rate very close to the rate of TT. Equation 9.1 shows the relationship of TAI to the dynamical timescales TDB and TDT as these scales were originally conceived.

The coordinate timescales can all be related mathematically as illustrated in Figure 9.2. That shows that, for practical applications, they can be related to TT and its convenient realization, TAI. The details of the formation of TAI and its relation to other timescales are discussed in following chapters. Figure 9.2 shows the constant rate difference between Geocentric Coordinate Time (TCG) and TT, and

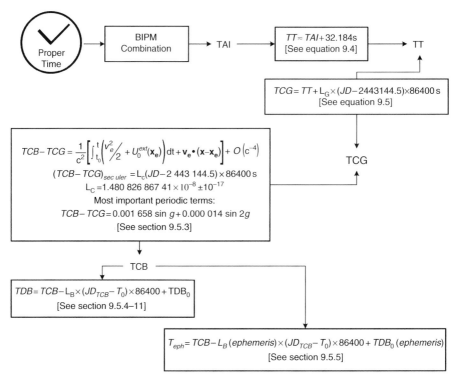

Figure 9.2 Coordinate timescales

both rate and periodic differences between TCB and TCG. The magnitudes of the differences between these different timescales are shown in Figure 9.3, originated by Seidelmann and Fukushima (1992).

Terrestrial Time is a scaled version of TCG, and TDB is a scaled version of TCB (Klioner, 2008). There is no physical significance to either scaling; they are both for the convenience of making the difference between the proper time of an observer and the coordinate times evaluated along his trajectory as small as possible. The relationships between proper and coordinate times, and between two coordinate times, are independent of the units. The SI second is the unit of proper time. If the theoretical formulas used relate proper times to coordinate times, with the proper time units of the SI second, the units of time of the coordinate time would also be the SI second. Scaled coordinate timescales are accompanied by scaled spatial coordinates and mass parameters in order to keep the equations of motion of celestial bodies and photons invariant (Klioner et al., 2009).

Therefore, Klioner et al. (2009) suggest that nomenclature such as "TDB units" and "TT units" be avoided, and that they should not be contrasted with "SI units." Instead the nomenclature "TCG-compatible," "TCB-compatible,"

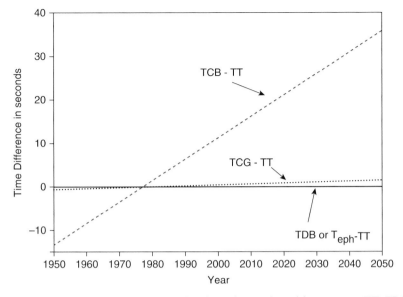

Figure 9.3 Differences in readings of various timescales with respect to TT, TDT, and ET equivalent to TT

"TT-compatible," or "TDB-compatible" should be used in referring to units in the respective reference systems. The names of the units second and meter for numerical values of all these quantities, they suggest, should be used without any adjectives.

References

Arias, E. F. (2013). Atomic Time Scales for the 21st Century. *RevMexAA(Serie de Conferencias*, **43**, 29.

The Astronomical Almanac. Washington, DC: US Government Printing Office.

Audoin, C. & Guinot, B. (2001). *The Measurement of Time*. Cambridge: Cambridge University Press.

Bureau International des Poids et Mesures (2014). *The International System of Units (SI)*.

Fairhead, L. & Bretagnon, P. (1990). An Analytical Formula for the Time Transformation TB-TT. *Astron. Astrophys.*, **229**, 240.

Fairhead, L. Bretagnon, P., & Lestrade, J. F. (1988).The Time Transformation TDB-TDT: An Analytical Formula and Related Problem of Convention. In A. K. Babcock & G. A. Wilkins, eds., *The Earth Rotation and Reference Frames for Geodesy and Geodynamics*. Dordrecht: Kluwer, 419.

Fricke, W., Schwan, H., & Lederle, T. (1988). *Fifth Fundamental Catalogue, Part I*. Heidelberg: Veröff. Astron. Rechen Inst.

Fukushima, T. (1995). Time Ephemeris, *Astron. Astrophys.* **294**, 895–890.

Guinot, B. & Seidelmann, P. K. (1988). Time Scales: Their History, Definition, and Interpretation. *Astron. Astrophys.* **194**, 304–308.

Hirayama, T., Kinoshita, H., Fujimoto, M.-K., & Fukushima, T. (1987). Analytical Expression of TDB-TDT. *Proc. IAG Symposia, IUGG XIX General Assembly*, Vancouver, August 10–33, 91.

Hughes, J. A., Smith, C. A., & Kaplan, G. H. (1991). *IAU Proc. 127th Colloquium, Reference Systems*. Washington, DC: US Naval Observatory.

Irwin, A. W. & Fukushima, T. (1999). A Numerical Time Ephemeris of the Earth. *Astron. Astrophys.*, **348**, 642.

Kaplan, G. H. (1981). The IAU Resolutions on Astronomical Reference System, Time Scales, and the Fundamental Reference Frames. *USNO Circular 163*. Washington, DC: US Naval Observatory.

Klioner, S. A. (2008). Relativistic Scaling of Astronomical Quantities and the System of Astronomical Units. *Astron. Astrophys.*, **478**, 951–958.

Klioner, S. A., Capitaine, N., Folkner, W. M., Guinot, B., Huang, T.-Y., Kopeikin, S., Pitjeva, E., Seidelmann, P. K., & Soffel, M. (2009). Units of Relativistic Time Scales and Associated Quantities. In S. A. Klioner, P. K. Seidelmann, & M. H. Soffel, eds., *Relativity in Fundamental Astronomy, Proceedings IAU Symposium No 261*. Cambridge: Cambridge University Press, pp. 79–84.

McCarthy, D. D. & Petit, G. P., eds. (2004). *IERS Conventions (2003), IERS Technical Note 32*, Frankfurt am Main: Verlag des Bundesamts für Kartographie und Geodäsie.

Moisson, X. & Bretagnon, P. (2001). Analytical Planetary Solution VSOP2000. *Celest. Mech. & Dynam. Astron*. **80**, 205–213.

Moyer, T. D. (1981). Transformations from Proper Time on Earth to Coordinate Time in Solar System Barycentric Space-Time Frame of Reference. *Celestial Mechanics* **23**, 33–56 and 57–68.

Nelson, R. A., McCarthy, D. D., Malys, S., Levine, J., Guinot, B., Fliegel, H. F., Beard, R. L., & Bartholomew, T. R. (2001). The Leap Second: Its History and Possible Future. *Metrologia*, **38**, 509–529.

Petit, G. & Arias, F.(2015). Long Term Stability of Atomic Time Scales. *Highlights of Astronomy*, **16**, 209.

Petit, G. & Luzum, B., eds. (2010). *IERS Conventions (2010)*. Verlag des Bundesamts für Kartographie und Geodäsie: Frankfurt am Main.

Seidelmann, P. K. & Fukushima, T. (1992). Why New Time Scales? *Astron. Astrophys.*, **265**, 833–838.

Soffel, M., Klioner, S. A., Petit, G., Wolf, P., Kopeikin, S. M., Bretagnon, P., Brumberg, V. A., Capitaine, N., Damour, T., Fukushima, T., Guinot, B., Huang, T., Lindegren, L., Ma, C., Nordtvedt, K., Ries, J., Seidelmann, P. K., Vokrouhlicky, D., Will, C., & Xu, Ch. (2003). The IAU 2000 Resolutions for Astrometry, Celestial Mechanics, and Metrology in the Relativistic Framework: Explanatory Supplement. *Astron. J.*, 126, 2687–2706.

Standish, E. M. (1998). Time Scales in the JPL and CfA Ephemerides. *Astron. Astrophys.*, 336, 381–384.

Trans. Int. Astron. Union, Vol. XVI B Proc. 16th General Assembly Grenoble, 1976 (1977) E. Müller and A. Jappel, eds. Washington, DC: Association of Universities for Research in Astronomy.

Trans. Int. Astron. Union, Vol. XVII B Proc. 17th General Assembly Montreal, 1979 (1980) P. Wayman, ed. Washington, DC: Association of Universities for Research in Astronomy.

Trans. Int. Astron. Union, Vol. XXI B Proc. 21st General Assembly Buenos Aires, 1991 (1992). J. Bergeron, ed. Dordrecht: Kluwer Academic Publishers.

Trans. Int. Astron. Union, Vol. XXIV B Proc. 24th General Assembly Manchester, 2000. (2001). H. Rickman, ed. San Francisco, CA: Astronomical Society of the Pacific.

Trans. Int. Astron. Union, Vol. XXVI B Proc. 26th General Assembly, Prague, 2006 (2007). K. A. van der Hucht, ed. Cambridge: Cambridge University Press.

10

Clock Developments

10.1 Introduction

In the English language, the word *clock* is related to the medieval Latin word *clocca* and to the French *cloche*. Both words mean "bell," and refer to the original purpose of clockwork, that is to ring bells at desired time intervals (Whitrow, 1988). Today we use the word *clock* to refer to any device used to measure the passage of time. These may range from the "clocks" of antiquity that provided measures of time by the flow of water to the most sophisticated clocks of recent times. Generally speaking, we can use the word to mean any manmade instrument used to measure time that does not depend on our ability to see the skies. That would then exclude those devices that make use of the length or direction of the Sun's shadow, or the apparent motion of the stars in the sky to tell time.

In the European Middle Ages, the word *horologium* was used to refer to any device used to keep time. The continued use of this word throughout the Middle Ages makes it difficult now to determine who was responsible for the first mechanical clock and when and where it came into existence, because writers of the early Middle Ages did not distinguish among the different timekeeping devices. Clock development reflects the overall growth in technological ability as well as society's requirements for accurate time to meet everyday needs.

10.2 Keeping Time in Antiquity

Early societies made use of the heavens to meet their need for timing. Measurement of the passage of time in terms of years and days was fairly straightforward. The direction of the Sun's shadow, or its length, are obvious ways to mark time's passage during the day. Instrumentation can be as simple as a stick stuck into the ground or as complex as the most sophisticated sundials. During the night, the motion of the stars can be used to indicate the passage of the seasons as well as the passage of time.

Egyptians were the first to divide the day into shorter time intervals. Their system was to divide the daylight into 10 equal parts, to allow an hour each for morning and evening twilight, and to divide the night into 12 equal parts. This provided for a day of 24 hours, but the length of the hours would vary depending on the seasons. In the summer, when the period of daylight would be longer, the daylight hours would be longer and the nighttime hours shorter. The opposite would occur in winter. Nighttime would end with the heliacal rising of a particular star or constellation. Heliacal rising refers to the rising of a celestial object at the same time as the Sun. The apparent motion of the Sun in the sky, due to the Earth's orbital motion, means that heliacal risings of these asterisms will change throughout the year. Egyptian priests chose to designate the heliacal rising of a different object at 10-day intervals to serve as the markers for the passage of time during the year. These asterisms are known as "decans," referring to their change at 10-day intervals.

The Egyptian calendar called for 365 days, and so there would be 36 decans with the last five days at the end of the year left over for festivals. The identity of all of the decans is not certain, but the star Sirius is known to be one of them. At the time of the heliacal rising of Sirius, or the beginning of the Egyptian year, there are 12 decans that pass during the night. As a result, the nighttime was divided into 12 hours, but with hours of unequal length called "seasonal" hours. This process resulted in our 24-hour day.

Hellenistic astronomers used a system of hours of equal length for scientific purposes, but the system of seasonal hours continued to be used in everyday practice well into the Middle Ages. The length of the equal hours was equivalent to the seasonal hours at the time of the equinoxes. These astronomers, following the Babylonian sexagesimal system of counting, further divided the hour into 60 "firsts," each containing 60 "seconds," resulting in our common division of the days into hours, minutes, and seconds (Whitrow, 1988). With the growth of civilizations, the need for time without reference to the sky became a concern, and the difference between the system of seasonal and equal hours posed a problem for the development of clocks through the ages.

10.2.1 Clepsydrae and Water "Clocks"

The measured flow of water, either into or out of a vessel, was an early means of dealing with the need to have a non-astronomical source of time. A clepsydra (pl.: clepsydrae) depends on the uniform flow of water to measure time. The word "clepsydra" is derived from the Greek words for "to steal" and "water" and literally means a water thief. It refers to the fact that early forms of clepsydrae were developed from simple siphon devices used to lift water liquids from one vessel to another. Clepsydrae were mentioned in Indian and Chinese texts of the first

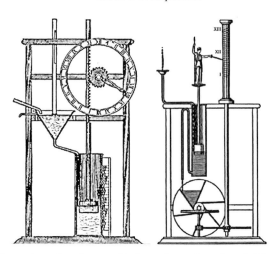

Figure 10.1 Examples of outflow and inflow clepsydrae from Milham (1923) with
permission of Oxford University Press

millennium BC as well as in Babylonian texts. An example of an Egyptian version
can be found at Karnak dating to about 1400 BC. This device was even able to
allow for the changing length of the hours of day and night during the year.
In China, clepsydrae were available from about 1500 BC, apparently making use
of technology from Babylon.

Greek versions appeared later, when simple outflow timers were used in
Athenian law courts to limit the length of speeches. In the third century BC, the
Greeks developed an inflow clepsydra with a means to regulate the water pressure
and thus improve its precision. Such a device was used in the Tower of the Winds in
Athens to provide public time (Dohrn-van Rossum, 1996). Figure 10.1 shows two
different types. Clepsydrae adapted for use as water "clocks" were used well into
the Middle Ages. Although they were used widely in monasteries, they also found
civil applications, and were even used in medical applications of the time.

10.2.2 Other Timekeeping Devices

The flow of water was not the only means used in early timekeeping. Burning
calibrated candles or incense sticks provided one means of keeping time without
access to the skies. Candles were popular in the Western and Islamic worlds, while
incense was used in Asia (Dohrn-van Rossum, 1996).

10.3 The First Mechanical Clocks

The first weight-driven mechanical clocks began to appear in Europe in the latter
part of the 13th century. There is no documented evidence of the first such clock or

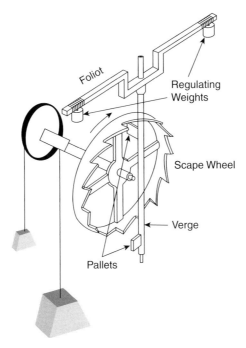

Figure 10.2 Schematic depiction of a verge-and-foliot clock

the person who built it. Water clocks, based on gear trains regulated by the flow of water, had been in use for some time, and it seems possible that a geared clock based on the regulated fall of a weight might have been a natural progression.

The mechanical clock came into existence with the invention of the verge escape-ment and foliot regulator (see Figure 10.2). An escapement is a device that controls the continuous motion of the clock's driving mechanism using the periodic motion of its regulator. A weight attached to a pulley would be expected to fall without stopping, but an escapement was introduced to limit the fall to small increments. An arm extending from the "verge" would then stop the fall and the wheel would apply a rotational impulse to the verge. That releases the wheel until it contacts the second arm of the verge. A crossbar on the verge, called a "foliot," controls the time it takes for the verge to rotate. The timing is changed by adjusting the positions of the regulating weights. Errors of clocks that used this mechanism were typically about 15 minutes per day. The advent of these verge-and-foliot clocks led to the gradual decline of water clocks and the development of more sophisticated variants for everyday use.

10.4 Pendulum Clocks

The next important step in clock development occurred with the adaptation of the pendulum to mechanical clockworks. This advance began with the work of Galileo

Galilei in the late 16th century, and while no longer used as the source of our most precise time, the pendulum clock did serve in that capacity well into the 20th century.

The period of a pendulum's swing is given by the equation

$$T = 2\pi\sqrt{\frac{\ell}{g}}\left(1 + \frac{1}{4}\sin^2\frac{\theta}{2} + \frac{9}{64}\sin^4\frac{\theta}{2} + \ldots\right), \qquad (10.1)$$

where ℓ is the length of the pendulum, θ is the maximum semi-amplitude of the pendulum swing, and g is the acceleration of gravity. The period then depends chiefly on the pendulum's length for small angles and does not depend on its mass. However, the period does depend on the strength of the local gravity vector, and so the period of the motion will depend on the pendulum's location.

10.4.1 Galileo

Galileo's secretary and earliest biographer, Vincenzo Viviani, tells us that Galileo's interest in pendulums began when he was a student in Pisa, where he noticed the motion of a suspended lamp in the cathedral. His notes on pendulums began in 1588, and in 1602, he wrote to a colleague about his observations, that the swings of a pendulum were isochronous and that pendulums of the same string length had the same period of swing. In practice, the pendulum was not used as a clock at that time, but it was used to measure heart rates and in scientific experiments. In using the pendulum, its oscillations would be made to continue by an assistant occasionally applying impulses. It was not until 1641, a year before his death, that Galileo began to consider using a pendulum to regulate a clock (see Figure 10.3). Viviani, writing in 1759, 17 years after Galileo's death, described the scene:

"One day in 1641, while I was living with him at his villa in Arcetri, I remember that the idea occurred to him that the pendulum could be adapted to clocks with weights or springs, serving in place of the usual *tempo*, he hoping that the very even and natural motions of the pendulum would correct all the defects in the art of clocks. But because his being deprived of sight prevented his making drawings and models to the desired effect, and his son Vincenzio coming one day from Florence to Arcetri, Galileo told him his idea and several discussions followed."

(Drake, 1978)

10.4.2 Huygens

Christiaan Huygens (1629–1695) receives the credit for the next significant advance in clock development with his design for the first successful operational pendulum clock in 1656. Huygens arranged for the patent for his clock to be assigned to a clockmaker named Salomon Coster who was able to manufacture

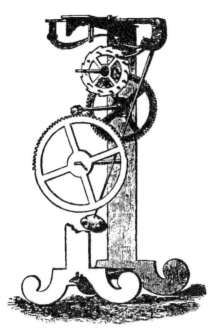

Figure 10.3 Viviani's drawing of Galileo's concept for a pendulum clock from Whitrow (1988)

a few clocks before his death in 1659. His initial effort had an error of less than one minute per day, but subsequent improvements reduced the error to about 10 seconds per day.

Johannes Hevelius, an astronomer who was in the process of building an observatory in Gdansk, had employed a clockmaker, Wolfgang Günther, to work on producing a pendulum clock, but he was unable to develop his clock before Huygens's success. In 1673 Hevelius wrote in his *Machinae Coelestis, Pars Prior* (van Leeuwen, 2007 transl.):

"Around this time, whilst the two pendulum clocks were being worked upon by the clockmaker but which were not completely finished (the clockmaker had little time available due to his work on the larger astronomic instruments), the very distinguished and very scholarly Christiaan Huygens invented similar clocks in 1657. This was also a very successful enterprise, and, a short time later, in 1658, he published an illustration of the pendulum clock, to the great advantage of (scientific) literature and for which I congratulate him. For, this prestigious invention offers an excellent remedy for all the ills of clocks built as yet, as well as solving the problems of inaccuracies which have crept into the escapements as well as the axles, pins and cog-wheels."

The 1658 illustration to which Hevelius refers is contained in a booklet entitled *Horologium*. Huygens pursued his clock development and went on to discover that the circular arc of the pendulum was not truly isochronous. However, if the

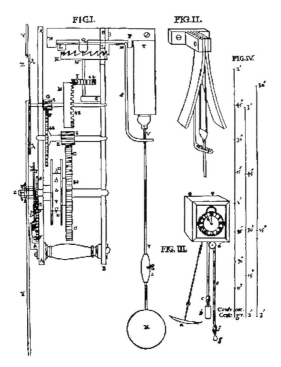

Figure 10.4 Diagram of Huygens's pendulum clock from his *Horologium Oscillatorium*

pendulum bob were constrained to follow the path of a cycloid, it would indeed provide an isochronous regulator. The word *cycloid* refers to the path of a point on the rim of a rolling wheel. In practice, Huygens did this by employing metal cheeks that constrained the pendulum to follow a cycloidal curve. This development was described in his 1673 publication *Horologium Oscillatorium*. Figure 10.4 reproduces the diagram for Huygens's clock as presented in that publication.

Huygens was also responsible for the first attempt to use time as a means to define a standard of length, when he suggested that the length of a pendulum that produced a particular time could be used as a standard length. However, since the period depends on local gravity, it could not be considered seriously in the definition of a conventional standard.

Huygens's clock still made use of a verge escapement that allowed the weight to drop about five centimeters in one hour. Until then, clocks generally used only an hour hand to indicate time. The improved accuracy of his clock allowed him to add minute and second hands. However, second hands did not become common until further improvements were made in the escapement.

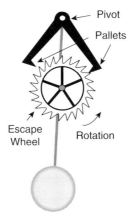

Figure 10.5 Anchor escapement after Whitrow (1988)

10.4.3 Pendulum Clock Developments

The first improvement in clock escapements came in the late 1660s, when the first "anchor" escapement appeared. Its invention is usually credited to William Clement (Whitrow, 1988), but Robert Hooke and Joseph Knibb are also mentioned as possible inventors. Although Huygens's design corrected for the fact that the circular swing of the pendulum was not isochronous, the invention of the anchor escapement allowed pendulum clocks to operate with a reduced swing, making the addition of the metal cheeks less important. The swing of the pendulum could be reduced from around 100° to about 5°. The reduced angle through which the pendulum must swing also resulted in the familiar shape of the longcase or "grandfather" clocks.

In the anchor escapement, the pallets look like a ship's anchor as seen in Figure 10.5. The swinging of the pendulum moves each pallet alternately between the escape wheel teeth. This wheel moves a little in the interval between the time one pallet moves away and the other engages another tooth.

The development of the pendulum did not stop there, however. A well-known family of clockmakers in England, the Fromanteels, sent Johannes Fromanteel to the workshop of Salomon Coster, Huygens's clockmaker, in 1657 to learn the details of the pendulum clock. On his return in 1658, the Fromanteels were able to build the first English pendulum clocks. They went on to introduce new features and to set the basic design for longcase clocks into the 18th century (Landes, 1983).

The development of pendulum clocks was so successful that by the end of the 17th century, they could be used for astronomical observations. Thomas Tompion (1639–1713) had finished making two 13-foot pendulums for the new Royal Observatory at Greenwich England in 1676. In 1715, George Graham, Tompion's partner, invented the "dead-beat" escapement, which provided a short impulse to the pendulum when it was nearly vertical. Other improvements over the following centuries dealt with compensating for the effects of temperature on the length of the pendulum, as well as changes in air pressure on the pendulum motion (Landes, 1983).

The most precise clocks were installed in observatories and were referred to as "astronomical regulators." These became the devices used to distribute time locally and eventually nationally. Astronomical observations would be made at the observatories in order to maintain the clock's long-term agreement with time determined from the Earth's rotation.

In 1889, the Riefler escapement was invented by Siegmund Riefler. Designed for use with regulators, it made use of a suspension spring, to keep the pendulum swinging. This spring not only suspended the pendulum but it was also able to provide impulses to the pendulum, made of invar and minimally entangled with the escape wheel (Milham, 1923; Sullivan, 2001). Riefler clocks (Figure 10.6) served as the basis for US time beginning in 1909.

In 1921, William H. Shortt invented a new pendulum clock that would soon replace the Riefler clocks for use as regulators. It actually made use of two pendulums (Figure 10.7), the master pendulum that actually kept the time and a slave pendulum that provided impulses to the master every 30 seconds (Hope-Jones, 1931). The pulses were delivered electrically to the master pendulum that was again made of invar, to reduce thermal effects, and housed in a partial vacuum. This form of clock, manufactured by the Synchronome Company, became the basis for US time in 1929 (Sullivan, 2001), and with an accuracy of one millisecond per day, was the first clock capable of measuring variations in the Earth's rotation.

10.4.4 Chronometers

In addition to the development of the regulator clocks, significant progress was also being made in the development of chronometers. The word *chronometer* is usually applied to high-precision, portable timekeeping devices. Such devices have been used historically to determine longitude at sea or for maintaining railroad schedules. The most significant development in this area of timekeeping was the work of John Harrison (1693–1776). The Longitude Act of Great Britain, passed in 1714, offered a first prize of £20,000 to anyone who could find longitude at sea with accuracy better than half a degree. Harrison proposed to do this by using a clock, but it meant that he would need to develop a portable clock that was accurate to better than three seconds in one day (Harrison, 1767). He succeeded with his Chronometer "H4" and eventually won the prize (Sobel, 1995). In France, Ferdinand Berthoud (1727–1807) also pioneered in producing marine chronometers (Berthoud, 1773).

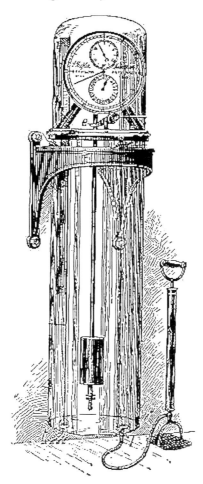

Figure 10.6 Riefler clock in airtight enclosure from Milham (1923) with permission of Oxford University Press

10.5 Quartz Crystal Clocks

The next significant advance in clock development occurred with the replacement of the Shortt clock by quartz crystal clocks. Timekeeping using these clocks is based on the oscillation of an electrical resonant circuit with current flowing through an inductor with inductance, L, and a condenser with capacitance C (Figure 10.8). The period T of the oscillation is given in hertz by the expression

$$T = 2\pi\sqrt{LC}, \tag{10.2}$$

where L is in henries and C is in farads.

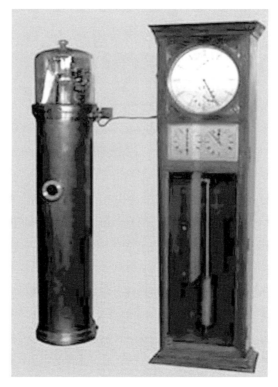

Figure 10.7 Shortt clock system. The master pendulum is on the left and the slave is on the right. National Institute of Standards and Technology (NIST).

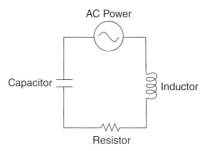

Figure 10.8 Electrical resonant circuit

The quartz crystal clock is the result of a long series of steps that began in 1857, when Jules Lissajous was able to sustain the mechanical motion of a tuning fork indefinitely using an electromagnet. A number of developments in electric oscillators occurred over the following years as outlined by Marrison (1948). Time and frequency standards using a tuning fork were gradually developed (Horton & Marrison, 1928; Dye & Essen, 1934), but during that time, quartz crystals were

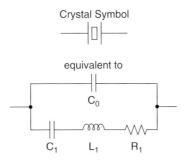

Figure 10.9 Electrical representation of a quartz crystal. The capacitance, C, the inductance, L, and the resistance, R, are inherent in the properties of the crystal itself, and are not separate components within the crystal. The capacitance C_0 is due to the wire connections across the quartz and can be measured physically.

also being investigated for possible application in oscillating circuits. The piezo-electric activity, mechanical and chemical stability of quartz, and the fact that it needed a relatively small amount of energy to maintain oscillations, made it particularly attractive (Katzir, 2015).

Piezoelectricity is the phenomenon describing the capability of some materials to generate electric potential when mechanically stressed. This ability is called the *direct piezoelectric effect*. The *converse piezoelectric effect* refers to the ability of a material to stress or strain when an electric potential is applied. The magnitude of the mechanical motion involved is quite small and only detected optically with high magnification. Some natural, as well as manmade materials, exhibit this capability, including quartz. Jacques and Pierre Curie first demonstrated the direct piezo-electric effect in 1880 (J. Curie & P. Curie, 1880, 1882) and through the following years, piezoelectricity has found many practical applications.

The natural piezoelectric quality of a quartz crystal means that it will resonate, or "ring," with a frequency depending on its dimensions, producing a voltage across points on its surface to which wires can be attached. Also, it means that the crystal will vibrate the most, if the voltage applied across the appropriate spots on its surface is applied with the proper resonant frequency. Quartz crystals can be produced, then, with the correct dimensions to achieve the desired frequencies in the circuits in which they are placed. In terms of electrical circuits the crystal can be drawn as in Figure 10.9.

Depending on its size and shape, each quartz crystal has its own natural frequency. If it is placed in an oscillating electric circuit that has nearly the same frequency, the crystal will vibrate at its natural frequency, and the frequency of the circuit will become the same as that of the crystal.

In 1917, Alexander McLean Nicolson first used a piezoelectric crystal to control the frequency of an oscillator (Marrison, 1948). Marrison also reports that:

"The first published quartz-controlled oscillator circuit is reproduced in Fig. 8A from Cady's 1922 article (Cady, 1922). In this oscillator the 'direct' and 'inverse' piezo-electric effects were employed separately, making use of two separate pairs of electrodes. The output of a three-stage amplifier was used to drive a rod-shaped crystal at its natural frequency through one pair of electrodes making use of the 'inverse' effect, while the input to the amplifier was provided through the 'direct' effect from the other pair. The feedback to sustain oscillations in the electrical circuit could be obtained only through the vibration of the quartz rod and hence was precisely controlled by it. Cady's results were received with widespread interest and were duplicated and continued in many laboratories, which soon resulted in many new discoveries and inventions."

The first quartz clock (Figure 10.10) was built in 1927 by Warren Marrison and J. W. Horton (Horton & Marrison, 1928; Marrison, 1948). Afterward, in the late 1930s, quartz crystal clocks began to replace the pendulum clocks as the time-keeping standards.

If the environment of the quartz oscillator is carefully controlled, its frequency will remain fairly stable. Carefully controlled laboratory quartz crystal clocks may accumulate errors of only a few thousandths of a second in a year.

10.6 Clock Performance

Clocks have been, and continue to be, used to provide the time of day in accordance with conventional standards. They also are used to provide time interval or the time between two events in conventional units. Time interval measurements need not depend on the time of day. To measure either time of day or time interval, clocks depend on the frequency of a repeating phenomenon. In many modern applications, the frequency, or time interval, provided by a standard may be more important than the time. For a repeating phenomenon with a period T measured in seconds, the frequency is given as $f = 1/T$ in units of hertz. Frequency comparisons often make use of a dimensionless fractional frequency given by $\Delta f/f$, where Δf is the measured frequency difference between two standards measured at frequency f. We refer to the process of setting two standards to read the same time as *synchronization*, and we refer to the process of setting two standards to the same frequency as *syntonization*.

The performance of a clock can be characterized by using a number of mathe-matical terms to describe how well it is, or remains, synchronized or syntonized. For some applications, it may not be necessary to relate a standard's time to a particular reference standard; the stability of its frequency may be the only characteristic of interest. In other applications, it may be important to characterize the clock's ability to reproduce the time provided by a conventionally accepted standard.

Figure 10.10 The first quartz clock on display at the International Watchmaking Museum, in La Chaux-De-Fonds, Switzerland

10.6.1 *Quality (Q) Factor*

A term used to describe oscillating systems is the "Q" or *quality factor*. It is defined as the ratio of the total energy in a system to the energy lost per cycle. In general, it refers to the comparison of the time constant describing the decay of an oscillation's amplitude to the period of the oscillation. This is equivalent to a comparison of the rate at which an oscillation dissipates energy to its oscillation frequency. A high value of Q would mean that an

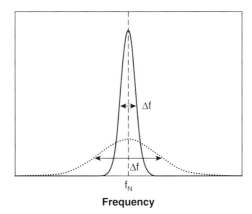

Figure 10.11 Response of a high-Q system (solid) compared to a low-Q system
(dotted)

oscillation would die out slowly. A pendulum with a high Q, for example,
would be expected to oscillate for a long time, but one with a low value
would die out relatively quickly.

For applications in modern timekeeping, the Q factor is related to the width of
the resonance phenomenon used to regulate the frequency of the timekeeping
device.

$$Q = \frac{f_N}{\Delta f}. \tag{10.3}$$

where f_N is the natural or resonant frequency and Δf is the bandwidth or the range
in frequencies that produces oscillation energy at least half of the peak value (see
Figure 10.11). A resonant system responds more strongly to a driving frequency
closer to its natural frequency. The response of a high-Q system decays much
more rapidly as the driving frequency moves away from its natural frequency, so
the system with the highest Q would be the most desirable as a frequency
standard.

10.6.2 Precision

Precision refers to the ability of repeated measurements to provide the same
result. For timekeeping applications, we can say that a device is precise if it
provides very repeatable measures of a chosen time interval, for example, the
length of a second. This can be measured statistically by the familiar standard
deviation, σ, of a sample of N measures of the quantity of interest x_i, with
mean $\bar{x}$.

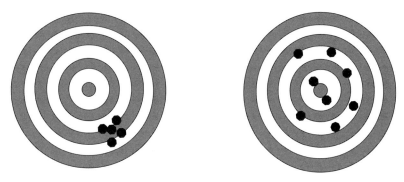

Figure 10.12 Examples of precision without accuracy (left) and accuracy without precision (right)

$$\sigma = \sqrt{\frac{\sum_{1}^{N} (x_i - \bar{x})^2}{N - 1}}. \tag{10.4}$$

10.6.3 Accuracy

In contrast to precision, *accuracy* describes the ability of measurements to conform to a standard value. Measurements could be accurate, but not precise, and vice versa (see Figure 10.12). For timekeeping applications, we might call a device accurate, if it reproduces the conventionally accepted standard value of a time interval after repeated measurements.

Statistically, this can be measured by the root mean square (RMS) of measures (x_i) of the difference in frequency, or time, between the clock in question and a conventionally adopted standard, or by determining the bias and standard deviation of those measures (see Figure 10.13).

$$\text{RMS} = \sqrt{\frac{\sum_{1}^{N} (x_i)^2}{N}}, \tag{10.5}$$

Figure 10.14 shows the improvement in timekeeping accuracy with time.

10.6.4 Stability

Stability refers to the ability of a standard to maintain its synchronization or syntonization over time. A stable clock would be one that would produce the same measures over a range of time intervals. It does not necessarily have to be accurate.

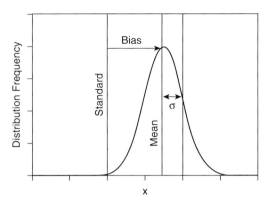

Figure 10.13 Representative distribution plot showing the relationship between the mean, bias, and standard deviation of measurements of the difference between a clock reading and the conventional standard

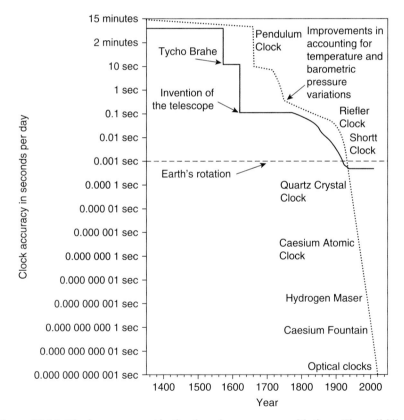

Figure 10.14 The improvement in timekeeping accuracy with time. The solid line shows the accuracy of astronomical optical measurements. The dotted line shows the accuracy of clock measurements. The dashed line shows the accuracy of the Earth rotation to provide an accurate timescale.

If we model the time produced by a clock to be:

$$t_c = t + x(t),$$ (10.6)

where t is the accurate "true" time and $x(t)$ is the time varying error in time (or phase of an output sinusoidal signal), then the error in the clock frequency is given by:

$$y(t) = \frac{dx(t)}{dt},$$ (10.7)

Statistically we can represent the variance of the errors in time and frequency, respectively, in terms of a Fourier representation by:

$$var(x) = \int_1^\infty S_x(f) \, df \text{ and } var(y) = \int_1^\infty S_y(f) \, df,$$ (10.8)

where f is the Fourier frequency and $S_x(f)$ and $S_y(f)$ are the spectral densities of the errors in time and frequency, respectively. The spectral densities of time and frequency are related by the expression:

$$S_y(f) = 4\pi^2 f^2 S_x(f).$$ (10.9)

Clock stability is modeled empirically considering five different types of noise: white phase noise, flicker phase noise, white frequency noise, flicker frequency noise, and random walk frequency noise. The power spectrum of clock noise can be considered, then, to be the sum of these components. For example,

$$S_y(f) = \sum_{\alpha=-2}^{\alpha=+2} h_\alpha f^\alpha,$$ (10.10)

where the individual contributions in the frequency domain are:

$$
\begin{aligned}
\text{white phase noise} &: h_2 f^2 \\
\text{flicker phase noise} &: h_1 f^1 \\
\text{white frequency noise} &: h_0 f^0 \\
\text{flicker frequency noise} &: h_{-1} f^{-1} \\
\text{random walk frequency noise} &: h_{-2} f^{-2}
\end{aligned}
$$ (10.11)

In the time domain, we can refer to Figure 10.15 for a schematic representation of the effects of these types of noise on the observed clock errors as a function of time. Figure 10.16 shows the spectral appearance of the different noise types.

Quantitatively, stability can be measured in practice by the two-sample or Allan variance, whose square root may be called the two-sample or Allan deviation,

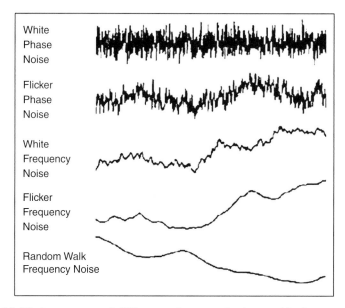

Figure 10.15 Appearance of different noise types in time residuals

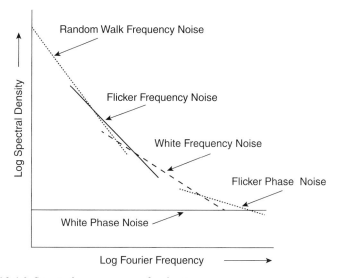

Figure 10.16 Spectral appearance of noise types

$$\sigma_y(\tau) = \sqrt{\frac{\sum_{i=1}^{M-1}\left(y_{i+1} - y_i\right)^2}{2(M-1)}} \text{ or } \sigma_y(\tau) = \sqrt{\frac{\sum_{i=1}^{N-2}\left(x_{i+2} - 2x_{i+1} - x_i\right)^2}{2(N-2)\tau^2}}, \quad (10.12)$$

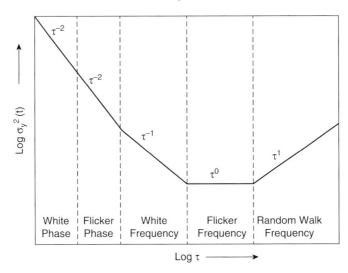

Figure 10.17 Representation of different noise processes in a plot of the Allan Variance as a function of sampling time

where y_i is a set of M frequency offset measurements, or x_i is a set of N time difference measurements, all the data being equally spaced at intervals of τ seconds. Instead of subtracting the mean from each data point in the summation, as is the case with the standard deviation, the Allan Variance subtracts the following data point. This eliminates the possible contribution of any systematic offset in frequency (Levine, 1999; Lombardi, 1999).

In terms of the five noise types, if we plot $\sigma_y^2(\tau)$ as a function of τ, then we find the following proportional relationships:

$$
\begin{aligned}
\text{white phase noise} &: \ H_2\tau^{-2} \\
\text{flicker phase noise} &: \ H_1\tau^{-2} \\
\text{white frequency noise} &: \ H_0\tau^{-1} \\
\text{flicker frequency noise} &: \ H_{-1}\tau^0 \\
\text{random walk frequency noise} &: \ H_{-2}\tau^1
\end{aligned}
\tag{10.13}
$$

A plot of $\sigma_y^2(\tau)$ versus the sampling interval would typically appear as in Figure 10.17, and Figure 10.18 shows how quartz crystal clocks would typically appear in a plot of Allan Deviation.

Fig. 10.17 shows that the Allan Variance is unable to distinguish between white phase noise and flicker phase noise because both appear with a τ^{-2} dependency. Another concern in using the Allan Variance in describing clock stability is the effect of a deterministic clock variation. The Allan Variance assumes the clock data to be without any coherent variation. The presence of a coherent signal in the data will affect the description of the noise process. To remedy the first problem, Allan

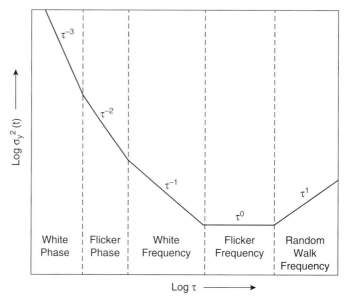

Figure 10.18 Representation of different noise processes in a plot of the Modified Allan Variance as a function of sampling time

devised the Modified Allan Variance given by Equation (10.14) for a set of phase data x_i sampled over N equal intervals of length τ_0 (Allan & Barnes, 1981).

$$\text{Mod } \sigma_y^2(\tau) = \frac{1}{2n^4\tau_0^2\left(N - 3n + 1\right)} \sum_{j=1}^{N-3n+1} \left[\sum_{i=j}^{n+j-1} (x_{i+2n} - 2x_{i+n} + x_i)\right]^2 . \quad (10.14)$$

This statistic is able to distinguish between white phase noise and flicker phase noise as shown in Figure 10.18. Essentially it replaces the individual sample in the expressions (10.12) with averages over adjacent intervals.

A related statistic called "Time Variation," or TVAR, is sometimes used to describe clock noise. It is given by the expression:

$$\text{TVAR} = \sigma_x^2 = \frac{\tau^2}{3}\text{mod } \sigma_y^2. \quad (10.15)$$

Other similar statistics have also been developed to characterize the time-varying behavior of clocks and frequency standards. These include Total Variance or Totvar (Howe & Greenhall, 1997; Greenhall et al., 1999; Howe & Vernotte, 1999) and the Hadamard Variance utilizing the second differences in frequency samples, or the third difference in phase samples (Baugh, 1971; Hutsell, 1995). This statistic has the advantage of being insensitive to frequency drift. Associated with the latter are the Modified Hadamard Variance (Bregni & Jmoda, 2006) and the Total Hadamard Variance (Howe et al., 2005).

References

Allan, D., Ashby, N., & Hodge, C. (1997). *The Science of Timekeeping*. Hewlett-Packard Application Note 1289. Hewlett-Packard Company.

Allan, D. & Barnes, J. A. (1981). A Modified "Allan Variance" with Increased Oscillator Characterization Ability. *Proc. 35th Ann. Freq. Control Symposium*, USAERADCOM, Ft. Monmouth, NJ. In D. B. Sullivan, D. W. Allan, D. A. Howe, & F. L. Walls, eds. (1990). *Characterization of Clocks and Oscillators*, Technical Note 1337. Boulder, CO: National Institute of Standards and Technology, pp. TN254–TN257.

Baugh, R. A. (1971). Frequency Modulation Analysis with the Hadamard Variance. *Proc. Ann. Symp. on Freq. Control.*, 222–225.

Berthoud, F. (1773). *Traité des horloges marines*. Paris: J. B. G. Musier fils.

Bregni, S., & Jmoda, L. (2006). Improved Estimation of the Hurst Parameter of Long-Range Dependent Traffic Using the Modified Hadamard Variance. *Proc. of the 2006 IEEE International Conference on Communications*, 566–572.

Cady, W. G. (1922). The Piezoelectric Resonator. *Proc. Inst. Radio Engineers*, **10**, 83–114.

Curie, J & Curie, P. (1880). Development par pression, de l'électricité polaire dans les cristaux hémièdres a faces inclinées. *Comptes Rendus*, **91**, 294.

Curie, J. & Curie, P. (1882). Déformations électrique du quarts. *Comptes Rendus*, **95**, 194–197.

Dohrn-van Rossum, G. (1996). *History of the Hour*, translated by Thomas Dunlap. Chicago, IL: University of Chicago Press.

Drake, S. (1978). *Galileo at Work: His Scientific Biography*. Chicago, IL: University of Chicago Press.

Dye, D. W. & Essen, L. (1934). The Valve Maintained Tuning Fork as a Primary Standard of Frequency. *Proc. Royal Soc. London Series A.*, **143**, 285–306.

Greenhall, C. A., Howe, D. A., & Percival, D. B. (1999). Total Variance, an Estimator of Long-Term Frequency Stability. *IEEE Trans. Ultrasonics, Ferroelectrics, and Freq. Control*, UFFC-46, 1183–1191.

Harrison, J. (1767). *The Principles of Mr. Harrison's Timekeeper*. London: W. Richardson and S. Clark and sold by J. Nourse, and Mess. Mount and Page.

Hevelius, J. (1673). *Machinae Coelestis, Pars Prior*. Gdansk.

Hope-Jones, F. (1931). *Electric Clocks*. London: N. A. G. Press.

Horton, J. W. & Marrison, W. A. (1928). Precision Determination of Frequency. *Proc. Inst. Radio Engineers*, **16**, 137–154.

Howe, D. A., Beard, R. L., Greenhall, C. A., Vernotte, F., Riley, W. J., & Peppler, T. K. (2005). Enhancements to GPS Operations and Clock Evaluations Using a "Total" Hadamard Deviation. *IEEE Trans. on Ultrasonics, Ferroelectrics, and Frequency Control*, **52**, 1253–1261.

Howe, D. A., & Greenhall, C. A. (1997). Total Variance: A Progress Report on a New Frequency Stability Characterization. *Proc. 29th Ann. PTTI Systems and Applications Meeting*, 39–48.

Howe, D. A. & Vernotte, F. (1999). Generalization of the Total Variance Approach to the Modified Allan Variance. *Proc. 31st PTTI Systems and Applications Meeting*, 267–276.

Hutsell, S. T. (1995). Relating the Hadamard Variance to MCS Kalman Filter Clock Estimation. *Proc. 27th PTTI Meeting*, 291–302.

Huygens, C. (1658). *Horologium*. Translated by E. L. Edwardes, 1970. Horologium, by Christiaan Huygens, 1658. *Antiquarian Horology*, **7**, 35–55.

Huygens, C. (1673). *Horologium oscillatorium sive de motu pendulorum ad horologia aptato demonstrationes geometricae*. Paris, 1673; English translation: *The Pendulum Clock or Geometrical Demonstrations Concerning the Motion of Pendula as Applied to Clocks*. Iowa State University Press, 1986. (available online at http://historical .library.cornell.edu/kmoddl/toc_huygens1.html and translation available online at www.17centurymaths.com/contents/huygens/horologiumpart5.pdf).

Katzir, S. (2015). Pursuing Frequency Standards and Control: The Invention of Quartz Clock Technologies. *Annals of Science*, **73**:1, 1–39.

Landes, D. S. (1983). *Revolution in Time, Clocks and the Making of the Modern World*. Cambridge, MA: Belknap Press of the Harvard University Press.

Levine, J. (1999). Introduction to Time and Frequency Metrology. *Review of Scientific Instruments*, **70**, 2567–2595.

Lombardi, M. A. (1999). Fundamentals of Time and Frequency. In Robert H. Bishop, ed., *The Mechatronics Handbook*. Boca Raton, FL: CRC Press, 17–1–17–18.

Marrison, W. A. (1948). The Evolution of the Quartz Crystal Clock. *Bell System Technical Journal*, **XXVII**, 510–588.

Milham, W. I. (1923). *Time & Timekeepers Including the History, Construction, Care, and Accuracy of Clocks and Watches*. New York, NY and London: Macmillan.

Sobel, D. (1995). *Longitude, the True Story of a Lone Genius Who Solved the Greatest Scientific Problem of His Time*. New York, NY: Walker and Company.

Sullivan, D. B. (2001). Time and Frequency Measurement at NIST: The First 100 Years. *2001 IEEE Int'l Frequency Control Symposium*. Gaithersburg, MD: National Institute of Standards and Technology.

van Leeuwen, P. (2007). www.antique-horology.org/_Editorial/Hevelius/.

Whitrow, G. J. (1988). *Time in History*. Oxford: Oxford University Press.

11

Microwave Atomic Clocks

11.1 Beyond Quartz-Crystal Oscillators

Although quartz-crystal oscillators had provided a major advance in timekeeping, it was apparent that there were limitations to that technology. These devices could provide frequency with a precision of about 10^{-10}, but going beyond that would be a challenge. Operationally, fundamental mode crystals could be made to provide frequencies up to 50 MHz. Higher frequencies capable of providing more precise time were possible using overtones but not commonly used. Aging and changes in environment, including temperature, humidity, pressure, and vibration, affected the frequency of the crystal, so systems were designed to attempt to compensate for these problems, including temperature-compensated crystal oscillators (TCXO), oven-controlled crystal oscillators (OCXO), and microcomputer-controlled crystal oscillators (MCXO). Today systems are commonly used that utilize signals from navigation satellite systems, such as the Global Positioning System (GPS), to discipline crystal oscillators in order to correct for these effects.

To make a significant advance in precision timekeeping of laboratory standards a fundamental change was required. In his autobiographical account, *Time for Reflection*, Louis Essen, the maker of the first operational Caesium (symbol Cs) atomic clock, writes that:

"The only possible alternative appeared to be a natural periodicity within the atom. The science of optical spectroscopy had given a picture of the atom as a miniature solar system having a central nucleus with a number of electrons revolving round it in permitted orbits. When an electron jumped from one orbit to another, light of a specific frequency was emitted or absorbed according to whether the jump was to a higher or lower energy state. Well known examples are the mercury and sodium lamps that illuminated the streets. The frequencies of these optical spectral lines are too high to be measured directly in terms of quartz clocks but are calculated from their wavelength which can be measured, and the velocity of light, velocity being the product of wavelength and frequency.

"The development of microwave techniques during the war provided spectroscopists with a powerful new tool and enabled them to study the response of atoms to electromagnetic waves covering a whole new band of frequencies. Atoms of the alkali metals were of particular interest because they have a single electron in the outermost orbit and, therefore, give the simplest spectrum. The results were brilliantly interpreted and led to the assumption that the outer electron and nucleus were spinning in either the same or opposite direction and that the two conditions represented states having slightly different energies. Transitions between them were accompanied by the emission or absorption of a (quantum of) radiation in the microwave region of the spectrum. The significance of this from our point of view is that the frequency can be measured in terms of the quartz standards and it becomes potentially possible to define the second as the time occupied by a certain number of cycles of an atomic spectral line. It was suggested by Rabi that a spectral line of an isotope of caesium might be suitable. Its frequency was near 10^{10} Hz in a band used for radar."

The story of atomic clocks doesn't begin with Essen, however. It actually goes back to the late 19th century, but to trace that history it is important to first understand the physics of using atoms to determine time and frequency.

11.2 Physics of Atomic Clocks

Elements are composed of atoms, each of which has an identical structure that characterizes that particular element. Isolated atoms of an element have identical energy levels, and changes of those energy levels occur when the atomic electrons interact with electromagnetic fields. Energy level transitions are related to the frequency of the relevant electromagnetic field causing a transition by the expression:

$$\nu = \frac{E_2 - E_1}{h}, \tag{11.1}$$

where E_1 and E_2 are the energies of the levels involved, ν is the frequency of the magnetic field, and h is Planck's constant. This expression shows that it is possible to provide a frequency from atomic energy level transitions that might be useful for timekeeping purposes.

Atomic energy levels take on discrete values and are classified according to a series of distinct physical states. The principal levels, which are characterized by the principal quantum number n, describe the radii of electron orbitals about the nucleus. These levels describe the largest atomic energy separations. The elements of principal interest for atomic clocks are those with just a single electron in their outer orbital shell. The alkali metals (including rubidium and caesium) all have one unpaired electron in the outer shell. Their inner shells are either empty or totally full. Similarly, ions with only one electron in the outer shell can be used in atomic timekeeping. The principal quantum numbers for the energy levels used in the most common clock transitions are $n = 1$ for hydrogen, $n = 5$ for rubidium, and $n = 6$ for caesium and singly ionized mercury.

While n characterizes the radii of the electron orbitals about the nucleus, the principal energy levels are subdivided further as a result of the quantization of the angular momentum of the electrons. These levels are characterized by the quantum number l. For the transitions of interest in atomic timekeeping $l = 0$ (the ground state) and $l = 1$. Electrons are further distinguished by another property called spin, characterized by the quantum number s. This number can only take on the values of $+\frac{1}{2}$ or $-\frac{1}{2}$. Consequently, the total electron angular momentum, which is the sum of the orbital and spin angular momentum of the electron, is characterized by the quantum number J given by $l+\frac{1}{2}$ or $|l-\frac{1}{2}|$. For the ground state, then, $J=\frac{1}{2}$, and in the first excited state $(l=1)$ $J=\frac{1}{2}$ or $J=\frac{3}{2}$. The nucleus of the atom also possesses spin with an associated quantum number I, which, for alkali metals, is given by an odd multiple of $\frac{1}{2}$. For the elements of interest for atomic clocks these numbers are $I=\frac{1}{2}$ for hydrogen and singly ionized mercury, $I=\frac{3}{2}$ for rubidium, and $I=\frac{7}{2}$ for caesium.

The total angular momentum of the atom, then, is the vector sum of the angular momenta of the electron and the nucleus and is characterized by the quantum number F. It can only take on integer values $|I - J| \leq F \leq I + J$. In the presence of a magnetic field the Zeeman effect splits each hyperfine level, characterized by the quantum level F, into 2 F + 1 sublevels characterized by the quantum number m_F, which can take on values $-F \leq m_F \leq F$. The levels characterized by $m_F = 0$ are referred to as "clock transitions" because they have very small variations in the values of their energy as the magnetic field varies (Audoin & Guinot, 2001). Figure 11.1 shows an energy level representation of the fine and hyperfine structure of the outer $(n = 6)$ electron of the caesium atom as an illustration. The hyperfine structure is further subdivided into Zeeman sublevels caused by the application of constant magnetic fields.

A common notation, also shown in Figure 11.1, is used to specify these levels. An upper-case letter is used to specify the orbital angular momentum number l. So for $l = 0$ the letter S is used and for $l = 1$ P is used. A superscript index preceding the uppercase letter specifies the quantity $2s + 1$, and a subscript index after the letter corresponds to the value of J.

The accuracy of the frequency corresponding to any transition is limited by the Heisenberg uncertainty principle:

$$\Delta E \; \Delta t \geq \frac{h}{4\pi}. \tag{11.2}$$

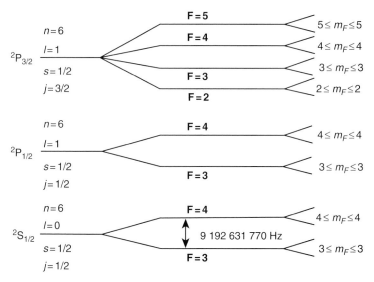

$^2P_{3/2}$ $n=6$ $l=1$ $s=1/2$ $j=3/2$

F=5 $5\le m_F\le5$
F=4 $4\le m_F\le4$
     $3\le m_F\le3$
F=3
F=2 $2\le m_F\le2$

$^2P_{1/2}$ $n=6$ $l=1$ $s=1/2$ $j=1/2$

F=4 $4\le m_F\le4$
F=3 $3\le m_F\le3$

$^2S_{1/2}$ $n=6$ $l=0$ $s=1/2$ $j=1/2$

F=4 $4\le m_F\le4$
9 192 631 770 Hz
F=3 $3\le m_F\le3$

Figure 11.1 Energy level representation of the fine (left) and hyperfine (right) structure of the outer ($n = 6$) electron of the caesium atom

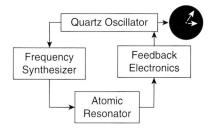

Figure 11.2 Atomic clock concept

Since E = hv, $\Delta\nu\,\Delta t \ge 1/4\pi$, implying that long observation times would lead to a smaller frequency uncertainty. In practice, the spectral lines corresponding to the energy level transitions are broadened for both physical reasons and because of the finite resolution of the electronics involved. The most significant physical causes are due to temperature and pressure broadening.

11.3 General Structure of Atomic Clocks

In general, atomic clocks are constructed by combining (1) an oscillator with excellent short-term stability such as a quartz oscillator, (2) an atomic resonator capable of being locked to the oscillator in an electronic circuit, and (3) the means to use the frequency of the electronic circuit to provide an indication of time. Figure 11.2 shows the concept schematically.

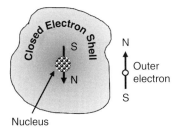

Figure 11.3 Hydrogen-like atom with one outer electron whose spin vector is oriented opposite to the nuclear spin vector

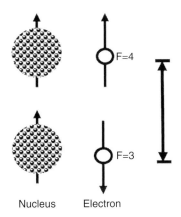

Figure 11.4 Representation of the energy level transition in the hydrogen-like atoms

Atomic frequency standards are commonly based on hyperfine transitions of hydrogen-like atoms, such as rubidium, caesium, and hydrogen. These atoms have a single unpaired electron in a symmetric orbit where there is no orbital angular momentum and no fine structure. Their fundamental state breaks down into only two hyperfine levels due to the interaction with the nucleus's magnetic spin. These provide transition frequencies that can be used conveniently in electronic circuitry, for example, 1.4 GHz for hydrogen, 6.8 GHz for rubidium, and 9.2 GHz for caesium. These hydrogen-like (or alkali) atoms have structures as shown in Figure 11.3. The energy level transitions then refer to a change in the quantum number F as shown in Figure 11.4. They correspond to the spin–spin interaction between the atomic nucleus and the outer electron in the ground state of the atom and are called the *ground state hyperfine transitions*. As mentioned earlier, the "clock transition" is the transition between the least magnetic-field-sensitive sub-levels. To minimize the possibility of other transitions occurring, a constant magnetic field, called the "C-field," is applied in the resonator.

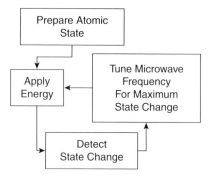

Figure 11.5 Schematic representation of an atomic resonator designed to provide the frequency corresponding to an energy level transition

An atomic resonator, designed to provide a frequency ν corresponding to that transition, is shown schematically in Figure 11.5. First, a maximum number of atoms in the ground state are prepared and exposed to microwave energy with a frequency close to the appropriate frequency corresponding to the desired transition. This frequency is then varied to achieve the maximum atoms in the higher energy level.

To be useful for timekeeping purposes, it is important to minimize the fractional frequency measurement uncertainty $\Delta f/f_0$ where Δf is the error in the frequency measurement and f_0 is the resonance frequency. This quantity is inversely proportional to the interrogation time, or the time during which the radiation is applied before measurements are made, and the square root of the number of atoms. Assuming no limitation to the detection efficiency, the Allan deviation of a frequency standard can be related to the clock transition frequency f_0, the interrogation time T, and the number of atoms N by the expression:

$$\sigma_y(\tau) = \frac{1}{f_0\sqrt{NT\tau}}. \tag{11.3}$$

So this fact makes it desirable for atomic clocks to make use of a high resonance frequency with a large number of atoms and a long interrogation time.

A number of physical effects can contribute to instability in atomic clocks. These include changes in temperature, humidity, atmospheric pressure, vibration, presence of outside electric and magnetic fields, noise in the circuit electronics, and change in the gravity field. Great care is exercised in the construction of atomic clocks to minimize these effects along with a series of other smaller effects. For example, extensive magnetic shielding is used to reduce the effect of the Earth's magnetic field.

Black body radiation due to the temperature of the surrounding environment contributes a systematic shift in the frequency that is measured by an atomic clock.

Because the atoms involved are in random motion, the observed frequency is broadened by the Doppler Effect. Restricting the space in which the atoms interact limits this problem. Nevertheless first-order Doppler remains a source of error. In addition to this effect there is a second-order Doppler Effect caused by relativistic time dilation effects for atoms moving with respect to the apparatus. Only slowing the speeds of the atoms can help with this problem. Collisions of the atoms with the confining chamber contribute to broadening of the observed frequency, and the signal-to-noise ratio contributes a random noise component. Tuning of the cavity in which the interaction occurs can affect the apparent frequency. This effect, called *cavity pulling*, can be helped by going to more narrow resonance peaks.

11.4 Development of Atomic Clocks

The story of the development of atomic clocks does not begin with an atomic resonator using the hydrogen-like atoms, however. The first atomic clock used the ammonia molecule instead. Work in the area of detecting spectral lines corresponding to energy level transitions had focused on the optical frequencies until 1934 when C. E. Cleaton and N. H. Williams were able to observe a change in state of the ammonia molecule that provided a radio frequency at 24 GHz (McCoubrey, 1996). Research in radar during World War II led to further development of microwave spectroscopy. In 1948, Harold Lyons of the US National Bureau of Standards began his work in developing this spectral line for use in timekeeping, and in 1949 he was able to report success (Lyons, 1949). This clock, shown in Figure 11.6, used a quartz-crystal oscillator to provide a frequency standard controlled by the vibrations of excited ammonia molecules. It reached a stability of the order of 1×10^{-7}, but was limited by Doppler (thermal) and pressure broadening of the spectral line. A second version that reached 2×10^{-8} was not able to outperform the best quartz crystals of the time (Lombardi et al., 2007), but these devices did point the way to future atomic clocks.

11.4.1 Caesium

The standard for modern timekeeping, the Système International second, is based on the caesium atom. Caesium is a nonradioactive element having atomic number 55 and atomic weight 133. It is part of the alkali metals (group I of the periodic table) and does react violently with water and oxygen. The development of atomic clocks is based on the work of Isaac Rabi and colleagues (1939), who pioneered the method of molecular beam magnetic resonance, eventually leading to the caesium beam frequency standard.

Figure 11.6 The first ammonia standard. The inventor, Harold Lyons, is on the right and the director of the National Bureau of Standards, Edward Condon, is on the left (NIST).

The first measurements of the hyperfine transition that has become the standard for timekeeping were reported by S. Millman and P. Kusch (1940). In his 1945 Richtmeyer lecture Rabi first mentioned the possibility of developing atomic clocks (Ramsey, 1993). A very significant development in the story is the invention in 1949 by Norman Ramsey (1949, 1950) of a method to use separated oscillatory fields to excite the atoms in an experiment involving the resonance of molecular hydrogen. Instead of distributing the energy over the entire transition region, Ramsey concentrated it in two coherently driven oscillating fields, one at the beginning and one at the end of the transition region. Using this method provides narrower resonance peaks that are not broadened by inhomogeneities in the field. It also enables the transition region

Figure 11.7 Parry (left) and Essen (eight) with the first operational caesium beam clock from Henderson (2005)
[Crown copyright 1960. Reproduced by permission of the Controller of HMSO and the Queen's Printer for Scotland.]

to be larger and permits improved sensitivity through the introduction of relative phase shifts between the two regions (Ramsey, 1983).

In 1952, Sherwood, Lyons, McCracken, and Kusch developed the concept of an atomic beam clock and proposed a plan for such a device. J. R. Zacharias in 1954 attempted unsuccessfully to use caesium in a fountain arrangement to provide a frequency source (Ramsey, 1983, 2005). Finally, in 1955, L. Essen and J. V. L. Parry at National Physical Laboratory (NPL) in Teddington, the United Kingdom, produced the first operational caesium beam atomic clock (Essen & Parry, 1957). Figure 11.7 shows Parry and Essen with the original clock.

In Essen's explanation in his *Time for Reflection* he writes that:

"Atoms leave the oven through a narrow slit and pass between the pole pieces of a powerful magnet which is shaped to give a non-uniform field. They follow various paths according to their initial direction, velocity and energy state. Only two paths are shown. A few of the atoms are selected by the slit half way along the path and continue through the pole pieces of a second magnet, which is the same as the first and increases the deflections in the same direction and away from the centre line. A weak radio field is applied in the space between the magnets and, when its frequency and strength are exactly right, the atoms jump to the other state, those initially in the low energy state absorbing energy from the field and, strangely enough, those in the high energy state being induced to emit energy, so that they are all reversed. Their deflections in the second magnet are also reversed and they are deflected back to the centre line where they strike the detector. This is a hot tungsten wire which imparts a charge to the atoms which boil off as charged particles and are attracted to electrodes and, after enormous amplification, measured as an electric current. Atoms which are not in the two states concerned are not deflected at all and give a steady signal which is useful for lining up the apparatus. The beam strength increases by ten per cent when transition occurs. The components are all contained in a metal pipe about 150 cm long evacuated as completely as possible."

Caesium was chosen as the element to be used in this effort because:

1. its fundamental state has only two hyperfine levels just as all the alkali atoms (see Figure 11.1), and at room temperature all caesium atoms are in that fundamental state;
2. in the (F = 3 to F = 4) transition, atoms in the F = 4 state stay there for a long time compared to the observation time;
3. the transition frequency is easily detectable using the electronic equipment in use at that time;
4. it is relatively insensitive to electric fields; and
5. it is less expensive than other alkali elements.

11.4.1.1 Calibration of the Caesium Frequency

In order to make practical use of the successful operation of the NPL caesium clock, it was necessary to establish a relation between the time derived from the caesium clock and the standard time in use at that time. A preliminary effort was accomplished in June 1955, when Essen and Parry (1955) calibrated the frequency of the transition in terms of the timescale maintained at the Royal Greenwich Observatory, which they could relate to the second of UT2. Concurrently the astronomical community was discussing the use of Ephemeris Time as a uniform timescale to replace UT2 as the timescale of choice for those needing a timescale independent of the Earth's variable rotational speed.

Essen described the concerns of astronomers again in his *Time for Reflection*:

"A few months after the atomic had been in operation The Astronomer Royal invited me to describe it at the meeting of the International Astronomical Union to be held in Dublin. One of the main subjects for discussion was the adoption of a new unit of time. Astronomers knew that the unit based on the rotation of the earth was no longer adequate and they were recommending a unit, the second of the ephemeris time, based on the revolution of the earth round the sun. Unfortunately, although this unit might be expected to be more constant than the mean solar second, it is much more difficult to measure, and the observations would have to be averaged over years to give the required accuracy. This rendered it useless as a unit of measurement which must be available immediately. I pointed out that whatever advantages this unit might have for the astronomer it was useless for the physicist and engineer, and suggested that since an atomic unit would be needed in the future it would be wise to defer a decision until agreement could be obtained on the definition of such a unit. There was no support for this suggestion and the second of ephemeris time was adopted and was later confirmed by the International Committee of Weights and Measures, showing how even scientific bodies can make ridiculous decisions. One useful outcome of the Dublin meeting was that with the help of Markowitz – I was not an official delegate myself – a resolution was passed to the effect that when the relationship between ephemeris time and atomic time was established the atomic clock could be used to make astronomical time available. This meant that we had international approval to introduce atomic time when the comparisons were completed without further international meetings. A detailed programme was arranged with Markowitz. The time interval between certain time signals was measured at the NPL in terms of the atomic clock and at the US naval observatory in terms of the ephemeris second. The comparisons took longer than anticipated because of the relative inaccuracy of the astronomical measurements but after three years it was decided that further averaging was not likely to improve the result. The value was, therefore, announced and was eventually accepted internationally as the unit of time."

As a result of this collaboration, the frequency of the transition between the two hyperfine states of caesium was determined to be 9 192 631 770 cycles per Ephemeris Time second (Markowitz et al., 1958). This continues to be the basis for the definition of the second using the world's primary frequency standards.

11.4.1.2 Caesium Beam Tubes

The first commercial version of a caesium standard called the "Atomichron" appeared in 1956 (Figure 11.8). It was manufactured by the National Company and developed by R. T. Daly, Jerrold Zacharias, and A. Orenberg (Forman, 1998). A period of technology development followed the introduction of the Atomichron, resulting in a caesium atomic beam tube that was about 30 cm long. This tube was eventually incorporated in the Hewlett Packard HP5060 Cesium Atomic Beam Frequency Standard, which quickly became a standard device for high-precision timekeeping.

The basic beam tube configuration is shown in Figure 11.9. Caesium atoms are heated to a temperature of about 90°C and exit the oven traveling with a speed of about 260 ms^{-1}. Magnetic state selection is used to prepare the proper atomic state. This technique makes use of the fact that atoms in states ($F = 3$, $m_F = 0$) and ($F = 4$, $m_F = 0$) follow different paths when subjected to magnetic fields. Atoms in the hyperfine state ($F = 3$, $m_F = 0$) are deflected toward the axis of the tube, while those in the state ($F = 4$, $m_F = 0$) are deflected off the axis. They go on to contact the walls or "getters" and are eliminated. Those atoms in the proper state then enter the C-field region where they encounter a uniform magnetic field shielding them from the Earth's or any other stray fields.

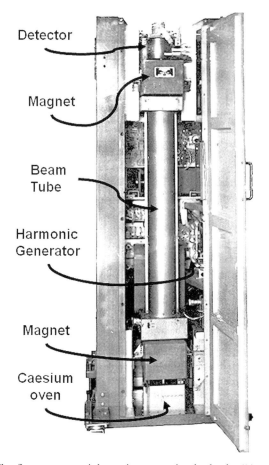

Figure 11.8 The first commercial caesium atomic clock, the "Atomichron"

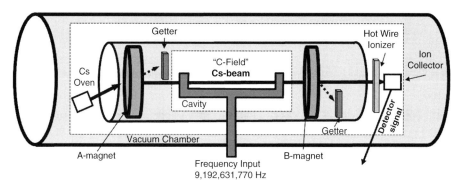

Figure 11.9 Schematic representation of a caesium beam tube. The typical cavity
length in commercial tubes is 10 to 20 cm; it is about 4 m in laboratory standards.

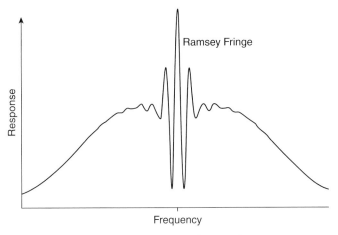

Figure 11.10 Frequency response of caesium beam tube

The caesium atoms then enter a region where they are exposed to two oscillatory fields. This method of using two oscillatory fields, invented by Ramsey (1950), allows atoms to pass through two equally long regions in which microwave magnetic induction is applied. A drift space separates the two regions. The phase of the microwave field in the second region is set to maximize the probability that the hyperfine transition occurs. In commercial caesium beam tubes, the two regions with the oscillatory fields are about 1.5 cm in length and the drift space is about 15 cm long. Laboratory standards may have extended drift regions of a few meters in length.

When the frequency (9,192,631,770 GHz) of the microwave signal is applied properly, the transition (F = 3, m_F = 0) to (F = 4, m_F = 0) occurs. A second set of magnets is then used to send those atoms in the (F = 4, m_F = 0) state in the direction of a hot wire detector made of a metal, such as tungsten, platinum, tantalum, or of an alloy such as platinum-iridium, where the atom stream is converted into an electric current, and amplified in an electron multiplier. This current is then monitored and maximized by adjusting the frequency of the microwave radiation to provide the maximum number of atoms in the proper state (Audoin, 1992).

The output current measured by the detector depends on the frequency of the microwave energy inserted into the resonator. The response as a function of frequency is shown in Figure 11.10. The central peak is called the *Ramsey fringe*.

A later development in using caesium beams was the introduction of optical pumping and optical detection of the resonance signal. The technique, proposed by Alfred Kastler (1950), would replace the magnet state selectors by optical interaction regions. One way to improve the signal-to-noise ratio is to ensure that the atomic states are modified to the greatest extent possible. This, in turn, means

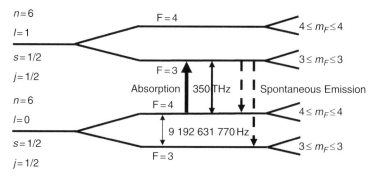

Figure 11.11 Optical pumping. An electron in the $l = 0$, $F = 4$ state is exposed to light of the proper frequency (approximately 350 THz) to excite it into the $l = 1$, $F = 3$ state from which it spontaneously emits a photon and decays to either $l = 0$, $F = 3$ or $l = 0$, $F = 4$. The net result is to increase the number of atoms with their outer electrons in the $l = 0$, $F = 3$ state while reducing the number in the $l = 0$, $F = 4$ state.

that the difference between the number of atoms in the ($F = 3$, $m_F = 0$) and the ($F = 4$, $m_F = 0$) states be as large as possible. Optical pumping provides an improvement in the signal-to-noise ratio by increasing the difference in the number of atoms in the states of interest. Figure 11.11 illustrates one possible application of the principle.

When the caesium atoms exit the oven, the ($F = 3$, $m_F = 0$) and the ($F = 4$, $m_F = 0$) levels of the ground state ($l = 0$) are approximately equally populated. In the case illustrated in Figure 11.11 the atoms absorb light with a wavelength of 0.85 µm (infrared) raising them from the $F = 4$ ground state to the $l = 1$, $F = 3$ level. From there they spontaneously decay within a few nanoseconds by emitting radiation and return to either the $F = 3$ or $F = 4$ levels of the ground state. This results in a net transfer of the atoms from the $F = 4$ to $F = 3$ levels of the ground state. The distribution of the atoms among the magnetic sublevels depends on the transition probabilities and on the polarization of the light (Piqué, 1977; Audoin & Guinot, 2001).

Just as in the case of the magnetic state selection technique, the atoms are then exposed to the microwave radiation to cause them to transition to the ground state $F = 4$ level. Lasers can also be used to replace the B magnets that are used to detect the atoms in this level. Here again, there are various possibilities. If, for example, light from the same laser is used to interact with (see Figure 11.12) the atoms exiting the microwave cavity, those in level $F = 4$ will be transferred back to the $F = 3$ level by emitting a photon. This fluorescence, which is proportional to the number of atoms in the $F = 4$ level, can be detected optically and used as a means to adjust the frequency of the microwave radiation to provide the maximum number of atoms in the proper state.

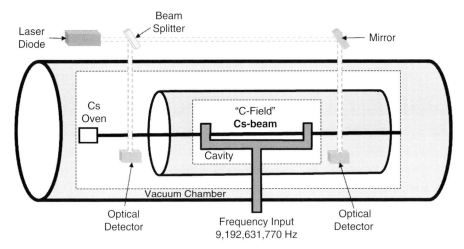

Figure 11.12 Optically pumped caesium beam tube using the light from a single laser

Figure 11.13 Doppler cooling

Two different lasers can also be used in optical pumping techniques. Referring to Figure 11.1, the first polarized laser, which is tuned to the $F = 3$ ground state transition to the $F = 4$ sublevel of the $^2P_{3/2}$ level, is used to pump atoms into the $F = 4$ level of the ground state in a manner similar to that described earlier. Making use of selection rules that describe some general rules about permitted transitions, another laser, tuned to the transition between the ground state $F = 4$ level and the $F = 4$ level of the $^2P_{3/2}$ level, removes atoms from the $F = 4$ level except for those in the $m_F = 0$ sublevel. The atoms prepared by this process are then sent to the resonance cavity where the frequency is tuned to the transition between the two hyperfine levels, causing stimulated recombination. The atoms in the $F = 3$ level in this case can then be detected optically.

The advantage of optical pumping is that it can enhance the number of atoms in the desired state rather than just reject the atoms in the undesired state. Using optical pumping for site selection and optical detection does not alter the source of the caesium beam or the microwave cavity. By eliminating the need for the state-selection magnet, a larger number of atoms contribute to the signal, which results in a superior signal-to-noise ratio. It also removes the need for strong magnets near the resonance cavity. The first operational optically pumped caesium device appeared in 1980 (Arditi & Picqué, 1980; Arditi, 1982).

11.4.1.3 Caesium Fountains

As mentioned in Section 11.4.1, one of the early attempts to use the caesium atom for timekeeping was that of Jerrold Zacharias, who in 1952 developed the concept of directing a thermal beam of atoms vertically through a microwave cavity. Some of the slower atoms would be expected to return to the source under the influence of gravity. This would provide a time of flight of the order of a second and permit Ramsey interrogation twice by the same microwave cavity. The implementation was not successful, however, because of scattering processes within the beam (Sullivan et al., 2001). Success would have to wait for the development of laser cooling of atoms.

Thermal broadening of the resonance line is one source of instability in a caesium beam atomic clock. It follows, then, that a possible improvement might be to reduce the speed of the atoms in the beam by cooling them to a very low temperature. The speed of the atoms when they exit the 100°C oven is about 260 m s^{-1}. In the 1970s, techniques were developed to make it possible to cool the atoms to the extent that they showed hardly any perceptible motion (Wineland & Itano, 1979). In the case of caesium this was made possible by the development of semi-conducting lasers that emitted light with a wavelength of 0.85 μm with high spectral purity. Doppler cooling of atoms is illustrated in Figure 11.13. Monochromatic light of the same intensity propagating in opposite directions is shown directed at an atom of velocity v. For caesium, the wavelength of the light is associated with the transition from the ground state to the $l = 1, j = \frac{3}{2}$ level and is tuned to have a frequency slightly less than the resonance frequency. The atom traveling toward the light propagating from the right, in this case, sees the frequency of the light in the direction it is traveling Doppler shifted toward its resonant frequency, and shifted away from its resonant frequency in the opposite direction. Consequently it absorbs more photons in the direction of travel, and the momentum exchange tends to slow its motion in that direction. When a cloud of atoms is exposed to light from three pairs of lasers in mutually orthogonal directions for a sufficient length of time to allow for tens of thousands of individual interactions, the atoms are slowed in three dimensions, resulting in the term *optical molasses*. This technique can be used to cool caesium atoms to a temperature close to 120 μK.

Even lower temperatures are possible using an additional cooling technique called the *Sisyphus effect* that makes use of spatial variations in the polarization of the light in two opposing beams of light. The interaction of atoms with the electromagnetic field of a laser light beam causes their energy levels to change. If, for example, two laser beams are oriented in opposite directions with equal amplitudes and perpendicular linear polarization, they create a standing wave

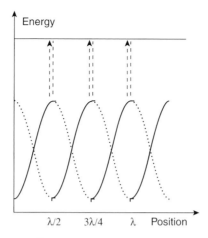

Figure 11.14 Sisyphus cooling

with polarization that varies with position and repeats spatially at intervals of $\lambda/2$. The energy of the sublevels of an atom interacting with these beams will likewise vary according to the position of the atom along the axis of the light beams. As the atom (assuming it has two energy levels) moves along this axis, it will experience a series of hills and valleys of potential energy.

When the atom reaches the top of a potential hill, conservation of energy requires that its kinetic energy be lowered. At that point, it transitions to a higher energy level and subsequently emits a photon removing the kinetic energy it gained in getting to the top of the potential hill. In emitting the photon, it transitions to the second level, which is now at a potential valley. This series of alternating potential highs and lows is called the *Sisyphus effect*, referring to the Greek mythological character Sisyphus who was endlessly doomed to push a rock up a hill only to have it roll down again (see Figure 11.14).

As the atom progresses along the axis, the atoms experience a series of potential hills and valleys until they reach a point where they lack the kinetic energy to climb another hill. This mechanism can result in atoms cooled to the level of a few μK. The possibility of cooling atoms to these temperatures made it possible to revisit the fountain concept.

The first demonstration of a laser-cooled fountain clock was developed in the late 1980s using sodium atoms (Kasevich, et al., 1989). The first caesium fountain was built at Laboratoire Primaire du Temps et des Fréquences in Paris in 1991 (Clairon et al., 1996). Caesium fountains typically cool about 10^7 atoms using six lasers tuned to the $l = 0, F = 4$ to $l = 1, F = 5$, transition with a wavelength of 852 nm. The resulting ball of cooled atoms in the $l = 0, F = 4$ state would then be launched vertically in a caesium fountain with a velocity of about 4 m s^{-1} (see Figure 11.15).

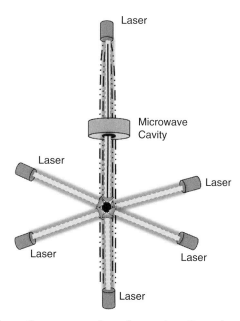

Figure 11.15 Schematic representation of a caesium fountain

This is done by introducing a change in the frequency between the laser beams in the vertical direction. The atoms then pass through a state preparation cavity, where they can be optically pumped to the $l = 0$, $F = 3$, $m_F = 0$ state. Next, they are irradiated by the clock radiation and continue drifting upward for a drift period of about one-half second, when gravity causes them to begin to fall back down. As with the caesium beam tube, they then experience the clock radiation a second time and pass through to an optical detection region where the number of atoms in the $l = 0$, $F = 4$ state is sensed. The cycle can then be repeated when another ball of atoms is cooled.

The advantages of caesium fountain technology over the caesium beam technology are 1) the laser cooling and preparation techniques make it possible to allow more atoms to be used since fewer are discarded in the selection process, 2) possible phase differences between the two arms of the clock radiation cavity in the beam tube are eliminated, and 3) a significantly longer interrogation time is possible. On the other hand, the beam tube provides a continuous beam of atoms, while the fountain operates with a series of pulses of atoms.

PTB has developed a second caesium fountain, which has a relative frequency instability of 22.5×10^{-13} and an uncertainty of realizing the SI second of 0.80×10^{-15} (Gerginov et al., 2010). The National Institute of Metrology (NIM) in China has a caesium primary frequency standard NIM5 operating more than 300 days a year with a fractional frequency instability of 3×10^{-13}. Comparisons with

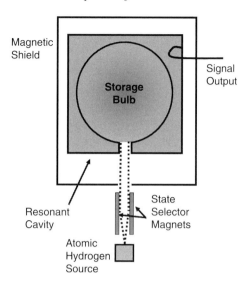

Figure 11.16 Schematic representation of a hydrogen maser

other fountain clocks showed good agreement within uncertainties (Fang et al., 2015).

11.4.2 Hydrogen

Hydrogen masers have found significant use in timekeeping since the 1960s. Early work using the single-electron hydrogen atom began in 1960 (Goldenberg et al., 1960) with a concept based on the ground state hyperfine transition $F = 0$ to $F = 1$ in the hydrogen atom and its corresponding frequency of 1 420 405 751.770 Hz (Audoin & Guinot, 2001). The principle relies on the coupling between the atoms and microwave energy field in a resonant cavity. The stimulated emission of radiation in the maser causes an amplification of the microwave field that sustains the required oscillation. The basic concept is shown in Figure 11.16.

An atomic hydrogen beam is created and passed through an area that magnetically selects those atoms in the $l = 0$, $F = 1$, $m_F = 1$ and $l = 0$, $F = 1$, $m_F = 0$ states to pass through to a quartz storage bulb whose inner wall is coated with a polymer containing fluorine such as Teflon. Approximately 10^{12} to 10^{13} atoms enter per second, where they undergo 10^4 to 10^5 collisions per second. The storage bulb is contained in a resonant cavity, which is contained behind a magnetic shield that maintains a constant magnetic field. Within the storage bulbs the hydrogen atoms undergo the transition from the $F = 1$ to the $F = 0$ level. The atoms entering the cavity amplify the applied frequency field provided that the frequency is tuned to be close to the frequency of the hyperfine atomic transition (Kleppner et al.,

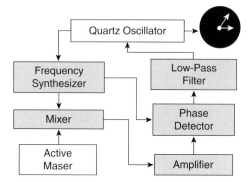

Figure 11.17 Block diagram of an active hydrogen maser

1962, 1965). Hydrogen masers differ from the caesium devices in that there is no direct measure of the change in the population of atoms in the different energy states, because there is no efficient means of detecting hydrogen atoms.

The stability of the hydrogen maser depends critically on the tuning of the cavity frequency. A variety of procedures can be used to tune that frequency automatically. In all cases an error signal is determined and that signal is then used to tune the cavity frequency. Sources of error in hydrogen masers include thermal noise, collisions between atoms, collisions with the storage bulb walls, and nonuniformity of the magnetic field (Audoin & Guinot, 2001). Operationally the maser can be used as an active or passive device for precise timekeeping.

11.4.2.1 Active Hydrogen Maser

An active hydrogen maser makes use of a quartz crystal phase locked to the output signal of the hydrogen maser. The maser operates spontaneously and its signal is detected by an antenna in the resonant cavity and synchronized to the crystal output as shown in Figure 11.17. The frequency produced by the quartz oscillator is mixed with the signal from the active maser to produce a signal that is used to steer the output of the oscillator. Some advantages of an active hydrogen maser are the narrow width of the resonant frequency due to the relatively long storage time, the unperturbed movement of the atoms due to its low pressure, and the low velocity of the atoms and its low noise level.

Efforts to develop a cryogenic hydrogen maser (Vessot et al., 1977; Crampton et al., 1979; Vessot et al., 1979; Hess et al., 1986; Hürlimann et al., 1986; Walsworth et al., 1986) began in the late 1970s. The concept is to reduce thermal noise in the atoms, which could lead to a possible improvement in stability of three orders of magnitude over that achieved by a hydrogen maser operating at room temperature. One such device has been operated at a temperature of 0.5°K, achieving results similar to that of an active hydrogen maser at room temperature

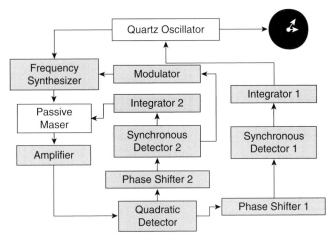

Figure 11.18 Block diagram of a passive hydrogen maser

for averaging times less than about 100 seconds. For averaging times greater than that, results have not matched those of a room-temperature maser (Vessot, 2005).

11.4.2.2 Passive Hydrogen Maser

In contrast with the active hydrogen maser, the passive device steers the quartz oscillator by locking the output of the crystal to the maser signal as is done with caesium oscillators. In the passive hydrogen maser, however, resonance is monitored by measuring the changes in amplitude of the electromagnetic field in the cavity. Both the frequency of the magnetic field in the cavity and the quartz oscillator frequency can be controlled following the process illustrated in Figure 11.18. The maser output signal is first amplified. The appropriate signals are detected and sent to two circuits: one is used to determine the difference between the interrogation signal carrier frequency and the atomic resonance frequency, and the second determines the difference between the carrier frequency and the cavity resonance. These are used to provide two error signals, one to control the quartz oscillator and the other the cavity resonance frequency.

Four hydrogen masers at the National Time Service Center (NTSC) are mutually referred and tested using various methods as described and documented (Song et al., 2016).

11.4.3 Rubidium

Another alkali atom, rubidium, is used extensively for precise timekeeping. It is a silvery-white metallic element having atomic number 37 that liquefies at 39.3°C and reacts vigorously in water. There are 24 isotopes of rubidium, but only two

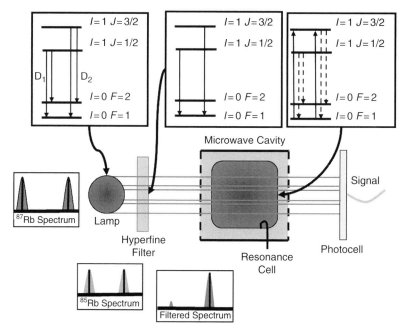

Figure 11.19 Rubidium cell frequency standard

occur naturally: Rb-85, which comprises 72.2% of the naturally occurring rubidium, and Rb-87, which is slightly radioactive with a half-life of 4.88×10^{10} years that comprises the remainder. Devices using rubidium make use of the ground state hyperfine clock transition of Rb-87 with a corresponding frequency of 6 834 682 610.904 Hz. Development of these standards began in the late 1950s (Arditi, 1958; Carpenter et al., 1960; Arditi & Carver, 1961, 1964; Packard & Schwartz, 1962; Davidovits, 1964; Davidovits & Novick, 1966). The CCTF (2012) has recommended a value for the frequency of rubidium of 6834 682 610.904312(9) with an uncertainty of 1.3×10^{-15}. Measurement values give 6834 682 610.904310(2.2) (Ovchinnikov et al., 2015).

11.4.3.1 Rubidium Cells

Rubidium cell clocks generally begin with a lamp operating at temperatures ranging between 65°C and 140°C that contains a small (about one mg) of Rb-87, or a mixture of Rb-85 and Rb-87, along with a noble gas such as krypton. A radio frequency generator creates a light discharge (see Figure 11.19) that is passed through a hyperfine filter that contains Rb-85 and a noble gas at high pressure (about 10^4 Pa). By a coincidence of the structure of rubidium, the atoms of Rb-85 in the filter absorb the components of the light emitted in the $I = 1$ to the ground state $F = 2$ level, resulting in a light with the spectrum shown in Figure 11.19. This is then passed to the resonance cell in a resonant cavity tuned to the resonance frequency of Rb-87 that contains Rb-87 along with nitrogen and rare gases. There the light is absorbed by the Rb-87 atoms in the $F = 1$ level of the ground state. They are then optically pumped to the $l = 1$ states, where they relax to either the $F = 1$ or $F = 2$ levels of the ground state. The net effect is to depopulate the $F = 1$ level and populate the $F = 2$ level. The resonance frequency in the cavity, however, stimulates the transition from the $F = 2$ to the $F = 1$ levels, thereby repopulating the $F=1$ level. These atoms, in turn, absorb part of the incident light, reducing the intensity detected by the photocell. This signal can be used to adjust the resonance frequency delivered to the cavity to achieve a minimum intensity (Audoin & Guinot, 2001). The buffer gases are used to keep the rubidium atoms away from the walls of the cell.

The additional gases in the resonance cell are used to damp the Doppler broadening of the resonance frequency. Various modifications can be made to this basic design. Frequently the filter is combined with the resonance cell to reduce the size of the standards, for example. These rubidium cell standards are not as accurate as caesium standards because the collisions with the molecules of the buffer gases introduce instabilities. Caesium standards are able to allow the caesium atoms to drift through an extended cavity and take advantage of the Ramsey separated fields to narrow the line width of the resonance frequency. Rubidium standards are also sensitive to environmental conditions. Optical pumping does, however, introduce a shift in the hyperfine structure of the atom, called *light shift*, which affects the accuracy of the device.

11.4.3.2 Rubidium Fountains

The largest source of instability in caesium fountains is the cold-collision shift, but this effect is much smaller for rubidium atoms. Also, because the clock frequency used for rubidium is smaller than that for caesium, the entrance and exit dimensions of the microwave cavity can be larger, allowing the use of more atoms. These considerations, along with the fact that rubidium has a less complicated series of magnetic states with which to contend, make the concept of a rubidium fountain attractive in spite of its lower resonant frequency (Fertig & Gibble, 2000). The principles of operation of a rubidium fountain are similar to those of a caesium fountain, with the exception of the substitution of rubidium for caesium.

The US Naval Observatory (USNO) has developed six operational, continuously running rubidium fountain clocks to supplement commercial caesium beams and hydrogen masers. Four located in Washington have been running for more than five years, and contribute to EAL. Two are at the USNO Alternate Master Clock Facility in Colorado. Their performance is consistent with an average stability of

5×10^{-17} and the relative drift between the fountains is zero at the level of 6×10^{-19} per day. A timescale based on these fountains shows no drift with respect to primary standards at a level of 1.2×10^{-18} per day (Peil et al., 2016). The National Physical Laboratory has also developed a Rb atomic fountain with a double magneto-optical arrangement (Ovchinnikov & Marra, 2011). Details concerning its accuracy are considered by Li and Gibble (2011). A portable rubidium fountain is under development in China (Du et al., 2013). A dual rubidium/caesium atomic fountain clock has also been developed for comparing the frequencies from the two elements (Guena et al., 2013).

11.4.3.3 Double-Bulb Rubidium Maser

A further application of rubidium in precise timekeeping is the double-bulb rubidium maser proposed in 1994 (Golding et al., 1994). To mitigate the effect of light shift in this design, optical pumping occurs in a bulb separated from the region in which the microwave interaction occurs. No buffer gas is required in this design, and the atoms are "recycled" in that they can be returned to the bulb, where the pumping occurs, to be "re-pumped." The lack of buffer gases in this design allows the atoms to effuse between the two areas, but this also makes the device sensitive to interactions of the atoms with the walls of the bulbs. This problem can be mitigated by a judicious choice of wall coatings. One of its advantages is the small size compared with a hydrogen maser.

11.5 Trapped Ion Clocks

The sources of error in all atomic clocks include perturbations due to confinement of the atoms. To eliminate the effect of wall collisions, ions trapped in an electro-magnetic field can be used to provide a timing source. The basic concept for applying stored ions in timekeeping is the same as that for atomic clocks. A local oscillator is used to provide a frequency near a resonant atomic frequency and the excited transitions between states provide an accurate frequency via a feedback chain. A number of singly ionized atoms with well-known hyperfine structure in the ground state are potential candidates for this application. Two basic kinds of ion traps can be used. These are the Paul trap and the Penning trap. Static electric and magnetic fields are used to confine ions in a Penning trap. A Paul trap makes use of a radio frequency field.

In a Penning trap, the plates on the ends (see Figure 11.20) are kept at the same potential with respect to the inner ring electrode. This field forces ions toward the center of the trap, if the ions are displaced in a direction toward either of the end plates. If they are displaced in any direction parallel to the planes of the end plates, however, a magnetic field $\vec{B}$ is required to force the atoms to the center of the trap.

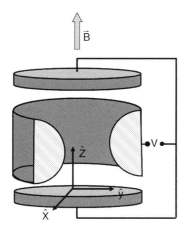

Figure 11.20 Schematic representation of a Penning ion trap

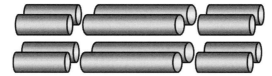

Figure 11.21 Linear Paul ion trap

As a result, a single ion will then undergo simple harmonic motion in the z direction, a circular cyclotron motion along with another lower frequency circular magnetron ($\vec{E} \times \vec{B}$) motion in the x-y plane (Blaum et al., 2010).

In a Paul trap, the electric potential between the inner element and the end plates oscillates at a high frequency that is determined by the geometry of the trap, and no magnetic field is required. In this configuration, an ion will undergo motion with a frequency equal to the frequency of the electric potential that is applied. This is called the *micromotion*. It also experiences motion with a much lower frequency, called the *secular motion* (Itano, 1991; Ludlow et al., 2015). A variation is the linear ion trap shown in Figure 11.21 (Prestage et al., 1989). In this design, an RF potential is applied between the rods. The phase at each rod differs by 180° from its neighbors. Such a trap attracts ions to a central axis where they have less micromotion in comparison with other types of traps. The rods can be connected in rings resulting in a "racetrack" trap (Church, 1969; Ludlow et al., 2015).

By trapping the ions in such traps, they can be allowed to interact with microwaves for several seconds. The longer they can interact without being disturbed, the more stable is the frequency standard. A number of ionized atoms with well-known ground state, hyperfine structure is available. These include helium, beryllium, magnesium, strontium, calcium, barium, ytterbium, cadmium, and mercury.

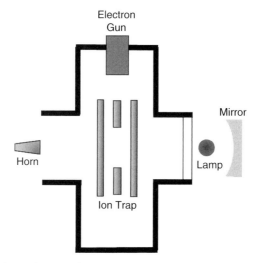

Figure 11.22 Linear ion trap frequency standard

11.5.1 Mercury

The first practical application of trapped ionized atoms for timekeeping made use of the element mercury (chemical symbol Hg). The transition of interest is the hyperfine transition of $^{199}Hg^+$, which has a very high frequency (40 507 347 996.841 Hz) and high Q. The measurement of the transition frequency was originally done in 1973 (Major & Werth, 1973), but the first frequency standard based on this technique was developed in the early 1980s (Jardino et al., 1981). Since that time there has been extensive development activity (Cutler et al., 1985, 1986; Tjoelker et al., 1996; Tjoelker et al., 2000; Prestage et al., 2001; Burt et al., 2008).

Mercury is used because its transition frequency is less affected by systematic effects related to magnetic fields and ion motions than would be the case in lighter atoms. This allows for better long-term stability while minimizing environmental issues. In an operational cycle, an electron gun ionizes the atoms of neutral ^{199}Hg introduced into the device (see Figure 11.22). A cloud of $^{199}Hg^+$ ions with a density of about 10^6 ions per cm^3 is used. State selection is done using optical pumping with a wavelength of 194 nm. This process takes advantage of a fortunate coincidence. The wavelength of the D$_1$ line of the spectrum of the $^{202}Hg^+$ ion of the mercury isotope of atomic weight 202 is very close to the wavelength corresponding to the transition between the $I = 0$, $F = 1$ ground state level and the $I = 1$, $F = 1$ level. A lamp providing the 194 nm-wavelength light using $^{202}Hg^+$ ions is used to pump the $^{199}Hg^+$ ions to the $I = 0$, $F = 0$ level. They can then be exposed to the 40.5 GHz hyperfine transition frequency that will repopulate the $I = 0$, $F = 1$ level.

When they decay back to the $l = 0$, $F = 0$ level, the fluorescence will intensify the 194 nm light and the intensity of this signal is monitored to determine the clock transition frequency. A buffer gas is mixed with the $^{199}Hg^+$ ions. A device following such a scheme called the *Linear Ion Trap Frequency Standard* (LITS) has been put in operation at the Jet Propulsion Laboratory (Tjoelker et al., 1996).

A variation of the LITS that makes use of two chambers is also being developed at the Jet Propulsion Laboratory (Prestage et al., 1995). It is called an *extended linear ion trap* (LITE) or *shuttle trap*. Two separate trap regions are used: a region for microwave interrogation, and a separate volume for loading and state preparation. Ions are shuttled between these two regions, hence the name "shuttle trap." This device provides for an interrogation region seven times longer than that of the LITS, and is less sensitive to environmental conditions. Improvement in stability by a factor of two can be expected over that provided by a single-chamber device. Mercury stored ion devices are limited in their stability, not only by detection noise, but by magnetic and thermal effects and collisions with the buffer gas molecules.

A laser-cooled linear ion trap using mercury ions has been developed (Berkeland et al., 1998). It uses the same Doppler cooling technique that was described for cooling atoms in a Paul trap to form a linear mercury crystal of seven ions. Light with wavelength of 198 nm is used to cool the ions without using a buffer gas.

11.5.2 Other Ions

Ions other than those of mercury have also been used in developmental frequency standards. These standards follow the same basic principles of the mercury ion trap, but have not enjoyed widespread use. The first laser-cooled ion trap used a Penning trap with $^9Be^+$ (beryllium) ions (Bollinger et al., 1985). Other ions involved include $^{173}Yb^+$ (ytterbium) (Münch et al., 1987), $^{171}Yb^+$ (ytterbium) (Fisk et al., 1995), $^{25}Mg^+$ (magnesium) (Itano & Wineland, 1981), $^{137}Ba^+$ (barium) (Blatt & Werth, 1982), $^{135}Ba^+$ (barium) (Becker & Werth, 1983).

11.6 PHARAO Laser-Cooled Microgravity Atomic Clock

Projet d'Horloge Atomique par Refroidissement d'Atomes en orbite (PHARAO) is the laser-cooled caesium clock developed under the Centre national d'études spatiales (CNES) for the European Space Agency mission ACES (Atomic Clock Ensemble in Space). In concept, it is similar to ground-based atomic fountains, but it is operated under microgravity conditions. Atoms are launched in free flight along the Pharao tube and cross a resonant cavity, where they interact two times with a microwave field tuned to the transition between the two hyperfine levels of

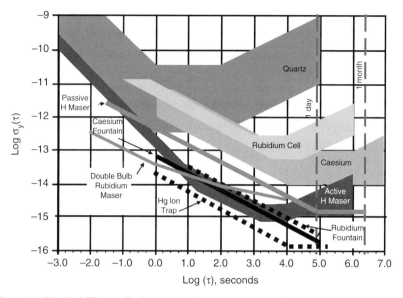

Figure 11.23 Stabilities of microwave timekeeping standards

caesium ground state. The interrogation method is the Ramsey scheme. In microgravity, the velocity of the atoms is constant and can be changed over almost two orders of magnitude, allowing detection of atomic signals with sub-Hertz linewidth. The primary feature of a microgravity laser-cooled caesium clock is long interrogation times, which change behaviors of some frequency shifts, including microwave lensing shift. ACES is planned to be on the International Space Station for 18–36 months, also with a hydrogen maser and time transfer systems. The mission is for fundamental physics tests of general relativity, precise geodesy, and Earth clock comparisons with fractional frequency precisions of 10^{-17} (Peterman et al., 2016).

11.7 Characterizing Atomic Clocks

The stabilities of the various microwave atomic timekeeping standards are shown in Figure 11.23. Standards that are still in development, such as the fountains or ion traps, may be expected to improve as research leads to new developments.

References

Arditi, M. (1958). L'influence des gaz tampons sur le déplacement de la fréquence et la largeur des raies des transitions hyperfines de l'état fondamental des atomes alcalins. *Le Journal de Physique et le Radium*, **19**, 873.

Arditi, M. (1982). A Caesium Beam Atomic Clock with Laser Optical Pumping, as a Potential Frequency Standard. *Metrologia*, **18**, 59–66.

Arditi, M. & Carver, T. R. (1961). Pressure, Light, and Temperature Shifts in Optical Detection of 0–0 Hyperfine Resonance of Alkali Metals. *Phys. Rev.*, **124**, 800–809.

Arditi, M. & Carver, T. R. (1964). Hyperfine Relaxation of Optically Pumped Rb^{87} Atoms in Buffer Gases. *Phys. Rev.*, **136**, 643–649.

Arditi, M. & Picqué, J.-L. (1980). Construction and Preliminary Tests of a Laser Optically Pumped Cesium Jet Atomic Clock. *Comptes Rendus, Série B – Sciences Physiques*, **290**, 461–464.

Audoin, C. (1992). Caesium Beam Frequency Standards: Classical and Optically Pumped. *Metrologia*, **29**, 113–134.

Audoin, C. & Guinot, B. (2001). *The Measurement of Time*. Cambridge: Cambridge University Press.

Becker, W. & Werth, G. (1983). Precise Determination of the Ground State Hyperfine Splitting of ^{135}Ba^{+}. *Zeitschrift für Physik A*, 311, 41–47.

Berkeland, D. J., Miller, J. D., Bergquist, J. C., Itano, W. M., & Wineland, D. J. (1998). Laser-Cooled Mercury Ion Frequency Standard. *Phys. Rev. Lett.*, **80**, 2089–2092.

Blatt, R. & Werth, G. (1982). Precision Determination of the Ground-State Hyperfine Splitting in ^{137}Ba^{+} Using the Ion-Storage Technique. *Phys. Rev. A.*, **25**, 1476–1482.

Blaum, K., Novokov, Yu, N., & Werth, G. (2010). Penning Traps as a Versatile Tool for Precise Experiments in Fundamental Physics. *Contemporary Physics*, **51**, 149–175.

Bollinger, J. J., Prestage, W. M., Itano, W. M., & Wineland, D. J. (1985). Laser-Cooled-Atomic Frequency Standard. *Phys. Rev. Lett.*, **54**, 1000–1003.

Burt, E. A., Diener, W. A., & Tjoelker, R. L. (2008). A Compensated Multi-Pole Linear Ion Trap Mercury Frequency Standard for Ultra-Stable Timekeeping. *IEEE Transactions on Ultrasonics, Ferroelectrics and Frequency Control*, 2586–2595.

Carpenter, R. J., Beaty, E. C., Bender, P. L., Saito, S., & Stone, R. O. (1960). A Prototype Rubidium Vapor Frequency Standard. *IRE Trans On Instrumentation*, **I-9**, 132–135.

Church, D. A. (1969). Storage-Ring Ion Trap Derived from the Linear Quadrupole Radio-Frequency Mass Filter. *J. Appl. Phys.*, **40**, 3127–3134.

Clairon, A., Ghezali, S., Santarelli, G., Laurent, Ph., Lea, S. N., Bahoura, M., Simon, E., Weyers, S., & Szymaniec, K. (1996). Preliminary Accuracy Evaluation of a Cesium Fountain Frequency Standard. In J. C. Bergquist, ed., *Proceedings of the Fifth Symposium on Frequency Standards and Metrology*. London: World Scientific, 49–59.

Crampton, S. B., Greytak, T. J., Kleppner, D., Phillips, W. D., Smith, D. A., & Weinrib, A. (1979). Hyperfine Resonance of Gaseous Atomic Hydrogen at 4.2 K. *Phys. Rev. Lett.*, **42**, 1039–1042.

Cutler, L. S., Flory, C. A., Giffard, R. P., & McGuire, M. D. (1986). Doppler Effects Due to Thermal Macromotion of Ions in an RF Quadrupole Trap. *Appl. Phys. B*, **39**, 251–259.

Cutler, L. S., Giffard, R. P., & McGuire, M. D. (1985). Thermalization of ^{199}Hg Ion Macromotion by a Light Background Gas in an RF Quadrupole Trap. *Appl. Phys. B*, **36**, 137–142.

Davidovits, P. (1964). An Optically Pumped Rb^{87} Maser Oscillator. *Appl. Phys. Letters*, **5**, 15–16.

Davidovits, P. & Novick, R. (1966). The Optically Pumped Rubidium Maser. *Proceedings of the IEEE*, **54**, 155–170.

Du Y., Wei R., Dong R., & Wang Y. (2013) Progress of the Portable Rubidium Atomic Fountain Clock in SIOM. In J. Sun, W., Jiao, H. Wu, & C. Shi, eds., *China Satellite Navigation Conference (CSNC) 2013 Proceedings. Lecture Notes in Electrical Engineering*, 245. Springer.

Essen, L., *Time for Reflection*, published privately and available at http://www.btinternet
.com/~time.lord/; also available in Henderson, D. (2005). Essen and the National
Physical Laboratory's Atomic Clock. *Metrologia*, **42**, S4–S9.

Essen, L. & Parry, J. V. L. (1955). An Atomic Standard of Frequency and Time Interval:
A Cæsium Resonator. *Nature*, **176**, 280–282.

Essen, L. & Parry, J. V. L. (1957). The Caesium Resonator as a Standard of Frequency and
Time. *Philos. Trans. Roy. Soc. London. Ser. A, Mathematical and Physical Sciences*,
250, 45–69.

Fang, F., Li, M., Lin, P., Chen, W., Liu, N., Lin, Y., Wang, P., Liu, K., Suo, R., & Li, T.
(2015). NIM5 Cs Fountain Clock and Its Evaluation. *Metrologia*, **52**, 454–468.

Fertig, C. & Gibble, K. (2000). Measurement and Cancellation of the Cold Collision
Frequency Shift in an ^{87}Rb Fountain Clock. *Phys. Rev. Lett.*, **85**, 1622–1625.

Fisk, P. T. H., Sellars, M. J., Lawn, M. A., Coles, C., Mann, A. G., & Blair, D. G. (1995).
Very High Q Microwave Spectroscopy on Trapped $^{171}Yb^{+}$ Ions: Application as
a Frequency Standard. *IEEE Transactions on Instrumentation and Measurement*, **44**,
113–116.

Forman, P. (1998). Atomichron: The Atomic Clock from Concept to Commercial Product.
Available at www.ieee-uffc.org/freqcontrol/atomichron/automichron.htm.

Gerginov, V., Nemitz, N., Weyers, S., Schröder, R., Griebsch, B., & Wynands, R. (2010).
Uncertainty Evaluation of the Caesium Fountain Clock PTB-CSF2. *Metrologia* **47**,
65–79.

Goldenberg, H. M., Kleppner, D., & Ramsey, N. F. (1960). Atomic Hydrogen Maser. *Phys.
Rev. Lett.*, **5**, 361–362.

Golding, W. M., Frank, A., Beard, R., White, J., Danzy, F., & Powers, E. (1994).
The Double Bulb Rubidium Maser. In *Proceedings of the 1994 IEEE International
Frequency Control Symposium*. Boston, MA: Institute of Electrical and Electronics
Engineers, 724–730.

Guena, J., Rosenbusch, P., Laurent, Ph., Abgrall, M., Rovera, D., Lours, M., Santarelli, G.,
Tobar, M. E., Bize, S., & Clairon, A. (2013). Demonstration of a Dual Alkali Rb/Cs
Atomic Fountain Clock. arXiv:1301.0483v1 3 Jan2013.

Hess, H. F., Kochanski, G. P., Doyle, J. M., Greytak, T. J., & Kleppner, D. (1986). Spin-
Polarized Hydrogen Maser. *Phys. Rev. A*, **34**, 1602–1604.

Hürlimann, M. D., Hardy, W. N., Berlinsky, A. J., & Cline, R. W. (1986). Recirculating
Cryogenic Hydrogen Maser. *Phys. Rev. A*, **34**, 1605–1608.

Itano, W. M. (1991). Atomic Ion Frequency Standards. *Proceedings of the IEEE*, **79**,
936–942.

Itano, W. M. & Wineland, D. J. (1981). Precision Measurement of the Ground-State
Hyperfine Constant of $^{25}Mg^{+}$. *Phys. Rev. A.*, **24**, 1364–1373.

Jardino, M., Desaintfuscien, M., Barillet, R., Viennet, J., Petit, P., & Audoin, C. (1981).
Frequency Stability of a Mercury Ion Frequency Standard. *Appl. Phys. A*, **24**,
107–112.

Kasevich, M., Riis, E., Chu, S., & DeVoe, R. G. (1989). RF Spectroscopy in an Atomic
Fountain. *Phys. Rev. Lett.*, **63**, 612–615.

Kastler, A. (1950). Quelques suggestions concernant la production optique et la détection
optique d'une inégalité de population des niveaux de quantification spatiale des
atomes. Application à l'expérience de Stern et Gerlach et à la résonance
magnétique. *Le Journal de Physique et le Radium*, **11**, 255–265.

Kleppner, D., Berg, H. C., Crampton, S. B., Ramsey, N. F., Vessot, R. F. C., Peters, H. E.,
& Vanier, J. (1965). Hydrogen Maser Principles and Techniques. *Phys. Rev.*, **138**,
A972–A983.

Kleppner, D., Goldberg, H. M., & Ramsey, N. F. (1962). Theory of the Hydrogen Maser. *Phys. Rev.*, **126**, 603–615.

Li, R. & Gibble, K. (2011). Comment on "Accurate Rubidium Atomic Fountain Frequency Standard." *Metrology* **48**, 446–447.

Lombardi, M. A., Heavner, T. P., & Jefferts, S. R. (2007). NIST Primary Frequency Standards and the Realization of the SI Second. *Measure*, **2**, 74–89.

Ludlow, A. D., Boyd, M. M., Ye, J., Peik, E., & Schmidt, P. O. (2015). Optical Atomic Clocks. *Rev. Mod. Phys.*, 87, 637–701.

Lyons, H. (1949). The Atomic Clock. *Instruments*, **22**, 133–135.

Major, F. G. & Werth, G. (1973). High-Resolution Magnetic Hyperfine Resonance in Harmonically Bound Ground State ^{199}Hg Ions. *Phys. Rev. Lett*. **30**, 1155–1158.

Markowitz, W., Hall, R. G., Essen, L., & Perry, J. V. L. (1958). Frequency of Cesium in Terms of Ephemeris Time. *Phys. Rev. Lett*. **1**, 105–107.

McCoubrey, A. O. (1996). History of Atomic Frequency Standards: A Trip through 20th Century Physics. In *Proceedings of the 1996 IEEE International Frequency Control Symposium*. IEEE, 1225–1241.

Millman, S. & Kusch, P. (1940). On the Radiofrequency Spectra of Sodium, Rubidium and Caesium. *Phys. Rev.*, **58**, 438–445.

Münch, A., Berkler, M., Gerz, Ch., Wilsdorf, D., & Werth, G. (1987). Precise Ground-State Hyperfine Splitting in ^{173}Yb II. *Phys. Rev. A.*, **35**, 4147–4150.

Ovchinnikov, Y. & Marra, G. (2011). Accurate Rubidium Atomic Fountain Frequency Standard. *Metrology*, **48**, 87–100.

Ovchinnikov, Y. B., Szymaniec, K., & Edris, S. (2015). Measurement of Rubidium Ground-State Hyperfine Transition Frequency Using Atomic Fountains. *Metrology*, **52**, 595–599.

Packard, M. E. & Swartz, B. E. (1962). The Optically Pumped Rubidium Vapor Frequency Standard. *IREE Trans. on Instrumentation*, **I-11**, 215–223.

Peil, S., Hanssen, J., Swanson, T. B., Taylor, J., & Ekstrom, C. R. (2016). The USNO Rubidium Fountains. *8th Symposium on Frequency Standards and Metrology 2015, Journal of Physics Conference Series*, **721**, 012004.

Peterman, P., Gibble, K., Laurent, Ph., & Salomon, C. (2016). Microwave Lensing Frequency Shift of the PHARAO Laser Cooled Microgravity Atomic Clock *Metrologia*, **53**, 899–907.

Picqué, J.-L. (1977). Hyperfine Optical Pumping of a Cesium Atomic Beam, and Applications. *Metrologia*, **13**, 115–119.

Prestage, J. D., Dick, G. J., & Maleki, L. (1989). New Ion Trap for Frequency Standard Applications. *J. Appl. Phys.*, **66**, 1013–1017.

Prestage, J. D., Tjoelker, R. J., Dick, G. J., & Maleki, L. (1995). Progress Report on the Improved Linear Ion Trap Physics Package. *Proc. 49th Ann. Symp. Freq. Control Symposium*, 82–85.

Prestage, J. D., Tjoelker, R. J., & Maleki, L. (2001). Recent Developments in Microwave Ion Clocks. In A. N. Luiten, ed., Topics in Applied Physics, Frequency Measurement and Control, **79**. Heidelberg: Springer-Verlag, 195–211.

Rabi, I. I., Millman, S., Kush P., & Zacharias J. R. (1939). The Molecular Beam Resonance Method for Measuring Nuclear Magnetic Moments, The Magnetic Moments of $_3$Li^6, $_3$Li^7 and $_9$F^{19}. *Phys. Rev.*, **55**, 526–535.

Ramsey, N. F. (1949). A New Molecular Beam Resonance Method. *Phys. Rev.*, **76**, 996.

Ramsey, N. F. (1950). A Molecular Beam Resonance Method with Separated Oscillating Fields. *Phys. Rev.*, **78**, 695–699.

Ramsey, N. F. (1983). History of Atomic Clocks. *Journal of Research of the National Bureau of Standards*, **88**, 301–320.

Ramsey, N. F. (1993). *I. I. Rabi 1898–1988, Biographical Memoir*. Washington, DC: National Academy of Sciences.

Ramsey, N. F. (2005). History of Early Atomic Clocks. *Metrologia*, **42**, S1–S3.

Sherwood, J., Lyons, H., McCracken, R., & Kusch, P. (1952). High Frequency Lines in the hfs Spectrum of Cesium. *Bulletin of the American Physical Society*, **27**, 43.

Song, H-j., Dong, S-w., & Wang, Z-m. (2016). An Analysis of NTSC's Timekeeping Hydrogen Masers. *Chinese Astronomy and Astrophysics*, **40**, 569–577.

Sullivan, D. B., Bergquist, J. C., Bollinger, J. J., Drullinger, R. E., Itano, W. M., Jefferts, S. R., Lee, W. D., Meekhof, D., Parker, T. E., Walls, F. L., & Wineland, D. J. (2001). Primary Atomic Frequency Standards at NIST. *Journal of Research of the National Institute of Standards and Technology*, **106**, 47–63.

Tjoelker, R. L., Bricker, C., Diener, W., Hamell, R. L., Kirk, A., Kuhnle, P., Maleki, L., Prestage, J. D., Santiago, D., Seidel, D., Stowers, D. A., Sydnor, R. L., Tucker, T. (1996). A Mercury Ion Frequency Standard Engineering Prototype for the NASA Deep Space Network. *Proceedings of the 1996 IEEE/EIA International Frequency Control Symposium and Exhibition*, 1073–1081.

Tjoelker, R. L., Chung, S., Diener, W., Kirk, A., Maleki, L., Prestage, J. D., & Young, B, (2000). Nitrogen Buffer Gas Experiments in Mercury Trapped Ion Frequency Standards. *Proceedings of the 2000 IEEE/EIA International Frequency Control Symposium and Exhibition*, 668–671.

Vessot, R. F. C. (2005). The Atomic Hydrogen Maser Oscillator. *Metrologia*, **42**, S80–S89.

Vessot, R. F. C., Levine, M. W., & Mattison, E. M. (1977). Comparison of Theoretical and Observed Maser Stability Limitation Due to Thermal Noise and the Prospect of Improvement by Low Temperature Operation. *Proceedings of the 9th Annual Precise Time and Time Interval (PTTI) Applications and Planning Meeting (NASA, Goddard Space Flight Center 29 November–1 December 1977)*, 549.

Vessot, R. F. C., Mattison, E. M., & Blomberg, E. L. (1979). Research with a Cold Atomic Hydrogen Maser. In *Annual Frequency Control Symposium, May 30–June 1, 1979, Proceedings*. Washington, DC: Electronic Industries Association, 511–514.

Walsworth, R. L., Silvera, I. F., Godfried, H. P., Agosta, C. C., Vessot, R. F. C., & Mattison, E. M. (1986). Hydrogen Maser at Temperatures below 1 K. *Phys. Rev. A*, **34**, 2550.

Wineland, D. J. & Itano, W. M. (1979). Laser Cooling of Atoms. *Phys. Rev. A.*, **20**, 1521–1540.

12

Optical Atomic Standards

12.1 Optical Transition Frequencies

As is the case with microwave standards, optical clocks consist of an oscillator regulated using the frequency of an atomic quantum-level transition. Optical clocks, however, realize this transition by exciting the atom with an electromagnetic wave in the optical region of the spectrum as opposed to the microwave region. The much higher optical transition frequencies provide an attractive alternative to the microwave frequencies, and, with continuing progress in laser technology, standards based on these frequencies are being realized. The interaction between atoms and light as a means of determining time and frequency is being developed by two systems. One uses a single ion with sufficient long transition lifetime when trapped by electric fields and then laser cooled. The other uses laser-cooled atoms confined in optical lattices. Sub-hertz linewidth lasers permit single ions to be probed with sufficient resolution for high-stability clock operation. The history and state of the art of the optical clock research over the past decade is given by Poli et al. (2013).

Table 12.1 lists optical frequencies that are being investigated for possible application in metrology. Figure 12.1 shows the electromagnetic spectrum with the microwave and optical frequencies along with the expected calibration errors of each of the transition frequencies suggested as possible secondary representations of the second. The high frequencies of the optical transitions pose a problem when applied to timekeeping. They need to be converted to frequencies that can be used in timekeeping circuitry before they can be useful for practical application. Radio frequencies up to about 100 GHz can be measured easily using electronic means, but measurement of frequencies higher than 100 GHz is problematic.

The answer to the problem is the development of frequency combs, which take advantage of developments in laser technology. When mode-locked lasers capable of producing light pulses with lengths of the order of femtoseconds (10^{-15} second)

Table 12.1 *Laser frequency standards recommended by the International Committee for Weights and Measures (CIPM) in 2015. SRS is a secondary representation of the second (Hong, 2017).*

Application	λ (nm)	Laser and reference	Spectroscopy	Frequency	Uncertainty
SRS	267	^{27}Al$^+$, 3s^2 ^{1}S$_0$–3s3p ^{3}P$_0$	Ion trap	1 121 015 393 207 857.3 Hz	1.9 × 10^{-15}
	282	^{199}Hg$^+$, 5d^{10}6s ^{2}S$_{1/2}$ ($F=0$)–5d^{9}6s^2 ^{2}D$_{5/2}$ ($F=2$)	Ion trap	1 064 721 609 899 145.3 Hz	1.9 × 10^{-15}
	436	^{171}Yb$^+$, 6s ^{2}S$_{1/2}$ ($F=0$) – 5d ^{2}D$_{3/2}$ ($F=2$)	Ion trap	688 358 979 309 308.3 Hz	6 × 10^{-16}
	467	^{171}Yb$^+$, 6s ^{2}S$_{1/2}$ ($F=0$)–4f^{13}6s^2 ^{2}F$_{7/2}$ ($F=3$)	Ion trap	642 121 496 772 645.0 Hz	6 × 10^{-16}
	578	^{171}Yb, 6s^2 ^{1}S$_0$ ($F=1/2$)–6s6p ^{3}P$_0$ ($F=1/2$)	Optical lattice	518 295 836 590 864.0 Hz	2 × 10^{-15}
	674	^{88}Sr$^+$, 5s ^{2}S$_{1/2}$–4d ^{2}D$_{5/2}$	Ion trap	444 779 044 095 486.6 Hz	1.6 × 10^{-15}
	698	^{87}Sr, 5s^2 ^{1}S$_0$ ($F=9/2$)–5s5p ^{3}P$_0$ ($F=9/2$)	Optical lattice	429 228 004 229 873.2 Hz	5 × 10^{-16}
Time	237	^{115}In$^+$, 5s^2 ^{1}S$_0$–5s5p ^{3}P$_0$	Ion trap	1267 402 452 899.92 kHz	3.6 × 10^{-13}
	243	^{1}H, 1S–2S, 2 photon	Atomic beam, two photons	1233 030 706 593 514 Hz	9 × 10^{-15}
	266	^{199}Hg, 6s^2 ^{1}S$_0$–6s6p ^{3}P$_0$	Optical lattice	1 128 575 290 808 154.8 Hz	6 × 10^{-16}
	657	^{40}Ca, ^{1}S$_0$–^{3}P$_1$, $\Delta m_J = 0$	Cold atoms	455 986 240 494 140 Hz	1.8 × 10^{-14}
	698	^{88}Sr, 5s^2 ^{1}S$_0$–5s5p ^{3}P$_0$	Optical lattice	429 228 066 418 012 Hz	1 × 10^{-14}
	729	^{40}Ca$^+$, 4s ^{2}S$_{1/2}$–3d ^{2}D$_{5/2}$	Ion trap	411 042 129 776 398.4 Hz	1.2 × 10^{-14}
Length and others	531	Diode laser, ^{127}I$_2$, R(36)32-0:a_1	Saturation absorption	564 074 632.42 MHz	1 × 10^{-10}
	532	Nd:YAG laser, ^{127}I$_2$, R(56)32-0:a_{10}	Saturation absorption	563 260 223 513 kHz	8.9 × 10^{-12}
	543	He-Ne laser, ^{127}I$_2$, R(106)28-8:b_{10}	Saturation absorption	551 580 162 400 kHz	4.5 × 10^{-11}
	633	He-Ne laser, ^{127}I$_2$, R(127)11-5:a_{16}	Saturation absorption	473 612 353 604 kHz	2.1 × 10^{-11}
	778	^{85}Rb, 5S$_{1/2}$($F=3$)–5D$_{5/2}$($F=5$), 2 photon	Two photons	385 285 142 375 kHz	1.3 × 10^{-11}
	780	^{87}Rb, 5S$_{1/2}$–5P$_{3/2}$, d/f crossover	Saturation absorption	384 227 981.9 MHz	5 × 10^{-10}
	1,540	^{13}C$_2$H$_2$, P(16)($v_1 + v_3$)	Saturation absorption	194 369 569 384 kHz	2.6 × 10^{-11}
	3,390	He-Ne laser, CH$_4$, n_3, P(7), F$_2^{(2)}$	Saturation absorption	88 376 181 600.18 kHz	3 × 10^{-12}

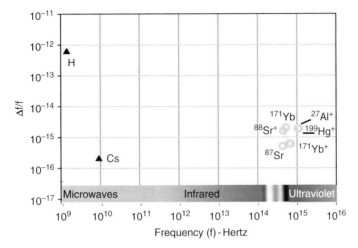

Figure 12.1 Spectrum of transition frequencies used to represent the second. The calibration error represented as *Δf/f* is shown for each transition. The microwave frequencies are represented by triangles and the optical frequencies currently suggested as possible secondary representations of the second for research purposes by the Consultative Committee for Time and Frequency of the Comité International des Poids et Mesures (CIPM) are shown as circles.

became available, frequency combs became possible. These mode-locked lasers are able to provide trains of extremely short optical pulses that are the result of a superposition of many continuous wave longitudinal cavity modes. The frequency of the nth mode is given by:

$$f_n = nf_R + f_0, \tag{12.1}$$

where f_R is the repetition frequency given by the reciprocal of the pulse repetition interval, and f_0 is a frequency offset. The appearance of the spectrum in the frequency domain is then extremely broad and comprised of a series of "spikes," spaced according to Equation (12.1), hence the name "frequency comb." The appearance of the spectrum then is directly related to the pulse rate of the laser. By matching the optical frequency of the laser linked to the energy level transition to the frequency comb by adjusting the repetition rate, we can relate the optical frequency to a signal useful for precise timekeeping (Udem et al., 1999; Holzworth et al., 2000; Diddams et al., 2002; Ma et al., 2004).

An optical clock, then, is quite similar to a microwave standard, but it uses a stabilized laser as the local oscillator, the output of which is used in a feedback loop to produce the desired energy-level transition. The feedback signal is provided by means of a cooling laser that is also used to detect the occurrence of energy-level transitions (Figure 12.2). When the light from the laser, acting as the local oscillator, causes transitions to occur, then no

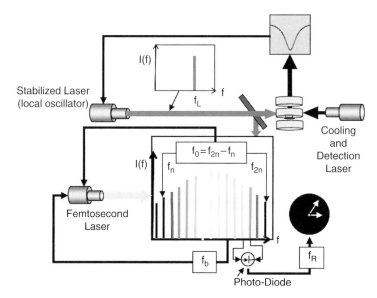

Figure 12.2 Schematic representation of an optical clock

fluorescence occurs from photons involved in the cooling transition. By monitoring the level of the fluorescence, it is possible to obtain the frequency of the energy-level transition.

The output signal of the clock is produced using a frequency comb based on another laser producing femtosecond pulses. That laser emits pulses at a nominal repetition rate f_R, and this, in turn, provides a comb of frequencies spaced at intervals f_n as given by Equation (12.1). If the comb spans one octave (a factor of two in frequency), the term f_0 can be derived by a self-referencing technique. In that process, the frequency of the nth infrared mode is doubled, producing a visible frequency that can be heterodyned with the nth mode frequency. The heterodyne signal provides a frequency $2(nf_R + f_0) - (2nf_R + f_0) = f_0$.

A second beat frequency, f_b, is also obtained between the mth frequency mode, $f_m = mf_R + f_0$, and the local oscillator frequency f_L. The clock then uses two feedback phase-locked loops to control f_0 and f_b. The first of these loops controls the femtosecond laser power so that f_0 maintains a constant relationship with f_R given by $f_0 = \beta f_R$. The second loop controls the cavity length of the femtosecond laser so that $f_b = \alpha f_R$. In this way, the frequencies of the two loops are phase referenced to the frequency of the local oscillator. A photo-diode is used to measure the frequency f_R, which is used as the clock frequency (Diddams et al., 2001). Figure 12.2 shows a schematic optical clock.

12.2 Optical Ion Clocks

The idea of a using a single ion as a frequency standard was first proposed by Hans Dehmelt (1982). The principles of an optical clock were realized practically in a mercury frequency standard based on a single laser-cooled ^{199}Hg$^+$ ion. This device made use of an energy transition that provided a frequency of 1.06×10^{15} Hz with a line width of 6.7 Hz (Rafac et al., 2000) from a single ion stored in a cryogenic Paul ion trap. This transition is in the ultraviolet (wavelength = 282 nm) and results in a Q value of 1.6×10^{14}. By comparing this standard to an optical standard based on neutral calcium atoms operating with a transition frequency of 456 THz (4.56×10^{14}), it was determined that the mercury device was capable of a 1-second stability of 1×10^{-15} (Diddams et al., 2001).

^{171}Yt$^+$ single-ion clocks (quadrupole transition) have been developed at Physikalisch-Technischen Bundesanstalt (PTB) and the National Physical Laboratory (NPL). The uncertainty of the CIPM-recommended frequency is 6×10^{-16}. The frequency uncertainty of these clocks can be as low as 1×10^{-16} and is limited by blackbody radiation shift and residual quadrupole shift (Godun, 2014). ^{171}Yb$^+$ single-ion clocks (octupole transition) are also being developed at NPL and PTB. The frequency uncertainty of these clocks can be as low as 3.2×10^{-18} and is limited by the quadratic Stark shift due to thermal radiation at room temperature (Huntemann et al., 2016).

^{88}Sr$^+$ single-ion clocks are being developed at NPL and the National Research Council of Canada, and the uncertainty of this CIPM-recommended frequency is 1.6×10^{-15}. The frequency uncertainty of these clocks can be as low as 5×10^{-17} and is limited by the blackbody radiation shift (Barwood et al., 2014).

^{199}Hg$^+$ and ^{27}Al$^+$ single-ion clocks are being developed at the National Institute for Standards and Technology (NIST). The CIPM-recommended frequency uncertainty for these clocks is 1.9×10^{-15}. The frequency uncertainty of ^{199}Hg$^+$ and ^{27}Al$^+$ single-ion clocks can be as low as 1.9×10^{-17} and 8.6×10^{-18}, respectively, and it is limited by residual quadrupole shift and excess micromotion, respectively. The uncertainty of the ^{27}Al$^+$ single-ion clock was validated at 2×10^{-17} using two clocks at NIST (Chou et al., 2010).

^{40}Ca$^+$ single-ion clocks have been developed by the Institut für Experimentalphysik, University of Innsbruck, National Institute of Information and Communications Technology (NICT), and the Wuhan Institute of Physics and Mathematics (WIPM). The uncertainty of the CIPM-recommended frequency is 1.2×10^{-14}. The frequency uncertainty of these clocks can be as low as 5×10^{-17} and is limited by shifts of excess micromotion and blackbody radiation (Huang et al., 2016). A compact transportable ^{40}Ca$^+$ optical clock has been engineered within a volume of 0.54 m^3. The systematic fractional uncertainty was evaluated to be 7.7×10^{-17} (Cao et al., 2016).

A ^{229}Th^{3+} single-ion optical clock is proposed to offer systematic shift suppression, which would allow clock performance with total fractional inaccuracy approaching 1×10^{-19} (Campbell et al., 2012).

12.3 Optical Neutral Atom Clocks

Neutral atoms can also be used in optical frequency standards. These clocks make use of optical standard technology and combine the advantages of using a trapped single ion with using a large number of free-falling neutral atoms. The performance of a single-ion clock is compromised for the reason that the signal-to-noise ratio is low because it only involves one ion. Using a cloud of neutral atoms in an optical clock is a possible means to get around the problem. Calcium has been used in this capacity because of a convenient wavelength at 657 nm and its insensitivity to external fields (Oates et al., 1999). Millions of calcium atoms are cooled in a magneto-optic trap, released, and then probed using laser light at 657 nm. A femtosecond mode-locked laser is used to provide a femtosecond comb, which, in turn, is used to provide the frequency of the clock transition.

Neutral atoms are also used in optical clocks by confining them in an optical lattice. To take advantage of some of the narrow line widths that are possible optically, it is necessary to increase the interrogation time of the atoms. Hidetoshi Katori et al. (2003) proposed using an optical lattice to confine the atoms so that the interrogation times could be increased. An optical lattice makes use of standing light waves to create a potential surface that is made up of a series of "valleys" in which the atoms are confined (Figure 12.3). The crystal-like structure in which the atoms are confined makes it possible to increase the interrogation time of the atoms. The interfering light beams that create the lattice must operate at a wavelength so that the light shifts that are exerted on the ground and upper states of the clock transition are exactly equal. This "magic frequency" prevents the optical field that creates the lattice from perturbing the clock-transition frequency.

Neutral atoms are first slowed and cooled and then loaded into the optical lattice. Atoms of strontium (^{87}Sr) (Ludlow et al., 2008) and ytterbium (^{171}Yb) (Barber et al., 2008) have been used in these standards. Sr optical lattice clocks are being developed in many institutes and are the most investigated optical clocks. The uncertainty of the CIPM-recommended frequency (5×10^{-16}) is the lowest of optical clocks. The frequency uncertainty of a Sr optical lattice clock can be as low as 10^{-18}, and is limited by the lattice light shift and density shift (Falke et al., 2011). Two Sr optical lattice clocks were compared intercontinentally with a baseline of 9,000 km between Germany and Japan with a two-way satellite time and frequency

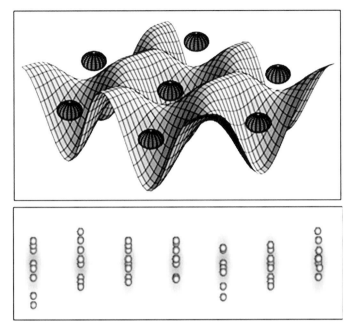

Figure 12.3 An optical lattice in three dimensions (top) and two dimensions (bottom)

transfer technique. A frequency comparison for 83640 s resulted in a fractional difference of $(1.1 \pm 1.6) \times 10^{-15}$ (Hachisu et al., 2014). The systematic uncertainty of a cryogenic clock is evaluated to be 7.2×10^{-18} by operating two such clocks synchronously. Statistical agreement between the clocks reached 2.0×10^{-18} after 11 measurements over a month (Ushijima et al., 2015). Two Sr optical lattice clocks at PTB and Systèmes de Référence Temps Espace (SYRTE) were compared and agreed at an uncertainty of 5×10^{-17} (Lisdat et al., 2016). The JILA has developed two ^{87}Sr optical lattice clocks and has achieved fractional stability of 2.2×10^{-16} at 1 s. Systematic uncertainties such as lattice AC Stark shift, atoms' thermal environment, and atomic response to room temperature blackbody radiation have been reduced. The total uncertainty of a clock is 2.1×10^{-18} (Bloom et al., 2014; Nicholson et al., 2015).

^{171}Yb optical lattice clocks have been developed in five institutes and are being actively investigated in the optical clock community. The frequency uncertainty of an Yb optical lattice clock can be as low as 10^{-17}, and is limited by the lattice light shift (Hong, 2017).

^{199}Hg optical lattice clocks are being developed by the RIKEN Center for Advanced Photonics/University of Tokyo and SYRTE. The uncertainty of the CIPM-recommended frequency is 6×10^{-16}. The frequency uncertainty of these clocks can be as low as 7.2×10^{-17} and is limited by the lattice light shift (Yamanaka et al., 2015). An optical lattice clock based on neutral mercury with a relative uncertainty of 1.7×10^{-16} was compared with ^{133}Cs and ^{87}Rb atomic fountains to determine the ratio between the ^{199}Hg clock transition and the ^{87}Rb ground state hyperfine transition. The main perturbations affecting the mercury clock frequency measurement are the trap AC Stark shift, the external magnetic field, the thermal radiation (black body radiation shift), the atomic density in the trap, and the pulsed interrogation.

12.4 Quantum Logic Clock

Quantum logic spectroscopy has opened another approach to a frequency standard. This technique uses a single ion of one element along with a single ion of an auxiliary element to handle the requirements for laser cooling and state detection. Such a standard has been developed using a ^{27}Al$^+$ ion and a ^{9}Be$^+$ ion (Rosenband et al., 2007). Both are loaded into a linear Paul trap where they can be expected to exist as an ion pair for several hours. A chemical interaction with background gas eventually removes one of the ions from the trap. The pair acts as a crystal within the trap. Lasers operating at a wavelength of 313 nm act on the beryllium to cool the ion pair. A second laser is used to get the aluminum ion to the proper state where the light from yet another laser at a wavelength of 267.4 nm is able to excite the clock transition. When the aluminum ion is in the excited state, the beryllium ion is constrained, by Coulomb interaction, to be in the level to which it had been previously pumped. The fluorescence of photons from this level can be monitored to provide the feedback signal to the excitation laser. The standard is operated by a series of pulses of this process. In each pulse seven photons are expected to be seen if the aluminum ion is in the proper state, and only one is expected to be seen, if the clock transition were not successful.

   The quantum logic standard receives this name because it employs techniques used in quantum computers that are based on quantum mechanics. There are two different logical states of the aluminum ion. The actual state of the ion is communicated to the beryllium ion, which then provides easily detected signals depending on its state. The optical frequency of the excitation laser can be translated to rf frequencies using a femtosecond comb as outlined previously. Aluminum is used in this process because it provides a stable frequency. Because it is difficult to detect, however, the beryllium ion is introduced.

12.5 Stabilized Lasers

Eight optical frequency standards that can be used for length and other applications are included in Table 12.1. Two of these, ^{127}I$_2$ and the 780 nm diode laser stabilized to ^{87}Rb, are newly included in the CIPM recommendations from 2015. The 633 nm iodine-stabilized He-Ne laser is the most popular standard for length applications, and the 3.99 μm CH$_4$-stabilized laser is being studied using new light sources. There is interest in development of compact, robust, efficient, and relatively inexpensive atomic frequency standards with frequency stabilities of 10^{-10} to 10^{-13} for the commercial and industrial markets.

A compact 531 nm iodine-stabilized diode laser has a frequency stability at the 10^{-12} level and a frequency uncertainty at the 10^{-11} level. The uncertainty of the CIPM-recommended frequency is 1×10^{-10}. This type of laser has recently been applied to long-range block measurements (Bitou et al., 2016).

Rubidium (Rb) 780 nm stabilized lasers are good for length applications because a commercially available laser and a compact spectroscopic configuration are used. The frequency uncertainty was 4.3×10^{-10}. The uncertainty of the CIPM-recommended frequency is 5×10^{-10}, based on two reported measurements (Ye et al., 1996; Hong et al., 2003).

A portable stand-alone optical frequency standard using a gas-filled hollow-core photonic crystal fiber was developed to stabilize a fiber laser to the ^{13}C$_2$H$_2$P (16) $(v_1 + v_3)$ transition at 1,542 nm using saturated absorption. The locked laser has a fractional frequency instability below 8×10^{-12} for an averaging time of 10^4 s. The system is portable and shows no change after shipment and return (Triches et al., 2015).

An optical standard is based on a two-photon, two-color Doppler-free transition of ^{87}Rb vapor within a HC-PCF, to which the sum-frequency of two lasers at 780 and 776 nm stabilizes. The standard has a fractional frequency stability of 9.8×10^{-12} at a 1.3 s integration time. A planned improvement should achieve $\sim 10^{-13}$ fractional frequency stability at 1 s integration times (Perrella et al., 2013).

12.6 Characterizing Optical Standards

At PTB a ^{87}Sr optical lattice clock, with time coverage of 46% over 25 days, is able to maintain a local timescale with a time error of less than 200 ps. At SYRTE a ^{87}Sr optical lattice clock operates reliably over several weeks, with time coverage of 80% (Hong, 2017).

Optical clocks are affected by systematic effects including the Doppler effect, interaction with magnetic fields (Zeeman effect), interaction with electric fields, interaction with blackbody radiation, collision and pressure shifts, gravitational effects, and locking errors due to technical effects (Poli et al., 2013).

A significant issue in characterizing optical clocks is the limitation posed by the precision achievable in comparing clocks at a distance. Figure 12.4 shows current capabilities graphically.

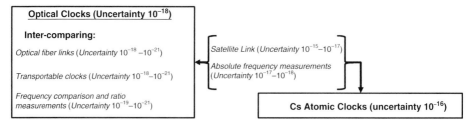

Figure 12.4 Standards and measurement tools involved in optical frequency metrology. When optical clocks are compared with microwave clocks or other optical clocks in the same laboratory, optical frequency combs are used to link different frequencies. When clocks in different laboratories are compared, optical fiber, satellites, or transportable clocks are used for a frequency link (Hong, 2017).

The uncertainty of caesium atomic clocks has been reduced to the 10^{-16} level with laser cooling. However, the optical clocks are offering uncertainties of 10^{-18}. They can be expected to improve with the goal of uncertainties $< 1 \times 10^{-18}$ by improving current optical clocks or by starting new clock schemes, but it may take some time due to the complications involved in their comparisons. For example, the geoid is not well defined at the 10^{-18} level.

References

Barber, Z. W., Stalnaker, J. E., Lemke, N. D., Poli, N., Oates, C. W., Fortier, T. M., Diddams, S. A., Hollberg, L., & Hoyt, C. W. (2008). Optical Lattice Induced Light Shifts in an Yb Atomic Clock. *Phys. Rev. Lett.*, **100**, 103002.

Barwood, G. P., Huang, G., Klein, H. A., Johnson, L. A. M., King, S. A., Margolis, H. S., Szymaniec, K., & Gill, P. (2014). Agreement between Two $^{88}Sr^+$ Optical Clocks to 4 Parts in 10^{17}. *Phys. Rev. A*, **89**, 050501.

Bitou, Y., Kobayashi, T., & Hong, F. L. (2016). Compact and Inexpensive Iodine-Stabilized Diode Laser System with an Output at 531 nm for Gauge Block Interferometers. *Precis. Eng.*, **47**, 528–531.

Bloom, B. J., Nicholson, T. L., Williams, J. R., Campbell, S. L., Bishof, M., Zhang, X., Zhang, W., Bromley, S. L., & Ye, J. (2014). An Optical Lattice Clock with Accuracy and Stability at the 10^{-18} Level. *Nature*, **506**, 71–75. doi:10.1038/nature 12941

Campbell, C. J., Radnaev, A. G., Kuzmich, A., Dzuba, V. A., Flambaum, V. V., & Derevianko, A. (2012). Single-Ion Nuclear Clock for Metrology at the 19th Decimal Place. *Phys. Rev. Lett.*, **108**, 120802.

Cao, J., Zhang, P., Shang, J., Cui, K., Yuan, J., Chao, S., Wang, S., Shu, H., & Huang, X. (2016). A Transportable $^{40}Ca^+$ Single-Ion Clock with 7.7 x 10^{-17} Systematic Uncertainty, arXiv:1607.03731C.

Chou, C. W., Hume, D. B., Koelemeij, J. C. J., Wineland, D. J., & Rosenband, T. (2010). Frequency Comparison of Two High-Accuracy Al^+ Optical Clocks. *Phys. Rev. Lett*, **104**, 070802.

Dehmelt, H. G. (1982). Mono-Ion Oscillator as Potential Ultimate Laser Frequency Standard. *IEEE Transactions on Instrumentation and Measurement*, **31**, 83–87.

Diddams, S. A., Hollberg, L., Ma, L.-S., & Robertsson, L. (2002). Femtosecond-Laser-Based Optical Clockwork with Instability ≤ Less Than or Equal to 6.3 X 10^{-16} in 1 s. *Opt. Lett.* **27**, 58–60.

Diddams, S. A., Udem, T., Bergquist, J. C., Curtis, E. A., Drullinger, R. E., Hollberg, L., Itano, W. M., Lee, W. D., Oates, C. W., Vogel, K. R., & Wineland, D. J. (2001). An Optical Clock Based on a Single Trapped ^{199}Hg$^+$ Ion. *Science*, **293**, 825–828.

Falke, S., Schnatz, H., Vellore Winfred, S. R., Middelmann, T., Vogt, S., Weyers, S., Lipphardt, B., Grosche, G., Riehle, F., Sterr, U., & Lisdat, C. (2011). The ^{87}Sr Optical Frequency Standard at PTB. *Metrologia*, **48**, 399–407.

Gill, P. (2005). Optical Frequency Standards. *Metrologia*, **42**, S125–S137.

Godun, R. M. (2014). Frequency Ratio of Two Optical Clock Transitions in ^{171}Yb and Constraints on the Time-Variation of Fundamental Constants. *Phys. Rev. Lett.*, **113**, 210801.

Hachisu, H., Fujieda, M., Nagano, S., Gotoh, T., Nogami, A., Ido, T., Falke, S., Huntemann, N., Grebing, C., Lipphardt, B., Lisdat, C., & Piester, D. (2014). Direct Comparison of Optical Lattice Clocks with an Intercontinental Baseline of 9000 km. *Optics Letters*, **39**, 4072–4075.

Holzwarth, R., Udem, T., Hänsch, T. W., Knight, J. C., Wadsworth, W. J., & Russell, P. S. J. (2000). Optical Frequency Synthesizer for Precision Spectroscopy. *Phys. Rev. Lett.*, **85**, 2264–2267.

Hong, F.-L. (2017). Optical Frequency Standards for Time and Length Applications. *Meas. Sci. Technol.*, **28**, 012002.

Hong, F.-L., Ishikawa, J., Sugiyama, K., Onae, A., Matsumoto, H., Ye, J., & Hall, J. L. (2003). Comparison of Independent Optical Frequency Measurement Using a Portable Iodine-Stabilized Nd:YAG Laser. *IEEE Trans. Instrum. Meas.*, **52**, 240–244.

Huang, Y., Guan, H., Liu, P., Bian, W., Ma, L., Liang, K., Li, T., & Gao, K. (2016). Frequency Comparison to Two ^{40}Ca$^+$ Optical Clocks with an Uncertainty at the 10^{-17} Level. *Phys. Rev. Lett.*, **116**, 013001.

Huntemann, N., Sanner, C., Lipphardt, B., Tamm, C., & Peik, E. (2016). Single-Ion Atomic Clock with 3 x 10^{-18} Systematic Uncertainty. *Phys. Rev. Lett.*, **116**, 063001.

Katori, H., Takamoto, M., Pal'chikov, V. G., & Ovsiannikov, V. D. (2003). Ultrastable Optical Clock with Neutral Atoms in an Engineered Light Shift Trap. *Phys. Rev. Lett.*, **91**, 173005–173008.

Lisdat, C., Grosche, G., Quintin, N., Shi, C., Raupach, S. M. F., Grebing, C., Nicolodi, D., Stefani, F., Al-Masoudi, A., Dörscher, S.,Häfner, S., Robyr, J.-L., Chiodo, N., Bilicki, S., Bookjans, E., Koczwara, A., Koke, S., Kuhl, A., Wiotte, F., Meynadier, F., Camisard, E., Abgrall, M., Lours, M., Legero, T., Schnatz, H., Sterr, U., Denker, J., Chardonnet, C., Le Coq, Y., Santarelli, G., Amy-Klein, A., Le Targat, R., Lodewyck, J., Lopez, O., & Pottie, P.-E. (2016). A Clock Network for Geodesy and Fundamental Science. *Nature Communications*, **7**, 12443.

Ludlow, A. D., Zelevinsky, T., Campbell, G. K., Blatt, S., Boyd, M. M., de Miranda, M. H. G., Martin, M. J., Thomsen, J. W., Foreman, S. M., Ye, J., Fortier, T. M., Stalnaker, J. E., Diddams, S. A., Le Coq, Y., Barber, Z. W., Poli, N., Lemke, N. D., Beck, K. M., & Oates, C. W. (2008). Sr Lattice Clock at 1×10^{-16} Fractional Uncertainty by Remote Optical Evaluation with a Ca Clock. *Science*, **319**, 1805–1808.

Ma, L.-S., Bi, Z., Bartels, A., Robertsson, L., Zucco, M., Windeler, R. S., Wilpers, G., Oates, C., Hollberg, L., & Diddams, S. A. (2004). Optical Frequency Synthesis and Comparison with Uncertainty at the 10–19 Level. *Science*, **303**, 1843–1845.

Nicholson, T. L., Campbell, S. L. Hutson, R. B., Marti, G. E., Bloom, B. J., McNally, R. L., Zhang, W., Barrett, M. D., Safronova, M. S., Strouse, G. F., Tew, W. L., & Ye, J.

(2015). Systematic Evaluation of an Atomic Clock at 2 x 10^{-18} Total Uncertainty. *Nature Communications*, **6**, 6896. doi:10.1038/ncomms 7896

Oates, C. W., Bondu, F., & Hollberg, L. (1999). A Diode-Laser Optical Frequency Reference Based on Laser-Cooled Ca Atoms. *Eur. Phys. J. D.*, **7**, 449–459.

Perrella, C., Light, P. S., Anstie, J. D., Baynes, F. N., Benabid, F., & Luiten, A. N. (2013). Two-Color Rubidium Fiber Frequency Standard. *Optics Letters*, **38**, 2122–2124.

Poli, N., Oates, C. W., Gill, P., & Tino, G. M. (2013). Optical Atomic Clocks, *Rivista del Nuovo Cimento*, **36**, 555–624.

Rafac, R. J., Young, B. C., Beall, J. A., Itano, W. M., Wineland, D. J., & Bergquist, J. C. (2000). Sub-Dekahertz Ultraviolet Spectroscopy of ^{199}Hg$^+$. *Phys. Rev. Lett.*, **85**, 2462–2465.

Rosenband, T., Schmidt, P. O., Hume, D. B., Itano, W. M., Fortier, T. M., Stalnaker, J. E., Kim, K., Diddams, S. A., Koelemeij, J. C. J., Bergquist, J. C., & Wineland, D. J. (2007). Observation of the $^1S_0 \rightarrow {}^3P_0$ Clock Transition in ^{27}Al$^+$. *Phys. Rev. Lett.*, **98**, 220801.

Triches, M., Brusch, A., & Hald, J. (2015). Portable Optical Frequency Standard Based on Sealed Gas-Filled Hollow Core Fiber Using a Novel Encapsulation Technique. *Appl. Phys. B*, **121**, 251–258.

Udem, T., Reichert, J., Holzwarth, R., & Hänsch, T. W. (1999). Accurate Measurement of Large Optical Frequency Differences with a Mode-Locked Laser. *Opt. Lett.*, **24**, 881–883.

Ushijima, I., Takamoto, M., Das, M., Ohkubo, T., & Katori, H. (2015). Cryogenic Optical Lattice Clocks. *Nature Photonics Letters*, **9**, 185–189.

Yamanaka, K., Ohmae, N., Ushijima, I., Takamoto, M., & Katori, H. (2015). Frequency Ratio of ^{199}Hg and ^{87}Sr Optical Lattice Clocks beyond the SI Limit. *Phys. Rev. Lett.*, **114**, 230801.

Ye, J., Swartz, S., Junger, P., & Hall, J. L. (1996). Hyperfine Structure and Absolute Frequency of the ^{87}Rb $5P_{3/2}$ State. *Opt. Lett.*, **21**, 1280–1282.

13

Definition and Role of a Second

13.1 The Historical Second

Early timekeepers had little need for time divisions finer than an hour or, at most, a simple fraction of an hour. In geometry, the circle was divided into 360 degrees in the last centuries BC by Babylonian astronomers, but they had developed the sexagesimal system much earlier for non-astronomical use (Neugebauer, 1969; Rochberg, 2016). The origin of the concept of 360 degrees in a circle is not clear. Some suspect that it is related to the length of the year in days. Another possibility is the fact that a hexagon composed of six equilateral triangles can be inscribed within a circle. Then, if each of the angles of the triangles could be described by 60 degrees, following the sexagesimal system, the circle would have 360 degrees. The first known Hellenistic geometer to make use of these divisions was Hupsikles around 150 BC (Irby-Massie & Keyser, 2002). Claudius Ptolemy (ca. 100–ca. 175) followed the custom and used finer subdivisions of the degree in his work *Mathematike Syntaxis*, now known as the *Almagest*, the title being a Latin form of the Arabic translation *al-kitabu-l-mijist*. The first Latin translation of Ptolemy's treatise did not become available until the 12th century, when the work by Gerard of Cremona was published (McCluskey, 1998). In it we find the Latin translation of the subdivisions of the degree used by Ptolemy as *partes minutae primae*, or first minutes, which became known simply as "minutes," and the subsequent further subdivision of that unit, *partes minutae secundae*, or "second minutes," which became known as "seconds." The use of these angular units was restricted mainly to theoretical and astronomical applications.

In everyday, common, timekeeping usage, the hour was essentially divided into halves, thirds, or quarters, or sometimes 12ths until the end of the 16th century, but not into 60 minutes. In dealing with time, Claudius Ptolemy had divided the day either into four parts, each containing six hours, or into 360 "chronoi." Consequently, each hour was made up of 15 chronoi. Venerable Bede divided the hour into 4 "puncti," 10 "minuta," 15 "partes," or 40 "momenta." He considered the punctum to be the smallest unit that could be measured with a sundial and the momentum as the smallest conceivable unit (Dohrn-van Rossum, 1996). In the medieval world, the word "minutum" was also used in various writings to indicate 1/15th of an hour, 1/10th of an hour, and 1/60th of a day. The word "ostentum" was used to indicate 1/60th of an hour (Holford-Strevens, 2005). Other units of time used were saeculum (century), lustrum (five years), annus (year), mensis (month), hebdomada (week), dies (day), hora (hour), quadrans (quarter of an hour), minutum (minute), momentum (the length of time needed to discern that time has moved on), ostentum (the time needed to take something in visually), and ictus oculi (twinkling of an eye), which was equated with the atomus (the unit of time that could not be divided further) (Fuhrman, 1986).

In the later Middle Ages, however, the concept of minutae primae, secundae, and even tertiae had arisen in the face of this non-standardization of time units. Such terminology had been used in referring to measures of arc, but was being applied to time units as well. Although the concept of an hour with 60 minutes of 60 seconds each was understood by the mid-14th century, the concept still did not find widespread common usage, because clocks were not reliably capable of providing the finer time units. Minutes are mentioned in the 14th century, and clocks that indicated minutes may have existed by the end of the 15th century, but there is no reliable evidence to support that supposition. Seconds were not considered in common practical timekeeping for at least another century. In the late 17th century, following the development of pendulum clocks and the anchor escapement (Chapter 9), clocks with minute hands began to appear (Milham, 1923), and clocks with second hands followed in the 18th century.

In the area of precise timekeeping for scientific usage, Tycho Brahe mentions the use of minutes in his journals only twice between 1563 and 1570. In 1577, he refers to new clocks showing minutes and in 1581, he refers to seconds. In 1587, he complained about the fact that his clocks could not be made to agree to better than four seconds (Landes, 1983). In 1579, a Swiss clockmaker, Jost Bürgi, at the court of William of Hesse is reported to have developed a clock that marked seconds as well as minutes. Further, this clock is said to have been precise at the level of a minute per day (Landes, 1983).

In the early 17th century, navigation at sea over large distances provided some incentive for more precise timekeeping. At that time, the determination of latitude was relatively well known, but accurate measurement of longitude remained a problem. Typically mariners would sail toward the parallel of latitude of their destination and sail along that until their port was in sight. Their progress in longitude was measured by dead reckoning, which was aided by estimates of their speed in the water. To do that they made use of a "log" attached to a knotted rope. The log was thrown overboard and the knots that passed through the mariner's hands during a specified time duration were counted in order to determine speed. That time interval might be measured by the length of time required to recite a prayer or a verse from a song, but in the Middle Ages, the sandglass began to be used to measure those intervals of time (Landes, 1983). Sandglasses came into being at about the same time as wheeled clocks in the late 13th century and were becoming familiar instruments to measure short time intervals in the 14th century (Dohrn-van Rossum, 1996). Although they are called "sandglasses," other materials were often used, including powdered rock or crushed egg shells (Lippincott, 1999).

The definition of the second was assumed to be the 60th part of a minute, which was the 60th part of an hour, and nothing more formal than that was required. Tito Livio Burattini (1617–1681) perhaps provides the first formal definition of the second (Burattini, 1675) as 1/86,400 of a solar day (Leschiutta, 2005). This appears in his work *Misura universale* [*Universal Measure*] (1675), where he suggests a standard length unit equivalent to the length of a pendulum with a period of one second. This definition of a second appears to have been sufficient for all practical purposes into the 20th century. No formal definition of the second beyond the appropriate fraction of a day appears to have been needed. The organization responsible for world metrology surprisingly never formally defined the second of mean solar time (Audoin & Guinot, 2001).

13.2 The "Ephemeris Second"

A definition of a second based on the Earth's variable rotation became impractical for precise timekeeping applications in the 20th century. Chapter 6 outlines the development of the concept of Ephemeris Time (ET) that led to the first modern formal definition of the second, which was proposed to address this problem. The 10th meeting of the Conférence Générale des Poids et Mesures (CGPM) in 1954, following the earlier recommendation of the Comite International des Poids et Mesures (CIPM), proposed the following formal definition of the second: "The second is the fraction 1/31 556 925.975 of the length of the tropical year for 1900.0" (*Trans. Int. Astron. Union*, 1957).

As seen in Chapter 6, this fraction was based on Newcomb's formula for the geometric mean longitude of the Sun for the epoch of January 0, 1900, 12h UT (Newcomb, 1895) given by:

$$L = 279° \ 41' \ 46.04'' + 129 \ 602 \ 768''.13T + 1''.089T^2, \qquad (13.1)$$

where *T* is the time reckoned in Julian centuries of 36,525 days.

In 1956, the CIPM adopted the slightly more precise value with the words: "La second est la fraction 1/31 556 925.9747 de l'année tropique pour 1900 janvier 0 a 12 heures de temps des ephemerides." It also created the Comite Consultatif pour la Définition de la Seconde (CCDS) to coordinate future work in the area (Procès Verbaux des Séances, 1957). The "ephemeris second," as a fraction of the tropical year, was formally adopted by the 11th CGPM in 1960.

13.3 The SI Second

Events leading to the definition of the second of the Système International (SI) began with the development of the microwave caesium frequency standard (Chapter 11), which took place concurrently with the discussion and adoption of the second of Ephemeris Time. Following the establishment of the possible viability of the caesium standard as a clock, it became necessary to calibrate the frequency of the atomic transition and, thus, establish its timescale unit in relation to the prevailing definition of the second. Essen and Parry (1957) report that values of the frequency of caesium were reported by Sherwood et al. (1952) and in 1956. The 1952 value was 9 192 632 000 ±2000 Hz, and the 1956 values were 9 192 631 970 ±90 Hz, 9 192 631 800 ±50 Hz, and 9 192 631 880 ±30 Hz. These were consistent with the Essen and Parry (1955) value of 9 192 631 830 ±10 Hz and the Essen and Parry (1957) value of 9 192 631 845 ±2 Hz, both of which were determined with respect to the second of Universal Time (UT2) determined using astronomical observations of the Royal Greenwich Observatory.

The actual definition used for the SI second, however, was the result of the collaboration of UK physicists L. Essen and J. V. L. Parry with US astronomers Wm. Markowitz and R. G. Hall (see Section 11.4.1.1 in this volume). From this collaboration, Markowitz et al. (1958) arrived at a final value of 9 192 631 770 ±20 Hz. The second in this determination was the second of Ephemeris Time that was determined from astronomical observations of the Moon with respect to a star background. Leschiutta (2005) states that:

"As a personal remark, taking into account the capabilities of the timing emissions at the moment, of the frequency standards available, of the inevitable scatter of the moon camera, and some other factors, not least the widespread use and abuse in "touching" the piezo-oscillator, it is almost impossible to explain the accuracy of the Markowitz determination. Similar events, i.e. results surpassing the capabilities of the moment, are not uncommon in the history of science that sometimes is prone to accepting the intervention of a serendipity principle. The other possible explanation calls for a first class understanding of physics, coupled with scientific integrity . . .

The question about the mental and experimental paths taken by Markowitz that led him to write (9192 631 770 ±10) Hz remains open, despite:

- using UT1 data with clocks corrected nearly every day, with a timing accuracy at the millisecond level at best, giving an accuracy in frequency of 1×10^{-8},
- taking data from the double-rate moon camera, having a resolution of around 0.5 s on ET and for UT from a PZT affected by a resolution of 10 ms,
- the need for a long chain of measurements and commands, composed of many laboratories belonging to different authorities, to be kept 'synchronized.'"

The Markowitz et al. (1958) publication does not provide extensive details of the process by which they arrived at this number that was destined to become the definition of the SI second. The frequency of the caesium hyperfine transition that defines the second is actually discussed in two papers by the collaborators (Essen et al., 1958, and Markowitz et al., 1958). The Essen et al. (1958) paper describes the calibration of the caesium transition in terms of time determined from the Earth's rotation. The second paper describes the correction to this frequency to put it in terms of the second of Ephemeris Time. Leschiutta (2005) attempts to fill in some of the details based on subsequent discussions.

Two piezoelectric quartz clocks were involved in the process. The first was the device used to provide the frequency required to produce the hyperfine transition in the caesium atoms of Essen's standard in the United Kingdom. So at the National Physical Laboratory (NPL) in the United Kingdom a quartz clock was adjusted in rate to the frequency provided by the caesium standard. The second quartz oscillator needed was one at the US Naval Observatory (USNO) in Washington, DC, which was regulated by astronomical observations of the Earth's rotation.

In the 1950s, the instruments used to make the astronomical observations for timekeeping at the USNO were the Photographic Zenith Tubes (PZT) at Washington, DC, and at Richmond (outside Miami), Florida. The PZT was a telescope having a lens with an eight-inch diameter and a very long focal length. It was constrained to look only at the zenith and designed to make use of a mercury pool to define the direction of the vertical. The pool was inserted midway in the light path and reflected the light to a photographic plate immediately underneath the lens, which recorded the stellar images. The telescope was capable of determining the difference in time between the astronomically determined time (Universal Time) and the clock time daily with an accuracy of a few milliseconds. See Figure 13.1.

The USNO clock, then, was regulated to provide a realization of UT2, which is the astronomically determined time corrected for the known seasonal variation in the Earth's rotation (see Chapter 15). Its frequency can be denoted as v_{UT2} and the clock reading after a time interval τ can be written $h_{UT2} = v_{UT2}\tau$. Similarly after the same interval the NPL clock, whose frequency v_{CS}, was regulated to provide time based on the frequency of the nominal caesium frequency, would read $h_{CS} = v_{CS}\tau$. To calibrate the caesium frequency, then, in terms of UT2 it was necessary to

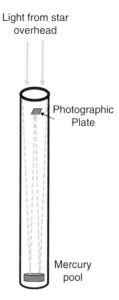

Light from star
overhead

Photographic
Plate

Mercury
pool

Figure 13.1 Photographic Zenith Tube

determine ν_{UT2} by comparing the frequencies of the two quartz clocks at opposite ends of the Atlantic Ocean. Each facility monitored the same timing signals broadcast by radio stations in order to determine the difference in time between the two clocks. The stations used for this were WWV in Washington, DC, USA, and GBR in Rugby, UK, 120 km away from NPL. In the Essen et al. (1958) analysis, smoothed monthly differences between the two clocks were used to determine the relative frequency difference. The smoothing of the astronomical observations also eliminated the effects of known tidal variations with periods of 27.6 and 13.6 days in the Earth's rotation. The radio signals were monitored near the beginning of each month from June 1955 through January 1958.

Mathematically, if each institution observes the same radio time signal at the same time at the beginning and end of the month, it is possible to determine the differences h_{UT2} and h_{CS}. Of course, corrections to the actual observations of the time differences must be made to allow for the times of transmission of the radio signals to the respective monitoring locations. Then, after making the systematic corrections, it is possible to write:

$$h_{CS} - h_{UT2} = (\nu_{CS} - \nu_{UT2})\tau, \qquad (13.2)$$

$$\text{but } \tau = {h_{UT2}}\big/{\nu_{UT2}}, \qquad (13.3)$$

$$\text{so } \Delta h = (\nu_{CS} - \nu_{UT2}) \frac{h_{UT2}}{\nu_{UT2}}, \text{ and} \qquad (13.4)$$

$$\nu_{UT2} = \nu_{CS} - \nu_{UT2} \frac{\Delta h}{h_{UT2}}. \qquad (13.5)$$

Following this process Markowitz et al. (1958) report that they had found the frequency of caesium in terms of the UT2 second to be $\nu_{UT2} = 9\,192\,631\,882$ cycles per second of UT2.

Observations made with the Markowitz Moon Camera (see Chapter 6) at Washington, DC, were then used to determine the difference between the Ephemeris Time second and the UT2 second. The same process that was used to determine ν_{UT2} could be used to determine ν_{ET}, the frequency of the caesium energy-level transition in terms of the "ephemeris second." A time series of the differences between Ephemeris Time and UT2 could be used in place of the time series of the differences between a clock reading UT2 and a clock operating with a nominal caesium frequency to estimate:

$$\nu_{ET} = \nu_{UT2} - \nu_{ET} \frac{\Delta h}{h_{ET}} \qquad (13.6)$$

Their result was (Markowitz et al., 1958):

$\nu_{ET} = 9\,192\,631\,770 \pm 20$ cycles per second of Ephemeris Time.

13.4 Adopting the SI Second

The SI brochure (BIPM, 2006) summarizes the transition to the atomic second defined by Markowitz, Hall, Essen, and Parry:

"In order to define the unit of time more precisely, the 11th CGPM (1960, <u>Resolution 9</u>) adopted a definition given by the International Astronomical Union based on the tropical year 1900. Experimental work, however, had already shown that an atomic standard of time, based on a transition between two energy levels of an atom or a molecule, could be realized and reproduced much more accurately. Considering that a very precise definition of the unit of time is indispensable for science and technology, the 13th CGPM (1967/68, Resolution 1) replaced the definition of the second by the following:

The second is the duration of 9 192 631 770 periods of the radiation corresponding to the transition between the two hyperfine levels of the ground state of the caesium 133 atom.

It follows that the hyperfine splitting in the ground state of the caesium 133 atom is exactly 9 192 631 770 hertz, ν(hfs Cs) = 9 192 631 770 Hz. [The symbol, (hfs Cs), is used to denote the frequency of the hyperfine transition in the ground state of the caesium atom.]"

At its 1997 meeting the CIPM affirmed that:

"This definition refers to a caesium atom at rest at a temperature of 0 K."

This note was intended to make it clear that the definition of the SI second is based on a Caesium atom unperturbed by black body radiation, that is, in an environment whose thermodynamic temperature is 0 K. The frequencies of all primary frequency standards should, therefore, be corrected for the shift due to ambient radiation, as stated at the meeting of the Consultative Committee for Time and Frequency in 1999.

The text of the 13th CGPM Resolution that defined the SI second is given in the brochure:

The 13th Conférence Générale des Poids et Mesures (CGPM),
    "**Considering**
    • that the definition of the second adopted by the Comité International des Poids et Mesures (CIPM) in 1956 (Resolution 1) and ratified by Resolution 9 of the 11th CGPM (1960), later upheld by Resolution 5 of the 12th CGPM (1964), is inadequate for the present needs of metrology,
    • that at its meeting of 1964 the CIPM, empowered by Resolution 5 of the 12th CGPM (1964), recommended, in order to fulfill these requirements, a caesium atomic frequency standard for temporary use,
    • that this frequency standard has now been sufficiently tested and found sufficiently accurate to provide a definition of the second fulfilling present requirements,
    • that the time has now come to replace the definition now in force of the unit of time of the Système International d'Unités by an atomic definition based on that standard,
    Decides
    1. The SI unit of time is the second defined as follows: "The second is the duration of 9 192 631 770 periods of the radiation corresponding to the transition between the two hyperfine levels of the ground state of the Caesium 133 atom";
    2. Resolution 1 adopted by the CIPM at its meeting of 1956 and Resolution 9 of the 11th CGPM are now abrogated."

In the world of international standards, the definition of the second directly affects the SI base unit of the meter which is now defined as:

The metre is the length of the path travelled by light in vacuum during a time interval of 1/299 792 458 of a second.

It also indirectly affects the definition of the SI base unit of the ampere, which is defined as:

The ampere is that constant current which, if maintained in two straight parallel conductors of infinite length, of negligible circular cross-section, and placed 1 metre apart in vacuum, would produce between these conductors a force equal to 2×10^{-7} newton per metre of length.

as well as the candela, which is defined as:

The candela is the luminous intensity, in a given direction, of a source that emits monochromatic radiation of frequency 540×10^{12} hertz and that has a radiant intensity in that direction of 1/683 watt per steradian.

Numerous other derived SI units also involve this definition of the second. Note also that the SI abbreviation for the unit of a second is "s".

In less than 30 years, from 1954 to 1983, we went from no formal definition of the second to the definition of the second being the most precise metrological unit and the basis for other units, including the definition of the meter in connection with the speed of light.

13.5 Toward the Redefinition of the Second

The emergence of optical atomic clocks with improved accuracies and the promise of even better accuracies compared to Caesium fountain clocks leads to possible consideration of a redefinition of the second based on optical frequencies. Although fountain standards are also improving in accuracy, the averaging time to achieve the current 10^{-16} uncertainty presents difficulties for some measurement applications.

In preparation for a possible redefinition of the second, the CIPM did introduce the concept of secondary representations of the second and has formed the Frequency Standards Working Group (WGFS) of the CCTF and the Consultative Committee for Length (CCL). Among its objectives are the maintenance, together with the BIPM, of a list of recommended frequency standard values and wavelength values for applications that include secondary representations of the second. The list includes optical frequency standards that utilize energy-level transitions in ions or atoms of aluminum, strontium, ytterbium, and mercury, as well as a rubidium microwave standard. These values are periodically updated and available on the International Bureau of Weights and Measures website (www .bipm.org/en/publications/mises-en-pratique/standard-frequencies.html) (Ido, 2014; Margolis & Gill, 2015; Gill, 2016). The choice of an appropriate transition or set of transitions is complicated currently by concerns about the limitations in comparing clocks at a distance and about modeling the effects of the Earth's gravitational field at ground level, which can cause frequency shifts of order 10^{-18} per cm of height difference. Future timekeeping standards are described in greater detail in Chapter 21, and techniques for distribution of time at the required improved accuracies are described in Section 17.3.

In addition to the transitions or transitions to be used, considerations regarding the redefinition of a second are when, and how, it might take place, the requirements for a redefinition, and the potential benefit from a redefinition.

A planned revision of the SI is planned for November 2018, when four of the SI base units, kilogram, ampere, kelvin, and mole, will be redefined. The new definitions will be based on fixed values of the Planck constant, elementary charge, Boltzmann constant, and Avogadro constant, respectively. This revision results in all base units, except the mole, being derived from the value of the SI second, but an improved value of the second will not affect the values of these base units (www .bipm.org/en/measurement-units/new-si/) (Riehle, 2015).

References

Audoin, C. & Guinot, B. (2001). *The Measurement of Time*. Cambridge: Cambridge University Press.

Burattini, T. L. (1675). *Misura universale*. Vilna: Stamperia dei Padri Francescani.

Bureau of Weights and Measures (BIPM) (2006). *The International System of Units (SI) Eighth Edition*. Paris: Bureau of Weights and Measures.

Dohrn-van Rossum, G. (1996). *History of the Hour*, translated by Thomas Dunlap. Chicago, IL: University of Chicago Press.

Essen, L., Parry, J. V. L., Markowitz, W., & Hall, R.G. (1958). Variation in the Speed of Rotation of the Earth since June 1955. *Nature*, **181**, 1054.

Fuhrman, H. (1986). *History of Germany in the High Middle Ages c. 1050–1200*, translated by Timothy Reuter. Cambridge: Cambridge University Press.

Gill, P. (2016). Is the Time Right for a Redefinition of the Second by Optical Atomic Clocks? *Journal of Physics Conference Series*, **723**, 012053.

Holford-Strevens, L. (2005). *The History of Time: A Very Short Introduction*. Oxford: University Press.

Ido, T. (2014). Frequency Comparison of Lattice Clocks toward the Redefinition of the Second. *Journal of Physics Conference Series*, **548**, 012057.

Irby-Massie, G. L. & Keyser, P. T. (2002). *Greek Science of the Hellenistic Era*. London: Routledge.

Landes, D. S. (1983). *Revolution in Time, Clocks and the Making of the Modern World*. Cambridge, MA: Belknap Press of the Harvard University Press.

Leschiutta, S. (2005). The Definition of the "Atomic" Second. *Metrologia*, **42**, S10–S19.

Lippincott, K. (1999). *The Story of Time*. London: Merrell Holberton.

Margolis, H. S. & Gill, P. (2015). Least Squares Analysis of Clock Frequency Comparison Data to Deduce Optimized Frequency and Frequency Ratio Values. *Metrology*, **52**, 628–634.

Markowitz, W., Hall, R. G., Essen, L., & Perry, J. V. L. (1958). Frequency of Cesium in Terms of Ephemeris Time. *Phys. Rev. Lett.*, **1**, 105–107.

McCluskey, S. C. (1998). *Astronomies and Cultures in Early Medieval Europe*. Cambridge: Cambridge University Press.

Milham, W. I. (1923). *Time & Timekeepers Including the History, Construction, Care, and Accuracy of Clocks and Watches*. New York and London: Macmillan.

Neugebauer, O. (1969). *The Exact Science in Antiquity* (2nd edn.). Providence, RI: Brown University Press.

Newcomb, S. (1895). *Astronomical Papers of the American Ephemeris and Nautical Almanac, vol. VI, Part I: Tables of the Sun*. Washington, DC: US Government Printing Office.

Procès-Verbaux des Séances, deuxième série (1957). **25**, 77.

Riehle, F. (2015). Towards a Re-Definition of the Second Based on Optical Atomic Clocks. arXiv: 1501.02068v2.

Rochberg, F. (2016). *Before Nature, Cuneiform Knowledge and the History of Science*, Chicago, IL: University of Chicago Press.

Sherwood, J. E., Lyons, H., McCracken, R. H., & Kusch, P. (1952). High Frequency Lines in the hfs Spectrum of Cesium. *Bull. Amer. Phys. Soc.*, **27**(I), 43.

Trans. Int. Astron. Union, **IX**, Proc. 9th General Assembly, Dublin, 1955 (1957). P. T. Oosterhoff, ed. New York, NY: Cambridge University Press, 451.

14

International Atomic Time (TAI)

14.1 Constructing an Atomic Timescale

The advent of a timescale based on atomic energy-level transitions soon followed the development of the new technology of atomic clocks. In his *Time for Reflection*, Louis Essen described the situation following his success with the first caesium standard:

"Our first task was to make every possible test to check to what extent the frequency could be varied by external conditions such as pressure, temperature, strength of the electric and magnetic fields, and so on. This could only be done by establishing a provisional atomic time scale, making use of the stability of our quartz clocks. They were set at intervals of a week by means of the atomic clock operating under standard conditions. These conditions were then varied and the effect measured by the quartz clocks. The test showed that with a very simple control of the conditions the atomic clock was enormously more accurate than astronomical time as well as having the advantages of being far simpler to use and being immediately available. It did not give the time of day, of course, but this is not required accurately. It is the length of a time interval and its inverse, frequency, that is needed ever more accurately for modern developments in navigation, computers and communication."

The earliest atomic timescales were constructed using a single atomic frequency standard to steer a quartz crystal clock. In 1955, shortly after the appearance of the first operational caesium frequency standard at the National Physical Laboratory (NPL) in the United Kingdom (Essen & Parry, 1957), the Royal Greenwich Observatory (RGO) established its timescale, called "Greenwich Atomic" (GA). It was formed using quartz crystal clocks whose frequencies were calibrated at periodic intervals with the NPL caesium frequency standard. On September 13, 1956, the US Naval Observatory began its atomic timescale, called "A.1," again using a quartz crystal clock calibrated daily with an Atomichron® (see Section 10.4.1.2) caesium beam standard located at the US Naval Research Laboratory (NRL) in Washington (*Time Service Notice No. 6*). In neither case were the caesium standard operated continuously (Nelson et al., 2001). The atomic

225

frequency standards were switched on periodically for a short duration in order to calibrate the frequency of the quartz crystal and then switched off again. Other laboratories and institutions followed quickly in constructing atomic timescales, including the US National Bureau of Standards (NBS) in Boulder, Colorado, which began an atomic timescale, NBS-A, on October 9, 1957. The epoch of this time-scale as well as that of A.1 were set equal to the astronomically determined time UT2 on January 1, 1958 (Barnes et al., 1965).

As more standards became available, constructing a timescale to take advantage of several standards became a challenge. In constructing such a timescale, some consideration needs to be given to the characteristics of the final product. These considerations include concerns about reliability, frequency stability, frequency accuracy, and accessibility (Guinot & Arias, 2005). The reliability of the scale depends critically on the reliability of the clocks whose data are used to produce the scale. To ensure this reliability it is desirable to have a large number of clocks contributing data, thereby reducing the dependency of the scale on a few contributors. Frequency stability refers to the ability of the timescale to maintain a scale interval with a constant relationship with respect to the conventionally adopted international standard second. To be stable the scale need not necessarily be accurate, but it must have a ratio of its scale interval to the standard second that is as constant as possible. Allan Variance is used to measure stability quantitatively. Frequency accuracy, on the other hand, requires that the scale interval be as close as possible to the standard second. Accessibility is provided by the clocks whose data are used in the formation of the timescale. In the process of computing the timescale, the arithmetic differences between the contributed clock time and the timescale can be provided to the contributors, who, in turn, can provide that to their client users of precise time.

Thought also must be given to the desired period over which the frequency of the resultant timescale is to be stabilized, as well as the period of time between the contributed comparisons of the clock data. Still other considerations include how often the timescale is to be computed and the time between the last contributed data and the time when the product is made available (Audoin & Guinot, 2001; Levine, 2012).

The USNO expanded its A.1 scale by making use of caesium standards at a number of cooperating institutions. By 1961, the A.1 scale was constructed using data from caesium frequency standards located at USNO, NRL, NBS, NPL, Harvard University, National Research Council of Canada in Ottawa, Centre National d'Études des Télécommunications (Bagneux, France), Observatoire de Neuchâtel (Switzerland), and the USNO Time Service Substation located at Richmond Heights, Florida, just south of Miami. The frequency comparisons were done by monitoring the phase of the signals from the very-low-frequency

(VLF) radio stations broadcasting with frequencies between 3 and 30 kHz (Markowitz, 1962; *Trans. Int. Astron. Union*, 1962).

14.2 History of TAI

In July 1955, the Bureau International de l'Heure (BIH) began an atomic timescale that has been continuous ever since. From 1955 through 1969, the BIH used VLF phase comparison data from distant atomic clocks and comparisons of local caesium standards to provide a mean timescale, called T_m or AM. All of these data were referenced to a quartz or rubidium clock at Observatoire de Paris where the BIH was located. Just as with A.1 this timescale was set equal to UT2 at 0h, January 1, 1958. Beginning in 1960, the BIH routinely published the difference between AM and astronomically observed UT2 in periodic bulletins.

In 1963, the BIH decided that, instead of giving all contributing clocks equal weight in the mean, it would be better to use as input only the three standards located at the metrology laboratories at the US National Bureau of Standards, the Swiss Laboratoire Suisse de Recherches Horlogères, and the National Physical Laboratory in the United Kingdom. The timescale then became known as A3 in recognition of the contributions of the three institutions. In 1966, the scale was expanded to make use of contributions from other laboratories, but the name A3 was still used (Guinot & Arias, 2005; Nelson et al., 2001).

Technological advances in time transfer, particularly the development of LORAN-C and time transfer via television signals, along with the advent of portable caesium clocks to improve calibrations, allowed the BIH to improve its timescale significantly. Consequently, it changed its procedure to average contributing independent local timescales, instead of the contributing clock data. The timescale created using this process began in 1969 and was designated TA(BIH). It was constrained to be continuous with A3. The initial timescales, designated TA(k) with k representing a laboratory identifier, that were included in TA(BIH) were those of the Physikalische Technische Bundesanstalt (PTB) of Germany, the Commission Nationale de l'Heure of France (F), and the US Naval Observatory (USNO). During 1969, the Royal Greenwich Observatory, the National Research Laboratory of Canada (NRC), the National Bureau of Standards (NBS), and the Observatoire de Neuchâtel (ON) in Switzerland were added to the list of contributors (Guinot & Arias, 2005).

The timekeeping community was able to use the BIH timescale by means of the published values of TA(BIH)−TA(k) that were available in monthly BIH circulars. Institution k could then determine the time-varying relation of its timescale TA(k) with respect to the standard TA(BIH) and make any appropriate changes in its scale to comply more closely with that of the BIH. The TA(BIH)−TA(k) data were available with a delay of one to two months and with uncertainties of 10 μs for those stations using VLF frequency comparisons and 1 μs for those using LORAN-C (Guinot & Arias, 2005).

Despite initial reluctance to accept time determined by means other than from astronomical observations of the Earth's rotation, atomic time began to be accepted gradually. Essen describes the gradual acceptance in the following remarks (Henderson, 2005):

"A sub-committee of the International Committee of Weights and Measures was set up to discuss atomic time and it is interesting to follow its gradual and reluctant acceptance by astronomers. The meeting in 1957 refused to accept the term atomic clock insisting that it was simply a frequency standard for the second: the second meeting in 1961 accepted that it was a standard of time interval but continued to stall by recommending that further work should be done: the third meeting in 1963 recommended the adoption of an atomic unit of time the value being that obtained at the NPL. No formal steps were taken to implement this recommendation and the International Scientific Radio Union, in which I was the chairman of the relevant section, had to stress the urgency of putting the resolution into effect. It was formally adopted as the unit of time in 1968 with only one abstention, the representative of the Greenwich Observatory, I regret to say."

The International Astronomical Union (IAU) recommended the unification of time through an atomic timescale in 1967. This was followed by similar recommendations of the International Union of Radio Sciences in 1969 and the International Radio Consultative Committee, now the International Telecommunication Union, in 1970 (Guinot & Arias, 2005). In June 1970, the Comité Consultatif pour la Définition de la Seconde (CCDS), discussed the need for an atomic scale and that the Bureau International des Poids et Mesures (BIPM) should deal not only with the definition of the timescale interval (the duration of the second), but also with timescales. Consequently they submitted two resolutions to the Comité International des Poids et Mesures (CIPM) on the subject of timescales. The first one pointed out the need for an international atomic timescale to coordinate time signals, serve as a uniform time reference especially for the dynamics of natural or artificial celestial objects, and to compare frequency standards operating in different places or times. The second suggested a definition of international atomic time (Terrien, 1970). These were then presented to the Conférence Générale des Poids et Mesures (CGPM), which met in October 1971 and passed the following resolutions (*Comptes Rendus de la 14ᵉ CGPM* (1971, 1972, 78):

"The 14th Conférence Générale des Poids et Mesures (CGPM),
  Considering

• that the second, unit of time of the Système International d'Unités, has since 1967 been defined in terms of a natural atomic frequency, and no longer in terms of the time scales provided by astronomical motions,

- that the need for an International Atomic Time (TAI) scale is a consequence of the atomic definition of the second,
- that several international organizations have ensured and are still successfully ensuring the establishment of the time scales based on astronomical motions, particularly thanks to the permanent services of the Bureau International de l'Heure (BIH),
- that the BIH has started to establish an atomic time scale of recognized quality and proven usefulness,
- that the atomic frequency standards for realizing the second have been considered and must continue to be considered by the Comité International des Poids et Mesures (CIPM) helped by a Consultative Committee, and that the unit interval of the International Atomic Time scale must be the second realized according to its atomic definition,
- that all the competent international scientific organizations and the national laboratories active in this field have expressed the wish that the CIPM and the CGPM should give a definition of International Atomic Time, and should contribute to the establishment of the International Atomic Time scale,
- that the usefulness of International Atomic Time entails close coordination with the time scales based on astronomical motions,

requests the CIPM

1. to give a definition of International Atomic Time,
   to take the necessary steps, in agreement with the international organizations concerned, to ensure that available scientific competence and existing facilities are used in the best possible way to realize the International Atomic Time scale and to satisfy the requirements of users of International Atomic Time."

"La 14e Conférence générale des poids et mesures,
   considérant

- qu'une échelle de Temps atomique international doit être mise à la disposition des utilisateurs,
- que le Bureau international de l'heure a prouvé qu'il est capable d'assurer ce service;

   rend hommage au Bureau international de l'heure pour l'œuvre qu'il a déjà accomplie;
   demande aux institutions nationales et internationales de bien vouloir continuer, et si possible augmenter, l'aide qu'elles donnent au Bureau international de l'heure, pour le bien de la communauté scientifique et technique internationale;
   autorise le Comité international des poids et mesures à conclure avec le Bureau international de l'heure les arrangements nécessaires pour la réalisation de l'échelle de Temps atomique international à définir par le Comité international.

They also endorsed the definition of International Atomic Time as: "The International Atomic Time is the time reference coordinate established by the Bureau International de l'Heure on the basis of the results given by the atomic clocks working in various establishments in accordance with the definition of the second, the time unit of the International System of Units" (Terrien, 1971). The abbreviation "TAI" first appears in a recommendation of the CGPM the following year (*Comptes Rendus de la 15e CGPM* (1975, 1976; Terrien, 1975).

"The 15th Conférence Générale des Poids et Mesures,

having examined the agreement between the Bureau International de l'Heure and the Bureau International des Poids et Mesures designed to meet the requirements of the users of the International Atomic Time (TAI),

notes with pleasure that TAI is made available to users under satisfactory conditions,

renews its request to the national and international organizations to continue, and if possible to increase, the help they provide to the Bureau International de l'Heure,

and asks the Comité International des Poids et Mesures to maintain their relationship with the Bureau International de l'Heure and its Directing Board with a view to improving the accuracy and the continuity of TAI."

In 1980, the Comité Consultatif pour la Définition de la Seconde (CCDS) discussed concerns related to TAI that resulted in formal recommendations. The first dealt with how TAI was to be considered in relativistic terms (Giacomo, 1981):

"The Comité Consultatif pour la Définition de la Seconde,

Considering

– that the 14th Conférence Générale has decided to establish an international reference time scale, TAI,
– that the CIPM at its 59th Session has defined TAI accordingly,
– that the BIH is charged with the determination of TAI, which is implemented according to the directives of CCDS ("Mise en pratique du Temps Atomique International," CCDS, 5th Session, 1970, p. S 22),
– that the use of TAI requires the application of transformations, commonly referred to as relativistic corrections, for measuring time differences between remote clocks,
– that these corrections require the adoption of a clearly-defined model,

States

– that TAI is a coordinate time scale defined in a geocentric reference frame with the SI second as realized on the rotating geoid as the scale unit,
– that accordingly it can be extended at the present state of the art with sufficient accuracy to any fixed or moving point in the vicinity of the geoid by applying first-order General Relativity corrections, *i.e.* corrections for gravitational potential and velocity differences and for the rotation of the Earth."

An explanatory note went on to provide details of the mathematical relativistic time transfer procedures to be used in the formation of TAI. The second recommendation thanked the BIH for its work in providing TAI (Giacomo, 1981):

"The Comité Consultatif pour la Définition de la Seconde,

Considering

– that the BIH with the support of the CIPM determines the international reference time scale TAI,
– that TAI meets present requirements of the users in applications requiring the highest precision,
– that the BIH continues to study ways of improving TAI,

– that the quality of this work and that of the published data meet the highest scientific standards and are at the limit of present day technological capabilities,

– that it is expected that additional applications of precision time technology will be forthcoming in the future,

Wishes to express its recognition and great appreciation for the excellent work of the BIH and encourages the Director of the BIH to continue and to follow the same guidelines which have so far proved remarkably successful."

In addition they made the following recommendations to the CIPM, which were endorsed at its 69th Session in 1980 (Giacomo, 1981).

"**Recommendation S 1 (1980)**

Algorithms for time-scale computation

The Comité Consultatif pour la Définition de la Seconde,

Considering

– that TAI should be as stable and as accurate as possible,

– that the many clocks and frequency standards available have various degrees of stability and accuracy,

– that the uncertainties in the current time comparisons can limit the quality of the computed time scales,

– that only a few primary standards are available to ensure the long-term stability of TAI and its conformity with the definition of the SI second, and

– that the algorithm employed can significantly affect the quality of the resulting time scale,

Recommends

that the development of time scale algorithms to ensure optimum use of the available data be actively pursued.

Recommendation S 2 (1980)

Primary frequency standards

The Comité Consultatif pour la Définition de la Seconde,

Taking note of

– the recommendation **S** 4 (1974) of the CCDS concerning primary cesium frequency standards, and

– the recommendation A.2 of URSI, August 1978, concerning primary cesium frequency standards, and

Considering

– that the needs are growing for a time scale, TAI, that is stable and accurate in rate,

– that the primary cesium frequency standards are the means to ensure accuracy of rate of TAI,

– that there are many laboratories contributing to the formation of TAI although there are still only three contributing to the accuracy of rate of TAI,

– that primary cesium standards operated as clocks can provide very stable and accurate time scales,

– that frequency standards other than cesium beam standards show significant promise,
– that there is currently insufficient research under way in the area of primary frequency standards,

Recommends

– that more laboratories pursue the development and operation of primary cesium frequency standards, both of conventional and new designs, and
– that laboratories undertake research and development on new frequency standards."

In 1988, the responsibility for the formation of TAI was passed from the BIH to the BIPM following the recommendation (*Comptes Rendus de la 18ᵉ CGPM* (1987), 1988, 98; Giacomo, 1988).

"The 18th Conférence Générale des Poids et Mesures
   considering the importance of measurements of time and in particular of the International Atomic Time scale, which has already been the subject of Resolution 2 of the 14th Conférence Générale des Poids et Mesures and of Resolutions 4 and 5 of the 15th Conférence Générale,
   having taken note of the resolutions adopted by the international Unions concerned – International Astronomical Union, International Union of Geodesy and Geophysics and International Union of Radio Science,
   pays tribute to the Bureau International de l'Heure and to its host organization, the Paris Observatory, for creating International Atomic Time and for the quality of the work carried out in order to establish it and diffuse it,
   approves the decisions of the Comité International which resulted in assumption by the Bureau International des Poids et Mesures of responsibility for establishing and diffusing International Atomic Time and
   recommends the national institutions concerned to pursue with the Bureau International des Poids et Mesures their collaboration for establishing and improving International Atomic Time."

Although TAI has been acknowledged as an atomic timescale, it has never been disseminated directly, nor has it been recognized as the international standard for timekeeping. That distinction is retained for Coordinated Universal Time (UTC) (Chapter 15). Since 1972, TAI and UTC differ by an integral number of seconds, but UTC is the timescale recognized as the international standard.

14.3 Formation of TAI

TAI is a timescale that incorporates the Système International (SI) second to provide users with reliability, frequency stability, frequency accuracy, and accessibility (Guinot & Arias, 2005). The operational concept is to make use of time comparisons of a large number of clocks that are contributed by cooperating institutions from around the world. The objective is to provide a timescale that is more stable than any of the clocks whose data are used in the process. These data are combined in a two-step process. The first step is to combine the data to create a preliminary timescale called Échelle Atomique Libre (EAL), translated into English as "Free Atomic Time Scale." In a separate step TAI is completed by steering the frequency of the EAL using data from primary and secondary frequency standards.

14.3.1 EAL

For the first step in the formation of TAI, the BIPM uses the comparison data from more than 500 individual contributing clocks from more than 70 timing laboratories worldwide. Some 87% of them are either commercial atomic clocks with high-performance caesium tubes or active hydrogen masers (Bureau International des Poids et Mesures, 2016). The formation of EAL is itself a two-step process. An algorithm to predict the frequency behavior of each contributing clock is first applied followed by an algorithm to determine the statistical weight to be used in forming the weighted mean of these clocks, which is EAL. The weighting procedure is designed for mid-term frequency stability of about three parts in 10^{16} over intervals of 20 to 40 days (Arias, 2014). To prevent domination of the scale by a small number of very stable clocks, a maximum relative weight is used each month that depends on the actual number of participating clocks during that month.

Since the first operational algorithm in 1973 called ALGOS (BIH, 1974 and 1975), the BIPM has made, and continues to make, improvements based on ongoing research in order to take advantage of improvements in clocks, frequency standards, and time comparison methods. The principles, however, have remained basically the same (Guinot & Arias, 2005).

If we assume that at time t, $h_i(t)$ is the reading of clock H_i, then the difference between that reading and EAL (t) can be given by:

$$x_i(t) = EAL(t) - h_i(t), \tag{14.1}$$

and, therefore, that the basic data reported to the BIPM in the form of time differences between clocks at time t are given by:

$$x_{ij}(t) = h_j(t) - h_i(t). \tag{14.2}$$

These quantities can be measured either within a given institution or between laboratories using appropriate techniques for precise time transfer. Currently these comparisons are made using the available time transfer techniques that include Two-Way Satellite Time and Frequency Transfer (TWSTFT) and methods that make use of the reception of signals from global navigation satellite systems (GNSS), particularly the Global Positioning System (GPS) and GLONASS. Other systems could be used in the future if they can provide the time comparison accuracy. Table 14.1 shows the institutions contributing to the formation of EAL in 2016. Figure 14.1 shows the network of participating laboratories as of mid-2016.

The explanatory note that accompanies the CCDS recommendation of 1980 dealing with the relativistic aspects of TAI (Giacomo, 1981) outlined the basic process of allowing for relativistic effects in the time comparisons. If time comparisons were to be done by portable clocks, the coordinate time accumulated during the transport of the portable clock from one point to another is given by:

$$\Delta t = \frac{2\omega}{c^2} A_E + \int_{\text{trajectory}} \left[1 - \frac{\Delta U(\vec{r})}{c^2} + \frac{\vec{v}}{2c^2} \right] ds, \qquad (14.3)$$

where $\vec{r}$ is a vector directed to the clock from the Earth's center; ds is the proper time element given by the clock; $\Delta U(\vec{r})$ is the difference in the gravity potential between the clock and the geoid (positive above the geoid); c is the speed of light; $\vec{v}$ is the velocity of the clock relative to the Earth; ω is the angular speed of the Earth's rotation; and A_E is the equatorial projection of the area swept by $\vec{r}$ in a coordinate system attached to the Earth.

$$A_E = \frac{1}{2} \int_{\text{trajectory}} \vec{r} \cdot \vec{v}_E \cos \varphi \, ds, \qquad (14.4)$$

$\vec{v}_E$ being the eastward component of $\vec{v}$ and φ the geocentric latitude. The element of area is considered positive when the projection of $\vec{r}$ rotates eastward. If the height of the clock above the surface of the Earth does not exceed 24 km, then it is possible to use $\Delta U(\vec{r}) = gh$, where g is the acceleration of gravity and h is the height of the clock above the geoid.

If the time comparison is done by means of an electromagnetic signal as opposed to a portable clock, the explanatory note called for the coordinate time elapsed between emission and reception of the signal to be given by:

$$\Delta t = \frac{2\omega}{c^2} + \frac{1}{c} \int_{\text{path}} d\sigma, \qquad (14.5)$$

where $d\sigma$ is an increment of the proper length for the transmission.

The uncertainty of these comparisons ranges between a nanosecond and a few tens of nanoseconds, and is made up of two components. Type A errors describe the statistical uncertainty of frequency measures, that is, without any regard for the possible effect of calibration errors in the measurement. Type B errors describe the systematic uncertainty of the calibration. Type B errors generally dominate with estimates ranging up to ± 20.0 ns.

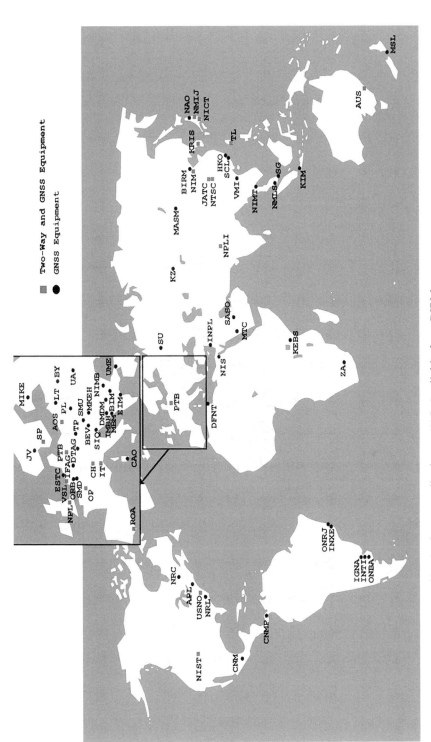

Figure 14.1 Distribution of institutions contributing to TAI. Available from BIPM.

The TWSTFT technique makes use of geostationary communications satellites to make time comparisons between clocks at two transmitting/receiving stations. Statistical uncertainty of time comparisons using this technique can be below ±1 ns (Arias, 2014). The other way in which time comparisons are made is by using GNSS. Timing receivers of GNSS signals have been developed to make these comparisons. One scheme is called the "common-view" method in which several receivers observe the same emitted signal from a navigational satellite. Another approach is the "all-in-view" method, which eliminates the need for simultaneous observations at the receiving sites. The availability of satellite clock products from the International GNSS Service (IGS) greatly facilitates this approach. Dual-frequency GPS receivers provide ionosphere-free data. Multipath and tropospheric delays contribute to the uncertainty of GNSS time comparisons, but the statistical errors can still be better than ±1 ns (Arias, 2014).

As for the Type B uncertainties, traveling time comparison equipment is used to determine internal delays of GNSS equipment at participating laboratories in campaigns organized by the BIPM. Stations that use TWSTFT procedures also organize calibrations.

The first step in forming EAL using the clock comparison data is to predict the behavior of each clock whose data are reported. Based on the past differences of each clock with respect to EAL clock reading corrections, $h_i'(t)$, are computed. A quadratic model is used to derive, for each clock, frequency corrections $\hat{y}_{i,Ik}$ that are treated as constants for monthly intervals, I_k (Panfilo et al., 2012). Each clock H_i is also assigned a statistical weight p_i based on its predictability.

EAL is calculated in monthly batches with each batch solution involving four iterations. First, the preliminary series $[EAL - h_i(t)]$ is determined using the weights from the previous iteration. For the first iteration, the weights of the previous monthly batch are used. After that, the absolute value of the difference between the frequency of H_i during the interval I_k designated $y_{i,Ik}$ and the predicted frequency, $\hat{y}_{i,Ik}$ designated $\varepsilon_{i,Ik} = |y_{i,Ik} - \hat{y}_{i,Ik}|$ is computed. One year of $\varepsilon^2_{i,Ik}$ is then filtered to give a more predominant role to more recent measurements with respect to older ones. The filtered values are used to determine a temporary weight for each clock based on its predictability. The new weight, p_i, of each clock is the same as the temporary weight, unless it exceeds a threshold higher limit for weights, or the clock has exhibited a frequency difference in excess of an adopted threshold. The weights and corrected clock readings are then used to compute EAL.

$$EAL(t) = \frac{\sum_{i=1}^{N} p_i[h_i(t) + h_i'(t)]}{\sum_{i=1}^{N} p_i}, \qquad (14.6)$$

where N is the number of clocks in the system.

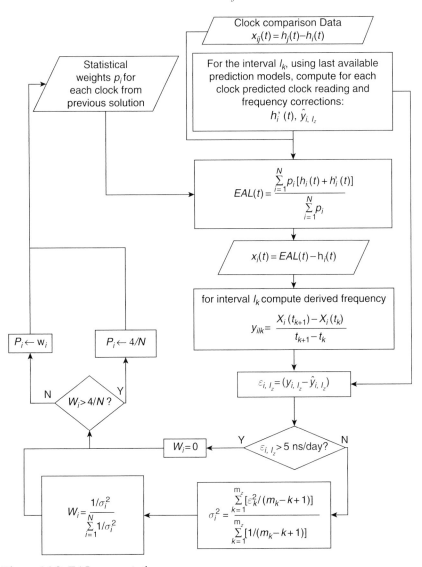

Figure 14.2 EAL computation process

These equations lead to the system (Tavella & Thomas, 1991):

$$x_{ij}(t) = x_i(t) - x_j(t),$$

$$\sum_{1}^{N} p_i x_i(t) = \sum p_i h'_i(t). \tag{14.7}$$

The iterative process is outlined in Figure 14.2 (Panfilo et al., 2014). In subsequent iterations the weights used are based on the statistics of the previous iteration.

In each iteration, new values are calculated for the clock reading correction terms (Thomas & Azoubib, 1996). Since its inception, the stability of EAL has improved mainly because of improvements in the contributing clocks and changes in weighting schemes (Arias, 2012, 2014; Petit et al., 2015).

Table 14.1 *Acronyms (k) and institutions that are possible contributors to the formation of EAL (http://www.bipm.org/en/scientific/tai/)*

AOS	Astrogeodynamical Observatory, Space Research Centre P.A.S., Borowiec, Poland
APL	Applied Physics Laboratory, Laurel, Maryland, USA
AUS	Consortium of laboratories in Australia
BEV	Bundesamt für Eich- und Vermessungswesen, Vienna, Austria
BIM	Bulgarian Institute of Metrology, Sofia, Bulgaria
BIRM	Beijing Institute of Radio Metrology and Measurement, Beijing, P. R. China
BY	Belarussian State Institute of Metrology, Minsk, Belarus
CAO	Stazione Astronomica di Cagliari (Cagliari Astronomical Observatory), Cagliari, Italy
CH	Federal Institute of Metrology (METAS), Bern-Wabern, Switzerland
CNM	Centro Nacional de Metrología, Querétaro (CENAM), Mexico
CNMP	Centro Nacional de Metrología de Panamá (CENAMEP), Panama
DFNT(1)	Laboratoire de Métrologie de la Direction Générale des Transmissions et de l'Informatique (DEF-NAT), Tunis, Tunisia
DMDM	Directorate of Measures and Precious Metals, Belgrade, Serbia
DTAG	Deutsche Telekom AG, Frankfurt/Main, Germany
EIM	Hellenic Institute of Metrology, Thessaloniki, Greece
ESTC	European Space Research and Technology Centre (ESA-ESTEC), Noordwijk, the Netherlands
HKO	Hong Kong Observatory, Hong Kong, China
IFAG	Bundesamt für Kartographie und Geodäsie (Federal Agency for Cartography and Geodesy), Fundamental Station, Wettzell, Kötzting, Germany
IGNA	Instituto Geográfico Nacional, Buenos Aires, Argentina
IMBH(2)	Institute of Metrology of Bosnia and Herzegovina, Sarajevo, Bosnia and Herzegovina
INPL	National Physical Laboratory, Jerusalem, Israel
INTI	Instituto Nacional de Tecnología Industrial, Buenos Aires, Argentina
INXE	National Institute for Metrology and Technology (INMETRO) – Time and Frequency Laboratory, Rio de Janeiro, Brazil
IT	Istituto Nazionale di Ricerca Metrologica (INRIM), Torino, Italy
JATC	Joint Atomic Time Commission, Lintong, P.R. China
JV	Justervesenet, Norwegian Metrology and Accreditation Service, Kjeller, Norway
KEBS	Kenya Bureau of Standards, Nairobi, Kenya
KIM	Research Centre for Calibration, Instrumentation and Metrology, The Indonesian Institute of Sciences, Serpong-Tangerang, Indonesia

Table 14.1 (*cont.*)

KRISS	Korea Research Institute of Standards and Science (KRISS), Daejeon, Rep. of Korea
KZ	Kazakhstan Institute of Metrology (KazInMetr), Astana, Kazakhstan
LT	Center for Physical Sciences and Technology (VMT/FTMC), Vilnius, Lithuania
MASM	Mongolian Agency for Standardization and Metrology, Bayanzurkh District, Mongolia
MBM(3)	Bureau of Metrology – Laboratory for Time and Frequency, Podgorica, Montenegro
MIKE	MIKES Metrology, VTT Technical Centre of Finland Ltd, Espoo, Finland
MKEH	Hungarian Trade Licensing Office, Budapest, Hungary
MSL	Measurement Standards Laboratory, Lower Hutt, New Zealand
MTC	MAKKAH Time Centre – King Abdulah Centre for Crescent Observations and Astronomy, Makkah, Saudi Arabia
NAO	National Astronomical Observatory, Misuzawa, Japan
NICT	National Institute of Information and Communications Technology, Tokyo, Japan
NIM	National Institute of Metrology, Beijing, P.R. China
NIMB	National Institute of Metrology, Bucharest, Romania
NIMT	National Institute of Metrology, Bangkok, Thailand
NIS	National Institute for Standards, Cairo, Egypt
NIST	National Institute of Standards and Technology, Boulder, Colorado, USA
NMIJ	National Metrology Institute of Japan, Tsukuba, Japan
NMLS	National Metrology Laboratory of SIRIM Berhad, Shah Alam, Malaysia
NPL	National Physical Laboratory, Teddington, United Kingdom
NPLI	National Physical Laboratory, New Delhi, India
NRC	National Research Council of Canada, Ottawa, Canada
NRL	US Naval Research Laboratory, Washington, DC, USA
NTSC	National Time Service Center of China, Lintong, P.R. China
ONBA	Observatorio Naval, Buenos Aires, Argentina
ONRJ	Observatório Nacional, Rio de Janeiro, Brazil
OP	Laboratoire national de métrologie et d'essais – Systèmes de références Temps-Espace, Observatoire de Paris (LNE-SYRTE), Paris, France
ORB	Observatoire Royal de Belgique, Brussels, Belgium
PL	Consortium of laboratories in Poland
PTB	Physikalisch-Technische Bundesanstalt, Braunschweig, Germany
ROA	Real Instituto y Observatorio de la Armada, San Fernando, Spain
SASO	Saudi Standards, Metrology and Quality Organization, Riyadh, Saudi Arabia
SCL	Standards and Calibration Laboratory, Hong Kong, China
SG	National Metrology Centre – Agency for Science, Technology and Research (A*STAR), Singapore

Table 14.1 (*cont.*)

SIQ	Slovenian Institute of Quality and Metrology, Ljubljana, Slovenia
SMD	Metrology Division of the Quality and Safety Department – Scientific Metrology, Brussels, Belgium
SMU	Slovenský Metrologický Ústav (Slovak Institute of Metrology), Bratislava, Slovakia
SP	Sveriges Provnings- och Forskningsinstitut (Swedish National Testing and Research Institute), Borås, Sweden
SU	Institute of Metrology for Time and Space (IMVP), NPO "VNIIFTRI" Mendeleevo, Moscow Region, Russian Federation
TL	Telecommunication Laboratories, Chung-Li, Chinese Taipei
TP	Institute of Photonics and Electronics, Czech Academy of Sciences (IPE/ASCR), Prague, Czech Republic
UA	National Science Center "Institute of Metrology" (NSC), Kharkhov, Ukraine
UME	Ulusai Metroloji Enstitüsü, Marmara Research Centre, (National Metrology Institute), Gebze Kocaeli, Turkey
USNO	US Naval Observatory, Washington, DC, USA
VMI	Vietnam Metrology Institute, Ha Noi, Viet Nam
VSL	VSL, Dutch Metrology Institute, Delft, the Netherlands

14.3.2 Steering EAL with Primary and Secondary Frequency Standards

The requirement for accuracy in a timescale is met by making use of data from specific primary and secondary laboratory standards to provide systematic corrections to the scale interval of EAL. Primary standards realize the transition in the caesium atom that defines the SI second directly. Secondary standards realize other transitions that have been recommended to be secondary representation to the SI second. These standards are frequency standards maintained at a few laboratories that realize the second with the highest accuracy. These institutions make corrections for various systematic effects and are consequently able to provide an accurate realization of the second with uncertainties on the order of parts in 10^{16}. Laboratories currently providing primary and secondary frequency standard data include IT, NIM, NIST, NPL, PTB, and SYRTE. The data supplied to the BIPM for each frequency comparison include the interval over which the comparison was done, the uncertainty of the measurement due to systematic effects in the primary frequency standard, the uncertainty due to the instability of the primary frequency standard, the uncertainty due to the link between the primary frequency standard and the clock whose data were used in the EAL calculation, and the uncertainty in the link between that clock and the EAL reference, along with a reference publication dealing with the evaluation of the systematic error (Petit, 2000). Corrections are made to account for the relativistic frequency shift caused by the distance of the standards above or below a conventional surface of equal gravity potential, very close to the rotating geoid. This correction is about 1×10^{-16} per meter. The total fractional frequency uncertainties of these data, including the errors due to the frequency transfers, range from ± 0.5 to $\pm 12 \times 10^{-15}$.

The BIPM uses these data in an algorithm created originally at the BIH to produce the steering corrections to the EAL (Azoubib et al., 1977). The combination of EAL with these systematic corrections provides the final product, TAI. Before January 1, 1977, TAI was equal to EAL. After that date, a systematic steering correction of 1×10^{-12} was applied to ensure agreement between the TAI scale unit and the SI second at sea level. In addition, EAL was modified using steps of 2×10^{-14} as needed to bring the scale unit into agreement with the SI second as defined by the primary frequency standards (Guinot, 1988). Following the CCDS recommendation S2 of 1996 (Quinn, 1997), the frequencies of the primary frequency standards were corrected systematically for blackbody radiation. From 1998 to 2004, steering corrections of $\pm 1 \times 10^{-15}$ were applied as needed, for intervals of two months at least. After July 2004, frequency corrections, up to 0.7×10^{-15}, were applied for intervals of one month, if the value of the difference between the TAI scale unit and the SI second provided by the primary frequency standards exceeded 2.5 times its uncertainty (Guinot & Arias, 2005).

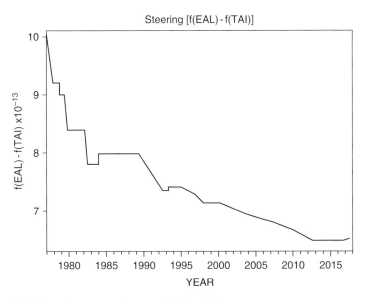

Figure 14.3 Steering corrections applied to EAL

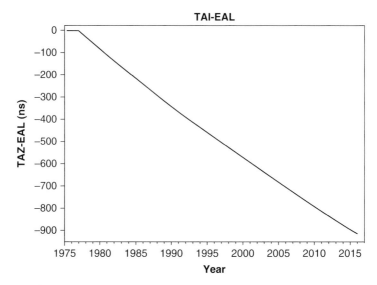

Figure 14.4 TAI-EAL

Steering corrections to TAI are reported in the monthly BIPM Circular T available at www.bipm.org/en/scientific/tai/. Figure 14.3 displays the steering of EAL and Figure 14.4 shows the difference between the EAL and TAI timescales since 1975.

14.4 Stability of TAI

The current estimate of the stability of TAI created using the process outlined earlier is 3×10^{-16}. The long-term stability is limited by the accuracy of the measurements of the primary frequency standards, which is currently at the level of 2×10^{-16} (Petit, 2010; Arias, 2014).

14.5 Distribution of TAI

Figure 14.5 displays the flow of data connected to the distribution of TAI.

TAI is available to users with a typical delay of a few weeks. However, it is not distributed directly, but it is made available through the use of Coordinated Universal Time (UTC) (see Chapter 15). UTC differs from TAI only by an integral number of seconds, so it is clear that UTC-$h_i(t)$ is equivalent to TAI $- h_i(t)$ modulo integral seconds, $h_i(t)$ being the contributed clock readings (see Equation [14.1]). UTC is used formally to provide this accessibility as it is the conventionally accepted international standard and the timescale in common use. Institutions that contribute to the formation of TAI by the BIPM maintain their local

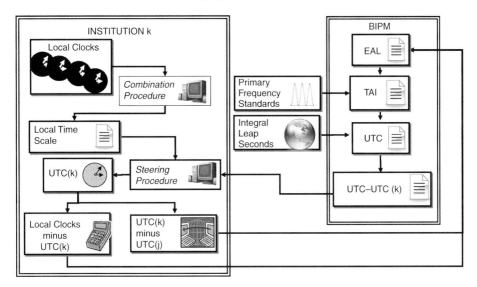

Figure 14.5 Distribution of TAI

approximation, designated UTC(k), of the international standard. Here the letter k designates an institutional abbreviation (see Table 14.1). The conventional nomenclature stipulates that UTC and TAI, appearing without a following set of parentheses, indicates the scale that is provided by the BIPM. Monthly the BIPM disseminates the time series UTC-UTC(k) at five-day intervals, and since 2005 it also provides the uncertainties of these offsets. From this information the contributing laboratories can then access TAI, since TAI and UTC differ only by integral seconds (BIPM, 2006; Arias, 2012).

This information appears in the monthly publication *Circular T*. This periodical bulletin also contains the data by which the contributing institutions can steer their respective clocks to match the frequency of TAI to meet their respective requirements. It provides access to the realization of the SI second by listing the deviation of the TAI scale interval with respect to the SI second, and the individual observations of each of the contributing primary frequency standards. The time links used in the calculation of TAI and their uncertainties are also shown. All data used to calculate TAI and *Circular T*, as well as related information, can be found on the ftp server of the BIPM time section (www.bipm.org).

14.6 Relationship of TAI to Terrestrial Time

Since its official beginning in 1972 with 56 clocks and 25 laboratories, TAI has been maintained as a continuous timescale that can be considered as a realization of Terrestrial Time (TT). TT was defined by the IAU in a 1991 resolution:

"The XXIst General Assembly of the International Astronomical Union,
  considering,

a) that the time scales used for dating events observed from the surface of the Earth and for terrestrial metrology should have as the unit of measurement the SI second, as realized by terrestrial time standards,
b) the definition of the International Atomic Time, TAI, approved by the 14th Conférence Générale des Poids et Mesures (1971) and completed by a declaration of the 9th session of the Comité Consultatif pour la Définition de la Seconde (1980),

recommends that,

1) the time reference for apparent geocentric ephemerides be Terrestrial Time, TT,
2) TT be a time scale differing from TCG [Geocentric Coordinate Time] of Recommendation III by a constant rate, the unit of measurement of TT being chosen so that it agrees with the SI second on the geoid,
3) at instant 1977 January 1, 0h 0 m 0s TAI exactly, TT have the reading 1977 January 1, 0h 0 m 32.184s exactly.

Notes for Recommendation IV

1. The basis of the measurement of time on the Earth is International Atomic Time (TAI) which is made available by the dissemination of corrections to be added to the readings of national time scales and clocks. The time scale TAI was defined by the 59th session of the Comité International des Poids et Mesures (1970) and approved by the 14th Conférence Générale des Poids et Mesures (1971) as a realized time scale. As the errors in the realization of TA are not always negligible, it has been found necessary to define an ideal form of TAI, apart from the 32.184s offset, now designated Terrestrial Time, TT.
2. The time scale TAI is established and disseminated according to the principle of coordinate synchronization, in the geocentric coordinate system, as explained in CCDS, 9th Session (1980) and in Reports of the CCIR, 1990, annex to Volume VII (1990).
3. In order to define TT it is necessary to define the coordinate system precisely, by the metric form, to which it belongs. To be consistent with the uncertainties of the frequency of the best standards, it is at present (1991) sufficient to use the relativistic metric given in Recommendation I.
4. For ensuring an approximate continuity with the previous time arguments of ephemerides, Ephemeris Time, ET, a time offset is introduced so that TT − TAI = 32.184s exactly at 1977 January 1, 0h TAI. This date corresponds to the implementation of a steering process of the TAI frequency, introduced so that the TAI unit of measurement remains in close agreement with the best realizations of the SI second on the geoid. TT can be considered as equivalent to TDT as defined by International Astronomical Union (IAU) Recommendation 5 (1976) of Commissions 4, 8 and 31, and Recommendation 5 (1979) of Commissions 4, 19 and 31.
5. The divergence between TAI and TT is a consequence of the physical defects of atomic time standards. In the interval 1977–1990, in addition to the constant offset of 32.184s, the deviation probably remained within the approximate limits of +/− 10 microseconds.

It is expected to increase more slowly in the future as a consequence of improvements in time standards. In many cases, especially for the publication of ephemerides, this deviation is negligible. In such cases, it can be stated that the argument of the ephemerides is TAI + 32.184s.

6. Terrestrial Time differs from TCG of Recommendation III by a scaling factor, in seconds:

$$TCG - TT = L_G \times (JD - 2443144.5) \times 86400.$$

The present estimate of the value of L_G is 6.969291×10^{-10} $(+/- 3 \times 10^{-16})$. The numerical value is derived from the latest estimate of gravitational potential on the geoid, $W = 62636860$ $(+/- 30)$ m^2/s^2 (Chovitz, *Bulletin Géodesique*, **62**, 359, 1988). The two time scales are distinguished by different names to avoid scaling errors. The relationship between L_B and L_C of Recommendation III, notes 1 and 2, and L_G is, $L_B = L_C + L_G$.

7. The unit of measurement of TT is the SI second on the geoid. The usual multiples, such as the TT day of 86400 SI seconds on the geoid and the TT Julian century of 36525 TT days, can be used provided that the reference to TT be clearly indicated whenever ambiguity may arise. Corresponding time intervals of TAI are in agreement with the TT intervals within the uncertainties of the primary atomic standards (e.g., within $+/- 2 \times 10^{-14}$ in relative value during 1990).

8. Markers of the TT scale can follow any date system based upon the second, e.g., the usual calendar date or the Julian Date, provided that the reference to TT be clearly indicated whenever ambiguity may arise.

9. It is suggested that realizations of TT be designated by TT (xxx) where xxx is an identifier. In most cases a convenient approximation is:

$$TT(TAI) = TAI + 32.184s.$$

However, in some applications it may be advantageous to use other realizations. The BIPM, for example, has issued time scales such as TT (BIPM90)."

In 2000, the IAU redefined TT with the following resolution (www.iau.org/static/resolutions/IAU1991_French.pdf):

"Resolution B1.9 Re-definition of Terrestrial Time TT
 The XXIVth International Astronomical Union General Assembly,
  Considering

1. that IAU Resolution A4 (1991) has defined Terrestrial Time (TT) in its Recommendation 4,
2. that the intricacy and temporal changes inherent to the definition and realisation of the geoid are a source of uncertainty in the definition and realisation of TT, which may become, in the near future, the dominant source of uncertainty in realising TT from atomic clocks,

Recommends
 that TT be a time scale differing from TCG by a constant rate: $dTT/dTCG = 1 - L_G$, where $L_G = 6.969290134 \times 10^{-10}$ is a defining constant.
 Note
 L_G was defined by the IAU Resolution A4 (1991) in its Recommendation 4 as equal to U_G/c^2 where U_G is the geopotential at the geoid. L_G is now used as a defining constant."

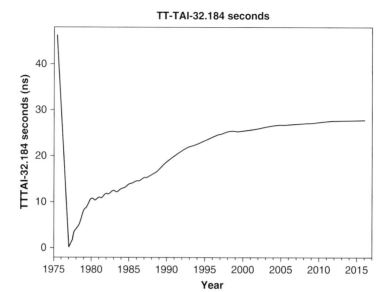

Figure 14.6 Difference between TT (BIPM15) and TAI

These resolutions make clear that for practical applications, TAI can be considered a near real-time realization of TT and that effectively TT = TAI + 32.184 s. In practice, TT is made available annually after a reanalysis of the data used to produce TAI. These versions of TT are designated TT(BIPMyy) where the yy stands for the last two digits of the last year of data used in the reanalysis. Thus TT (BIPM15), for example, was computed in January 2016, using all of the data contributed through 2015.

TT(BIPM15) is the most recent version available from the BIPM at ftp://ftp2 .bipm.org/pub/tai/scale/TTBIPM/ttbipm.15. It is identical to all past realizations since TT (BIPM99) for all dates before January 2, 1993 and is within 1 ns of the previous realization TT (BIPM06) for dates before September 2006. The difference between these two versions of TT reached 2 ns in December 2006. Figure 14.6 shows a plot of the difference between TT (BIPM07) and TAI.

References

Arias, E. F. (2012). Perspective: Time Scales and Clocks: Invited Review Article: The Statistical Modeling of Atomic Clocks and the Design of Time Scales. *Review of Scientific Instruments*, **83**, 020901.

Arias, E. F. (2014). Atomic Time Scales for the 21st Century. *Rev Mex AA(Serie de Conferencias)*, **43**, 29–34.

Audoin, C. & Guinot, B. (2001). *The Measurement of Time*. Cambridge: Cambridge University Press.

Azoubib, J., Granveaud, M., & Guinot, B. (1977). Estimation of the Scale Unit Duration of Time Scales. *Metrologia*, **13**, 87–93.

Barnes J. A., Andrews D. H., & Allan D. W. (1965). The NBS-A Time Scale: Its Generation and Dissemination. *IEEE Trans. Instrum. Meas., IM*, **14**, 228–232.

Bureau International de l'Heure (1974). *BIH Annual Report for 1973*. Observatoire de Paris (in French; English version appears in *BIH Annual Report for 1974*, published in 1975).

Bureau International des Poids et Mesures (2006). *The International System of Units (SI). Eighth Edition.* Paris: Bureau International des Poids et Mesures.

Bureau International des Poids et Mesures (2016). *BIPM Annual Report on Time Activities Volume 10 2015*. Paris: Bureau International des Poids et Mesures.

Comptes Rendus de la 14^e CGPM (1971) (1972). 78, available at www1.bipm.org/en/convention/cgpm/resolutions.html.

Comptes Rendus de la 15^e CGPM (1975) (1976). 104, available at www1.bipm.org/en/convention/cgpm/resolutions.html.

Comptes Rendus de la 18^e CGPM (1987) (1988). 98, available at www1.bipm.org/en/convention/cgpm/resolutions.html.

Essen, L. *Time for Reflection*. Published privately and available at www.btinternet.com /~time.lord/ also available in Henderson, D. (2005). Essen and the National Physical Laboratory's Atomic Clock. *Metrologia*, 42 S4–S9.

Essen, L. & Parry, J. V. L. (1957). The Caesium Resonator as a Standard of Frequency and Time. *Philos. Trans. Roy. Soc. London. Ser. A, Mathematical and Physical Sciences*, **250**, 45–69.

Giacomo, P. (1981). News from the BIPM. *Metrologia*, **17**, 69–74.

Giacomo, P. (1988). News from the BIPM. *Metrologia*, **25**,113–119.

Guinot, B. (1988). Atomic Time Scales for Pulsar Studies and Other Demanding Applications. *Astron. Astrophys*, **192**, 370–373.

Guinot, B. & Arias, E. F. (2005). Atomic Time-Keeping from 1955 to the Present. *Metrologia*, **42**, S20–S30.

Levine, J. (2012). The Statistical Modeling of Atomic Clocks and the Design of Time Scales. *Review of Scientific Instruments*, **83**, 021101.

Markowitz, W. (1962). *IRE Trans. Instrum.*, **I-11**, 239–242.

Nelson, R. A., McCarthy, D. D., Malys, S., Levine, J., Guinot, B., Fliegel, H. F., Beard, R. L., & Bartholomew, E. R. (2001). The Leap Second: Its History and Possible Future. *Metrologia*, **38**, 509–529.

Panfilo, G., Harmegnies, A., & Tisserand, L. (2012). A New Prediction Algorithm for the Generation of International Atomic Time. *Metrologia*, **49**, 49–56.

Panfilo, G., Harmegnies, A., & Tisserand, L. (2014). A New Weighting Procedure for UTC. *Metrologia*, **51**, 285–292.

Petit, G. (2000). Use of Primary Frequency Standards for Estimating the Duration of the Scale Unit of TAI. 31st Annual Precise Time and Time Interval (PTTI) Meeting, 297–304.

Petit, G. (2010). Atomic Time Scales TAI and TT (BIPM): Present Performances and Prospects. *Highlights of Astronomy*, **15**, XXVIIth IAU General Assembly, 220–221.

Petit, G. & Arias, E. F. (2015). Long Term Stability of Atomic Time Scales. *Highlights of Astronomy*, **16**, 209–210.

Quinn, T. J. (1997). News from the BIPM. *Metrologia*, **34**, 187–194.

Tavella, P. & Thomas, C. (1991). Comparative Study of Time Scale Algorithms. *Metrologia*, **28**, 57–63.

Terrien, J. (1970). News from the BIPM. *Metrologia*, **7**, 43–44.
Terrien, J. (1971). News from the BIPM. *Metrologia*, **8**, 32–36.
Terrien, J. (1975). News from the BIPM. *Metrologia*, **11**, 179–183.
Thomas, C. & Azoubib, J. (1996). TAI Computation: Study of an Alternative Choice for Implementing an Upper Limit of Clock Weights. *Metrologia*, **33**, 227–240.
Time Service Notice No. 6 (1959, January 1). Washington, DC: US Naval Observatory.
Trans. Int. Astron. Union (1962), Vol. **XI A**, Reports on Astronomy. Edited by D. H. Sadler. New York, NY: Academic Press, 362–363.
Trans. Int. Astron. Union (1991), Vol. **XXI B**, Reports on Astronomy. Edited by D. McNally. New York, NY: Academic Press, 362–363.

15

Coordinated Universal Time (UTC)

15.1 Universal Time before 1972

In the 19th century, the words "universal time" were used typically to refer to the concept of time that would read the same everywhere in the world and be used as a conventional, or "universal," time standard. This is in contrast with the common practice involving many "local" times referred to local meridians. The phrase did not necessarily refer to a particular timescale such as Greenwich Mean Time (GMT), the mean solar time of the Greenwich meridian in England. (See Chapter 2 for the definitions and history of GMT.) The most precise determinations of time in the 19th century were accomplished using astronomical observations of star transits of the observers' meridians. Consequently, the development of the concept of universal time is related directly to the acceptance of the idea of a standard meridian to which those astronomical observations could be referred.

The commencement of the British *Nautical Almanac* in 1767 had predisposed users of maritime charts to the Greenwich meridian. However, charts based on many other "standard" meridians were available, including Christiania (Oslo), Copenhagen, Naples, Paris, and Stockholm. In August 1871, the first International Geographical Congress met in Antwerp and passed a resolution expressing the participants' opinion that the Greenwich meridian should be used as the zero of longitude for all passage charts and that this should be obligatory within 15 years (Howse, 1997). The second International Geographical Congress met in Rome in 1875, producing further discussion without definitive results. At this meeting, however, the proposition was first suggested that France might consider adopting the Greenwich meridian, if Great Britain were to adopt the metric system.

In 1876, Sandford Fleming, the engineer-in-chief of the Canadian Pacific Railway published an article that promoted the concept of a universal time. This was followed in 1879 by two papers outlining his ideas regarding time (Fleming, 1879a, 1879b). In these works, he proposed what he first called "cosmopolitan"

Figure 15.1 Cosmopolitan Time clock face

time. However, he states that "[f]or this purpose either of the designations, 'common,' 'universal,' 'non-local,' 'uniform,' 'absolute,' 'all world,' 'terrestrial,' or 'cosmopolitan,' might be employed." (Fleming, 1879a). The words "cosmic time" were also used. The globe would be divided into 24 separate lunes, each corresponding to a chosen meridian of longitude designated by a letter of the alphabet, and cosmopolitan time would correspond to the time of the initial or prime meridian. To distinguish cosmopolitan time from a locally realized time, he suggested the use of a 24-hour system where the hours would be distinguished by letters (Figure 15.1). Although Fleming suggested the use of a prime meridian, he did not propose the use of Greenwich, because he apparently felt that this would be too politically sensitive (Blaise, 2002). He eventually favored the adoption of a meridian situated 180 degrees from that of Greenwich corresponding loosely to the current "date line."

In 1880, GMT did become the legal time in Great Britain and in 1883, the US and Canadian railways adopted a system of time zones based on the Greenwich meridian to facilitate scheduling. The US government did not implement a time zone system officially until 1918 (Bartky, 2000). Meanwhile, the Third International Geographical Congress met in 1881 in Venice to discuss the zero meridian and a standard time, among other issues. The participants voted to appoint an international commission to consider the problem, but no action was taken (Wheeler, 1885). In 1883, the issues were taken up at the Seventh General Conference of the International Association of Geodesy held in Rome. There, the delegates adopted resolutions that, among other things, (1) suggested Greenwich as the initial meridian, (2) recommended that longitude be measured from west to

Figure 15.2 Greenwich meridian

east, (3) recognized the usefulness of adopting "une heure universelle" in addition to "heures locales," (4) recommended that Greenwich noon, which corresponds with midnight on the meridian situated 12 hours from Greenwich in longitude, be the beginning of the cosmopolitan date, and (5) noted the convenience of measuring time from 0h to 24h. They also noted the special conference the US government had proposed regarding the standardization of longitude and time (Hirsch & von Oppolzer, 1883).

The International Meridian Conference held in Washington, DC, in October 1884 settled the matter by proposing "the meridian passing through the center of the transit instrument at the Observatory of Greenwich as the initial meridian for longitude" (see Figure 15.2). Participants in that conference also took on the issue of an international convention for time by proposing "the adoption of a universal day for all purposes for which it may be found convenient, and which shall not interfere with the use of local or other standard time where desirable." They further proposed that "this universal day is to be a mean solar day; is to begin at the moment of mean midnight of the international meridian, coinciding with the beginning of the civil day and date of that meridian; and is to be counted from zero up to twenty-four hours" (*International Conference Held at Washington for the Purpose of Fixing a Prime Meridian and a Universal Day, October 1884 – Protocols of the Proceedings*).

Despite the recommendations of the 1884 International Meridian Conference, astronomers continued to measure days from noon to noon. Following that tradition, the mean solar time measured from mean noon at Greenwich was designated as Greenwich Mean Time. In 1919, the Bureau International de l'Heure (BIH), the international service for time, began to coordinate the emission of time signals by radio stations based on Greenwich Civil Time (GCT), which is GMT plus 12 hours, following the recommendation of the International Meridian Conference. In 1925, however, the situation was changed in the astronomical almanacs, by introducing a 12-hour discontinuity whereby the date previously referred to as 31.5 December 1924 was now to be known as 1.0 January 1925. The British *Nautical Almanac* continued to call this time Greenwich Mean Time, but *The American Ephemeris* referred to this new timescale, measured from midnight to midnight, as Greenwich Civil Time. To avoid confusion, the name "Greenwich Mean Astronomical Time" was used to designate the time measured from noon to noon.

In 1928, the IAU recommended using the name "Universal Time" to replace GMT or GCT in astronomical almanacs. This was the first "official" designation of Universal Time. The International Research Council had established the Bureau International de l'Heure at the Paris Observatory in 1919 to coordinate the transmission of radio time signals. It published routinely the difference between the broadcast radio signal and the astronomically determined time.

The actual determination of this time continued to rely on astronomical observations of star transits that were used to set mechanical, and later, electronic clocks. Beginning in 1956, the IAU recognized three versions of Universal Time. The Greenwich Mean Solar Time as observed at any location on the Earth, without regard for the location of the Earth's rotation axis with respect to the observing site, was designated "UT0." If we also know the position of the pole with respect to the observing location, we can apply small corrections (on the order of tens of milliseconds) to produce a timescale, "UT1," that is free of the local effects of the station's geography. Finally, a third version was designated "UT2," that was obtained by applying a conventionally adopted seasonal variation to UT1 to account for the observed seasonal variation in the Earth's rotational speed. This time was generally regarded in the early 1950s as being the best representation of a uniform timescale, and radio time signals of that time were based on UT2.

In 1944, quartz crystal clocks began to be used to broadcast time signals. These devices kept time with a uniform rate and were adjusted as needed to keep pace with time determined astronomically. Atomic clocks based on the frequency of an atomic transition in the Caesium atom became available in 1955. A radio station broadcasting the national time standard for the United Kingdom began sending time signals, determined using an atomic clock based on a provisional calibration of the frequency of the atomic transition. In the United States, the US Naval Observatory and the National Bureau of Standards (NBS) (now the National Institute of Standards and Technology) also developed timescales based on caesium atomic clocks. This work was done from 1956 to 1957 and, as a result, the NBS radio station WWV began broadcasting time signals based on atomic clocks that were adjusted in rate and offsets to match the UT2 that was determined from star transits.

The World Administrative Radio Congress of 1959 recognized that different countries were sending inconsistent time signals, and they asked the International Radio Consultative Committee, abbreviated "CCIR," to study the problem. The United Kingdom and the United States had decided in 1957 to combine Nautical Almanacs beginning with the 1960 edition, and in 1959, they also agreed to coordinate their time and frequency transmissions by making the same adjustments, at the same time, to their caesium-based timescales to stay close to UT2. In 1959, the Royal Greenwich Observatory, the National Physical Laboratory in England, and the US Naval Observatory agreed to coordinate their time and frequency transmissions, which were based on UT2 and the atomic frequency. Based on a comparison of UT2 and the rotation rate of the Earth during the previous year, a factor S was determined and the actual frequency of transmission would be $F_0 (1 + S)$, where F_0 is the nominal atomic frequency. The time between pulses was 9192631770 (1-S) cycles of the caesium resonance. When the rotation of the Earth departed unpredictably from this offset atomic scale, step adjustments were introduced in the timescale in multiples of 50 milliseconds. The purpose of this cooperation was to avoid diverse timescales and to provide the same time and frequency from multiple sources. This coordination began on January 1, 1960, and the resulting timescale began to be called informally "Coordinated Universal Time." Timing laboratories from other countries also began to participate over time, and in 1961, the Bureau International de l'Heure at Paris Observatory began to coordinate the process internationally.

The original form of UTC was formalized in CCIR Recommendation 374 in 1963. However, the NBS continued to refer to its time signals as GMT. In 1965 the BIH started calculating UTC based on the atomic timescale A3 that would eventually evolve into TAI. Each year, the BIH would, after consulting other observatories, announce an offset in the atomic frequency in order to match UT2 as closely as possible. They would also announce 100 ms adjustments in UTC as required in order to maintain UTC with 0.1 s of UT2. In 1967, the CCIR adopted the names *Coordinated Universal Time* and *Temps Universel Coordonne* for the English and French names with the acronym UTC to be used in both languages. The name "Coordinated Universal Time (UTC)" was approved by a resolution of IAU Commissions 4 and 31 at the 13th General Assembly in 1967 (*Trans. IAU*,

1968). For provisional limited use, the CCIR in 1966 approved "Stepped Atomic Time," which used the atomic second with frequent 200 ms adjustments made in order to be within 0.1 s of UT2.

The resulting UTC timescale broadcast worldwide, with its seconds of variable length and potential "jumps" in time, began to cause concerns among users that needed stable timescales. There was an increasing need for precise frequencies for both military and civilian applications. Radio and television stations needed precise frequency standards to calibrate their transmitters so they could stay within their assigned places in the overcrowded frequency spectrum. Precise calibration of oscillators was also required for navigation systems such as Loran, Loran C, and Omega. Thus, the changing offset frequency was becoming a nuisance, and an attempt was made to maintain the same frequency for several years at a time. The proposed introduction of an air collision avoidance system in the early 1970s, based on precise frequency, made the use of frequency offsets intolerable. These concerns drove the acceptance of a new UTC, adopted in 1970 and implemented in 1972.

15.2 Coordinated Universal Time after 1972

The current system of Coordinated Universal Time can be traced back to a meeting of the International Union of Radio Science (URSI) in 1966 when participants noted the need for a uniform atomic frequency. At the 1967 meeting of URSI, participants agreed that all adjustments to atomic time should be eliminated, and that UT2 information could be distributed in tables or in radio transmissions. In May 1968, the idea of the current practice of introducing leap second adjustments in the UTC timescale was introduced independently by Louis Essen and Gernot Winkler, at a meeting of a commission organized by the International Committee for Weights and Measures (CIPM) to discuss the issue. In that same year, CCIR Study Group 7, meeting in Boulder, Colorado, discussed possible changes in the definition of UTC (Nelson et al., 2001).

They formed an "Interim Working Party" to provide proposals for a possible new definition of UTC. The options considered were (1) steps in UTC of 0.1 or 0.2 seconds to keep UTC close to UT2, (2) replacing UTC with a timescale with no adjustments, and (3) one-second adjustments.

Study Group 7 then formulated specific proposals that were approved in January 1970 at the CCIR XIIth Plenary Assembly in New Delhi. The recommendation adopted there provides the current definition of the world's civil time. It specified that (a) radio carrier frequencies and time intervals should correspond to the atomic second based on the caesium atom; (b) step adjustments should be exactly one second to maintain approximate agreement with UT; and (c) standard time signals should contain information on the difference between UTC and UT. The new system was to begin on January 1, 1972. In February 1971, Study Group 7 specified more details regarding the implementation of the 1970 recommendation 460. The predicted difference DUT1 = UT1 − UTC was to be coded into the broadcast time signals and DUT1 was not to exceed 0.7 s. A special offset of −0.1077580 second was given to UTC at the end of 1971, so that TAI − UTC was exactly 10 seconds. Since then, the UTC scale has been based on TAI with leap seconds added to keep UTC within less than a second of UT1.

The CCIR failed to send an official letter concerning the change to the IAU in time for its 1970 General Assembly. Hence, the IAU could not respond until the 1973 General Assembly, which was after the introduction of the change. In 1973, the IAU recognized that UTC provided mean solar time, recommended it for civil time, and suggested modifications to the leap second rules. In 1974 the CCIR revised recommendation 460–1 based on the input from the IAU, and raised the maximum difference between UTC and UT1 to 0.9 second. In 1975, the CGPM stated that UTC provided both atomic frequency and UT and endorsed it for civil time. On 1979 January 1 the rate of TAI was reduced by one part in 10^{12}, to better approximate the SI second. Thus the UTC rate was also changed. The CCTF determined that the TAI second was longer than the SI second, because the time standards were not being corrected for the effects of blackbody radiation. So, from 1996 to 1998, the TAI was steered to reduce the length of the second by two parts in 10^{14}. This also had a corresponding effect on UTC.

In 1988, the responsibility for TAI was transferred to the BIPM from the BIH, and the responsibility for determining the rotation of the Earth, and UT1, was transferred to the IERS. Thus, both the BIH and the International Latitude Service (ILS) ceased to exist at that time. In a 1992 reorganization, the International Telecommunications Union-Radiocommunications Sector (ITU-R) replaced the CCIR, and the UTC recommendation became ITU-R TF 460.

The current UTC system is defined by ITU-R (formerly CCIR) Recommendation ITU-R TF.460–5:

"UTC is the time scale maintained by the BIPM, with assistance from the IERS, which forms the basis of a coordinated dissemination of standard frequencies and time signals. It corresponds exactly in rate with TAI but differs from it by an integral number of seconds. The UTC scale is adjusted by the insertion or deletion of seconds (positive or negative leap seconds) to ensure approximate agreement with UT1."

15.3 Leap Seconds

To maintain the tolerance of 0.9 s between UTC and UT1, positive leap seconds have been introduced as needed since 1972. A positive, or negative, leap second could be the last second of any UTC month, but first preference should be given to the end of December and June, and second preference to the end of March and September. A positive leap second begins at 23h 59 m 60s and ends at 0h 0 m 0s of the first day of the following month. In that case the progression of seconds would read:

23h 59 m 58s
23h 59 m 59s
23h 59 m 60s
24h 00 m 00s

In the case of a negative leap second, 23h 59 m 58s will be followed one second later by 0h 0 m 0s of the first day of the following month. The IERS should decide upon and announce the introduction of a leap second at least eight weeks in advance. The first leap second was introduced on June 30, 1972. On average, leap seconds would be expected to occur approximately every 18 months at the present time, because of the secular deceleration of the Earth. However, due to other changes in the Earth's rotation, these events only can be predicted about a year in advance. Astronomical observations of the Earth's rotation angle are required to determine when leap seconds should be inserted. Table 15.1 provides the mathematical relationship between UTC and TAI since 1961.

15.4 DUT1

With the introduction of the new definition of UTC, a new variable, DUT1, was introduced as the expected difference between UT1 and UTC accurate at the level of 0.1 s. DUT1 was designed to provide a low-accuracy estimate of UT1−UTC primarily for celestial navigators. DUT1 is made available by the IERS and changes in its value are announced as required. Based on astronomical observations, the IERS decides upon the value of DUT1 and its date of introduction, and circulates the information one month in advance. In exceptional cases of a sudden change in the rate of rotation of the Earth, the IERS may issue a correction not later than two weeks in advance of the date of introduction. It is coded into some radio time signals and can be considered as a very coarse correction to UTC to give an approximation to UT1. The magnitude of DUT1 should not exceed 0.8 s, the difference between UTC + DUT1 and UT1 should not exceed 0.1 s, and the difference between UTC and UT1 should not exceed 0.9 s. A code was specified for the transmission of DUT1 with the radio time signals (Recommendation ITU-R TF.460).

15.5 UTC Worldwide

The availability of UTC as an international timescale approximating Greenwich mean solar time, with the precision of the SI second, and matching UT1 to within one second, led to its increasing acceptance by countries around the world. Over the years, UTC has become either the basis for legal time of many countries, or accepted as the de facto basis for standard civil time. In 2007, UTC became the official time of the United States, replacing mean solar time. UTC is now generally accepted internationally as the worldwide standard for time. The use of time zones, with offsets mostly of hour increments, provides the civil time based on UTC for all the zones around the world.

15.6 Time Distribution

The distribution of time via radio time signals has been replaced largely by other means of time distribution depending on the user's accuracy requirements. The Global Positioning System (GPS) and the Internet are now commonly used to obtain UTC. Chapter 17 describes these along with other methods of time transfer.

The IERS maintains updates to this table at www.iers.org. Figure 15.3 plots the difference between TAI and UT1 along with the difference between TAI and UTC to show the relationships among these scales since 1961.

15.7 The Future of UTC: Leap Seconds or Not?

The ITU-R and the IAU continue to discuss possible changes in the definition of UTC. These discussions are largely concerned with possibly eliminating the leap seconds. As the Earth's rotational speed slows, leap seconds will be required more frequently. If the rate of change of the length of the day is on average about 1.7 milliseconds per century, by the end of the 21st century, the length of the day would be about 86400.004 SI seconds, which would require inserting a leap second every 250 days. So, in the middle of the 22nd century, two leap seconds would be required every year. In the 25th century, four leap seconds would be required every year. In 2,000 years, having leap seconds only at the ends of the months would not be adequate. However, there have only been four leap seconds in the period from 1999 through 2016, so during that time, the Earth's rotational speed has not been slowing at the rate that would be expected from historical observations.

Communication and navigation systems exist that, for convenience of operations, use internal timescales that are independent of step changes, such as leap seconds. GPS is an example of one such system with worldwide availability. There

Table 15.1 *TAI–UTC*

From				To			TAI-UTC					
1961	Jan.	1	–	1961	Aug.	1	1.422818s	+	(MJD-37300)	×	0.001296s
	Aug.	1	–	1962	Jan.	1	1.372818s	+	(MJD-37300)	×	0.001296s
1962	Jan.	1	–	1963	Nov.	1	1.845858s	+	(MJD-37665)	×	0.0011232s
1963	Nov.	1	–	1964	Jan.	1	1.945858s	+	(MJD-37665)	×	0.0011232s
1964	Jan.	1	–		April	1	3.240130s	+	(MJD-38761)	×	0.001296s
	April	1	–		Sept.	1	3.340130s	+	(MJD-38761)	×	0.001296s
	Sept.	1	–	1965	Jan.	1	3.440130s	+	(MJD-38761)	×	0.001296s
1965	Jan.	1	–		Mar.	1	3.540130s	+	(MJD-38761)	×	0.001296s
	Mar.	1	–		Jul.	1	3.640130s	+	(MJD-38761)	×	0.001296s
	Jul.	1	–		Sept.	1	3.740130s	+	(MJD-38761)	×	0.001296s
	Sept.	1	–	1966	Jan.	1	3.840130s	+	(MJD-38761)	×	0.001296s
1966	Jan.	1	–	1968	Feb.	1	4.313170s	+	(MJD-39126)	×	0.002592s
1968	Feb.	1	–	1972	Jan.	1	4.213170s	+	(MJD-39126)	×	0.002592s
1972	Jan.	1	–		Jul.	1	10s				
	Jul.	1	–	1973	Jan.	1	11s				
1973	Jan.	1	–	1974	Jan.	1	12s				
1974	Jan.	1	–	1975	Jan.	1	13s				
1975	Jan.	1	–	1976	Jan.	1	14s				
1976	Jan.	1	–	1977	Jan.	1	15s				
1977	Jan.	1	–	1978	Jan.	1	16s				
1978	Jan.	1	–	1979	Jan.	1	17s				
1979	Jan.	1	–	1980	Jan.	1	18s				
1980	Jan.	1	–	1981	Jul.	1	19s				
1981	Jul.	1	–	1982	Jul.	1	20s				
1982	Jul.	1	–	1983	Jul.	1	21s				
1983	Jul.	1	–	1985	Jul.	1	22s				
1985	Jul.	1	–	1988	Jan.	1	23s				
1988	Jan.	1	–	1990	Jan.	1	24s				
1990	Jan.	1	–	1991	Jan.	1	25s				
1991	Jan.	1	–	1992	Jul.	1	26s				
1992	Jul.	1	–	1993	Jul	1	27s				
1993	Jul.	1	–	1994	Jul.	1	28s				
1994	Jul.	1	–	1996	Jan.	1	29s				
1996	Jan.	1	–	1997	Jul.	1	30s				
1997	Jul.	1	–	1999	Jan	1	31s				
1999	Jan.	1	–	2006	Jan	1	32s				
2006	Jan	1	–	2009	Jan	1	33s				
2009	Jan	1	–	2012	Jul	1	34s				
2012	Jul	1	–	2015	Jul	1	35s				
2015	Jul	1	–	2017	Jan	1	36s				
2017	Jan	1	–				37s				

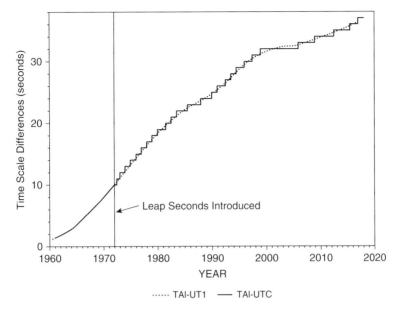

Figure 15.3 TAI-UT1 and TAI-UTC

are also a number of systems that can safely choose to ignore the need for UT1 information knowing that, with the current definition, UT1 cannot differ from UTC by more than 0.9 seconds. The relative costs of introducing leap seconds at unpredictable intervals into the world timing infrastructure versus the costs of making a change to an existing process are not well determined. So, at this time, the discussion continues concerning the question of changing the definition of UTC, and the procedure and date of such a possible change.

For more than 15 years, discussions concerning a possible redefinition of UTC in order either to eliminate the need to account for the variability of the Earth's rotational speed in a civil timescale, or to improve the current process have been continuing with no clear consensus. The World Radio Conference in 2015 discussed the issue of whether this matter should continue to be decided by the ITU-R or if it were more appropriate to have it decided by an international standards organization. A study involving the ITU-R, the BIPM, and other international organizations was requested, and further consideration of the issue at a World Radio Conference was postponed until 2023.

During this time period, some relevant meetings have occurred and resulting proceedings are available (Seago et al., 2011; Seago et al., 2013; Arias et al., 2017). Discussions have revolved around applications of civil timescales, the legal descriptions of civil timescales in different countries, and scientific, business, legal, cultural, religious, and public requirements. Suggested possibilities include (1) retaining the status quo, (2) eliminating leap seconds while retaining the name "UTC," (3) eliminating leap seconds and introducing a new name, and (4) retaining UTC as is, and officially recognizing a new civil timescale without leap seconds (Nelson et al., 2001).

Arguments in favor of retaining the status quo are the 45 years of experience with the system and the existing software and legal codes, the ability to ignore UT1 for low accuracy applications, and the close approximation to mean solar time. Arguments for eliminating the one-second jumps in the timescale are the availability of a uniform timescale, the unpredictability of the instants when leap seconds are to be inserted, the cost and consequences of potential errors in their introduction, and the fact that some electronics and software are unable to cope with a second labeled "60." Concern also remains about the name of a redefined UTC. Although some might favor retaining the UTC name in order to avoid potential confusion, others are concerned about the possibility of a violation of international naming standards.

Official recognition of a new timescale free of one-second jumps (such as from GPS and Galileo) would provide timescales for both those who need UTC within 0.9 s of UT1 and those who need a timescale without leap seconds. However, there is concern that two civil timescales might cause confusion among users regarding the actual time scales they might be using.

References

Arias, E. F., Combrinck, L., Gabor, P., Hohenkerk, C. Y., & Seidelmann, P. K., eds. (2017). *The Science of Time 2016*. Springer.

Bartky, I. R. (2000). *Selling the True Time, Nineteenth-Century Timekeeping in America*. Stanford, CA: Stanford University Press.

Blaise, C. (2002). *Time Lord, Sir Stanford Fleming and the Creation of Standard Time*. New York, NY: Random House.

Fleming, S. (1876). *Uniform Non-Local Time (Terrestrial Time)*. Ottawa.

Fleming, S. (1879a). Time-Reckoning. *Proceedings of the Canadian Institute*, **1**, 97–137.

Fleming, S. (1879b). Longitude and Time-Reckoning. *Proceedings of the Canadian Institute*, **1**, 138–149.

Hirsch, A. & Von Oppolzer, T. (1883). Unification des Longitudes par l'Adoption d'un Méridien initial unique et Introduction d'une Heure Universelle, *Bureau central de l'Association géodésique internationale*.

Howse, D. (1997). *Greenwich Time and the Longitude*. London: Philip Wilson Publishers.

International Conference Held at Washington for the Purpose of Fixing a Prime Meridian and a Universal Day, October 1884 – Protocols of the Proceedings.

Nelson, R. A., McCarthy, D. D., Malys, S., Levine, J., Guinot, B., Fliegel, H. F., Beard, R. L., & Bartholomew, E. R. (2001). The Leap Second: Its History and Possible Future. *Metrologia*, **38**, 509–529.

Seago, J. H., Seaman, R. L., & Allen, S. L., eds. (2011). Decoupling Civil Timekeeping and Earth Rotation. A Colloquium Exploring Implications of Redefining Coordinated

Universal Time (UTC). *AAS Science and Technology Series*, **113**, San Diego, CA: Univelt, Inc.

Seago, J. H., Seaman, R. L., Seidelmann, P. K., & Allen, S. L., eds. (2013). Requirements for UTC and Civil Timekeeping on Earth. A Colloquium Addressing a Continuous Time Standard. *AAS Science and Technology Series*, **115**, San Diego: Univelt, Inc.

Trans. Int. Astron. Union (1968). Vol. **XIII B**, *Proc. 13th General Assembly, Prague, 1967*, L. Perek, ed., Dordrecht: Reidel, 181.

Wheeler, G. M. (1885). *Report upon the Third International Geographic Congress and Exhibition at Venice, Italy, 1881*, Washington, DC: US Government Printing Office.

16

Time in the Solar System

16.1 The Solar System

Our solar system is composed of four terrestrial planets; four giant planets; minor planets mostly between Mars and Jupiter; Kuiper belt objects beyond Neptune, including Pluto, which can be called a planet, or dwarf planet, according to your taste; the Oort comet cloud; and comets, dust, and satellites around all of the planets except Mercury and Venus, and around Pluto, minor planets, and Kuiper belt objects. Our solar system is part of the Milky Way galaxy and in motion around its center. Our galaxy is a small part of the universe composed of billions of stars, galaxies, star clusters, pulsars, quasars, nebulae, and other objects. It is approximately 13.5 billion years old and expanding. The distances are large and measured in large numbers of light-years, the distance light travels in a year.

While knowledge of the solar system and the universe is continually improving, there is also much yet to be understood. Even in our solar system new objects are being discovered. The masses and compositions of the various bodies are not all well known. The interactive forces are not all well determined. The kinematics of the Earth is being determined with increased accuracy and complexity.

The observations that contribute to improved understanding of the solar system are made in different wavelengths of the electromagnetic spectrum from the Earth's surface, from satellites in orbit around the Earth, and from space probes traveling through the solar system. It is frequently essential to relate the timing of these observations in a space-time coordinate system that is fully relativistic and consistent with the gravitational potential and motion of the observational platform. As we consider time in a four-dimensional relativistic reference system, we must be aware of the presence of objects of large mass, their motions, and their effects on electromagnetic waves, and the motions of the objects of interest.

Table 16.1 *Uniformity of time*

Apparent Solar Time	10^{-4}	10 s/day
Mean Solar Time	10^{-8}	0.001 s /day
Ephemeris Time	10^{-10}	0.000 01 s/day
Atomic Time 1956	10^{-12}	0.000 000 1 s/day
Atomic Time 2007	10^{-15}	0.000 000 000 1 s/day
TCB	10^{-30}	0.000 000 000 000 000 000 000 000 1 s/day

16.2 Pursuit of Uniformity

The calculations of positions and motions of solar system objects are based on the independent variable, time. It is assumed that time is a uniform time in that the unit of time measure is the same at all times. A measure of the lack of uniformity of time, denoted by u, is provided by the variations of rate over specified intervals. It is also assumed that the theoretically defined time can be realized operationally by some physical source that can be used in recording the observations. To determine improved ephemerides, the observations of bodies must be compared to predicted positions at the same time, and the differences analyzed.

The uniformity of the timescale has been challenged repeatedly over the years. A progression of different timescales has been used for ephemerides. As accuracies have improved there has been a growing recognition of the need for improved timescales and an increase in the complexity of the definitions and relationships. We have progressed from time solely dependent on the rotation and orbital motion of the Earth to timescales based on atomic physics and the principles of the theory of general relativity.

Table 16.1 shows the uniformity of the timescales used to compute ephemerides of solar system bodies from Ptolemy in AD 150 to the present time. The values of u for the different timescales, except for ET and TCB, are the realized values for the scales at those times when they were used in the calculation of ephemerides. For Ephemeris Time (ET) and Barycentric Coordinate Time (TCB), the values are those of the definitions. The comparison of the accuracies for Ephemeris Time and Atomic Time in 1956 indicates a reason for the short lifetime of Ephemeris Time, and the emergence of the timescales based on atomic physics.

16.3 Pursuit of Accuracy

In parallel with the pursuit of a uniform timescale, the quest for accuracy, not only in the second of time but also in the positions and motions of the solar system bodies, is necessary for improving the understanding of our solar system.

The developments in time accuracy have been described in other chapters. The progression in the accuracy of observations and ephemerides will only be briefly outlined here and can be found in more detail in other sources (*Explanatory Supplement*, 1992, 2012). The observational techniques have ranged from the unaided eye, through the telescope, photography, charge coupled devices (CCDs), radar and laser ranging, to spacecraft visits. The accuracies have gone from arcminutes to milliarcseconds, from thousands of kilometers to millimeters.

The mathematical bases of ephemerides have progressed from epicycles to Kepler's laws, from Newton's universal law of gravity to Einstein's theory of relativity, from "computers" that were people with pencil, paper, and logarithmic tables, to calculators, to punched card equipment, to modern high-speed computers. Here, the accuracies have gone from arcminutes to microarcseconds, from thousands of kilometers to millimeters.

The continuing advances in the accuracy of the observational techniques, mathematical modeling, and computational capability indeed drive the need for improved accuracy in the ephemerides.

16.4 Time and Phenomena

A variety of solar system phenomena depend on the positions of the bodies and their motions. In most cases, observations of these phenomena and their timings depend on the observer's location. In many cases, useful observations require knowledge of the Earth's orientation in space, precise a priori ephemerides of the solar system, and an accurate, uniform clock time.

16.4.1 Eclipses, Occultations, Transits

Three types of phenomena are of particular importance. An eclipse takes place when one body passes into the shadow of another body. An occultation takes place when a large body passes in front of a smaller body. Transits take place when a smaller body passes in front of a larger body.

A solar eclipse occurs when the Moon comes between the Sun and the Earth and blocks the sunlight from reaching a specific location on the Earth, so it is really a kind of occultation (Figure 16.1). Due to the rapid motion of the Moon's shadow, the locations of solar eclipses are in restricted paths and very sensitive to the Earth's orientation at the time of the eclipse. The types and the geometry of eclipses are illustrated in Figure 16.2.

Lunar eclipses take place when the Earth comes between the Sun and the Moon. This means the Moon is in the Earth's shadow and wherever on Earth the Moon is above the horizon, the lunar eclipse can be observed. Thus, the circumstances of

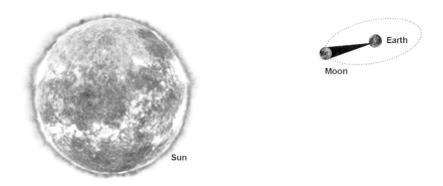

Figure 16.1 Solar eclipse

Figure 16.2 Types of solar eclipses

lunar eclipses are sensitive to the ephemerides of the Sun, Moon, and Earth, but not to the Earth orientation.

Lunar occultations take place when the Moon passes in front of another celestial body. The circumstances of these phenomena are very sensitive to Earth orientation, star catalog positions, ephemerides of the occulted body, lunar limb corrections, and time. Occultations can provide improved knowledge concerning all those quantities. They can also provide information about the occulted body, whether it is a double star, has an atmosphere or satellite, etc. An occultation of Jupiter by the Moon is shown in Figure 16.3.

There can be transits of Mercury and Venus in front of the Sun and of satellites of Jupiter and Saturn in front of those bodies. Transits of extrasolar planets in front of stars are currently the means of detecting the existence of the smallest extrasolar planets. The timings of transits can be used to determine relative distances. Thus, in the past, observations of the transit of Venus were made from many locations around the world in an attempt to determine an accurate value for the distance between the Earth and the Sun, also called the astronomical unit. Transits of Jupiter's satellites were used to try to determine the speed of light.

Solar eclipses are the most sensitive to variations in Earth rotation, and as has been seen, the locations of observers of historic solar eclipses are the basis for the determination of the long-term rate of rotation of the Earth. They also provide

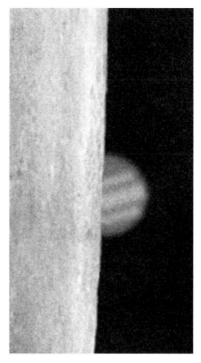

Figure 16.3 Occultation of Jupiter by the Moon

a means of determining the long-term parameters of the ephemerides of the Earth and Moon (Stephenson et al., 2016).

16.4.2 Sunrises and Sunsets

The daily phenomena of sunrises and sunsets are the most visible evidence of the rotation and orbital motion of the Earth, as the times and locations of the phenomena change during the year. From sundials the difference between apparent and mean solar time, or the equation of time, can be detected. The atmospheric refraction effect causes differences of up to minutes in the real times of rises and sets. Hence, while the times are dependent on the Earth's orientation, the variations in Earth rotation do not have a significant effect on the times.

16.4.3 Moonrises and Moonsets

The rapid motion of the Moon is most obvious from its monthly cycle through its phases. Observations of the crescent moon immediately after new Moon are often the bases for lunar calendars, and these depend critically on the separations of the

Sun and Moon and the rotation of the Earth. It is now possible to predict accurately the times for sighting the crescent Moon. Moon rises and sets are also predictable and dependent on ephemerides and Earth orientation, but the accuracies are such that variations in Earth orientation do not have a significant effect on the times.

16.5 Tropical Year

Observations of the Sun are not only used to provide solar time of day; they also provide the length of the year. The tropical year is conceptually the time between successive passages of the Sun through the same point on the ecliptic. The currently accepted definition is the time for the Sun's mean longitude to increase by 360 degrees (Danjon, 1959, Meeus & Savoie, 1992). The name comes from the return of the Sun to the same tropics (i.e., Tropic of Cancer or Tropic of Capricorn). The use of the word "tropic" is derived from the Greek *tropikós*, referring to "turn." This definition of the tropical year differs from a definition as the period of time between equinoxes. Actually, the times between the vernal equinoxes, the autumnal equinoxes, and the two solstices are different. The values for these periods for years 0 and 2000 are shown in Table 16.2 (Meeus & Savoie, 1992). The average of these values gives the value for the mean tropical year.

Table 16.2 *Length of the year determined by length of time between successive passages of the Sun through the equinoxes and solstices*

	Year 0	Year 2000
March equinoxes	365.242137 days	365.242374 days
June solstices	365.241726 days	365.241626 days
September equinoxes	365.242496 days	365.242202 days
December solstices	365.242883 days	365.242740 days

The tropical year can vary by several minutes from year to year due to the motion of the Earth's perihelion, the secular increase in the rate of precession, and the periodic actions of the Moon and planets on the Earth's orbit. Averaging over time gives us a specified value for a mean tropical year. The precession rate is increasing, so the length of the tropical year is decreasing by 0.53 second per century.

The conventionally accepted value of the duration of the tropical year is 365.2421897 days, or 365 days 5 hours, 48 minutes, and 45.19 seconds. An accurate expression for calculating its length in days in the distant past (Laskar, 1986) is:

$$365.2421896698 - 0.00000615359 \, T - 7.29 \times 10^{-10} \, T^2$$
$$+ 2.64 \times 10^{-10} \, T^3, \tag{16.1}$$

where T is in Julian centuries of 36,525 days measured from 2000 January 1 Terrestrial Time (TT) (see Chapter 9). Different ephemerides may give different values for the expression for the length of the tropical year.

16.6 Time and Distance

Historically the units of time, length, and mass were independent with separate standards for each. However, atomic clocks have provided a measure of the duration of the second that is significantly more accurate than the standards for length or mass. In 1975, the 15th Conférence Générale des Poids et Mesures adopted the standard value for the speed of light based on physical measurements (SI Units) (Comptes Rendus of the Conférence Générale des Poids et Mesures [CGPM] 1975):

"The 15th Conférence Générale des Poids et Mesures,
   considering the excellent agreement among the results of wavelength measurements on the radiations of lasers locked on a molecular absorption line in the visible or infrared region, with an uncertainty estimated at $\pm 4 \times 10^{-9}$ which corresponds to the uncertainty of the realization of the metre,
   considering also the concordant measurements of the frequencies of several of these radiations,
   recommends the use of the resulting value for the speed of propagation of electromagnetic waves in vacuum c = 299 792 458 metres per second."

The combination of the knowledge of the speed of light and the standard for the second meant that the meter was better determined from those values, than from any artifact in an environmental chamber. In addition, the use of radar and laser signals provided the means to measure distances from observatories to satellites, the Moon, and planets with new precision. These measures were in terms of the round trip time for the signal up to the body to be reflected and observed as a return signal. Consequently the solar system distance measurements are known more precisely than ever before using time measures and our knowledge of the speed of light. See Section 3.14 for the definition of the au.

16.6.1 Meter Definition

The meter, the fundamental unit of length in the International System of units (SI), is now defined by the speed of light in a vacuum and the SI second. It was originally defined by a prototype platinum-iridium object meant to represent $^1/_{10,000,000}$ of the length of the quarter meridian between the poles and the equator. Since 1983 the meter is defined as the length of the path traveled by light in vacuum during a time interval of 1/299 792 458 of a second (SI Units).

In practice, the meter can be determined from the simple equation, $\lambda = c/f$, where λ is the wavelength corresponding to the frequency f of an energy-level transition in a particular atom or molecule in free space, and c is the speed of light in free space = 299 792 458 m/s. If f is taken as the frequency of the hyperfine structure of caesium 133, namely $f_{Cs} = 9\ 192\ 631\ 770$ Hz, then λ is fixed for this transition at λ_{Cs} = 32.6122557174941 mm, effectively defining the meter.

16.6.2 Radar Ranging

Radar ranging is the timing of the round trip of a radar signal from the Earth to the planet (uplink) and back from the planet to the Earth (downlink). The timing must be done by an accurate clock. To make an accurate determination of distance, analysis of the ranging time must account for the fact that the Earth and planet are moving and rotating during the process. (See *Explanatory Supplement*, 1992, 2012, and Section 3.8.1.)

16.6.3 Laser Ranging

Similar to radar ranging, laser ranging is the actual timing of the round trip light signal between the Earth and the Moon, or an artificial satellite. Since, in this case, light is being used, the equipment is very different. Telescopes are used to send and receive the signals, and retroreflectors are required to provide a return signal. Retroreflectors were placed on the Moon during the *Apollo* missions (Figure 16.4) and on Russian lunar landers. The McDonald Laser Ranging Station, of the University of Texas, which was one of the first continuing lunar laser ranging operations, is shown in Figure 16.5. Since the wavelengths are much shorter than those of the radar signals, better accuracies are possible. From lunar laser ranging observations secular tidal changes in the lunar orbit and Earth rotation can be determined (Williams & Boggs, 2016). Since 2015 infrared wavelength lunar laser ranging has been used in France to improve constraints on the internal structure of the Moon modeled in the planetary ephemeris, INPOP (Viswanathan et al., 2016). Currently lunar laser ranging accuracies of millimeters can be achieved (Murphy, 2013). The next generation of Lunar Retroreflectors is being developed currently (Currie, 2014). See Section 3.8.2.

16.6.4 Navigation Systems

Determining the length of time it takes an electromagnetic signal to travel between locations and, assuming a standard value for the speed of light, allows us to determine the linear distance between the two sites. Knowledge of the distance to multiple reference sites then allows us to determine a position within the reference

Figure 16.4 Retroreflectors placed on the Moon during the *Apollo* missions. Courtesy of NASA.

frame of the known locations. An early application of this principle occurred during World War II with the development of the British GEE system and the American LORAN system. These systems are called hyperbolic navigation systems because they make use of signals from two time-synchronized stations to determine the difference in the arrival times of the two signals, and the knowledge of that time difference places the observer on a particular hyperbolic curve between the two stations. The concept of operations is to use transmitters in "chains" of at least three stations to permit the user to determine the intersection of two hyperbolic curves and thus find his/her location (Figure 16.6).

In the 1960s, the US Navy Timation system expanded the concept of using times of arrival of signals from precisely synchronized clocks with a limited artificial satellite system for navigation. That concept was further developed into the Global Positioning System (GPS) (Figure 16.7). Similar systems have been, or are being, developed, such as GLONASS, Galileo, and Beidou (Compass). The satellites each carry accurate clocks and transmit a timed signal. Receivers monitor multiple satellites, and from the timed signals, determine their distance from each satellite.

Figure 16.5 McDonald Laser Ranging Station of the University of Texas. Reproduced by permission of McDonald Observatory.

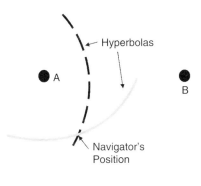

Figure 16.6 Hyperbolic navigation system. The solid line represents the hyperbolic path defined by the difference in the time of arrival of signals from stations A and C while the dashed line represents the hyperbolic path defined by the difference in the time of arrival of signals from stations A and B. Their intersection is the navigator's position.

The multiple distance measures determine the position of the receiver in three dimensions. These measurements can be used to determine the kinematics of the Earth at fixed positions on Earth, as well as determine the motions of moving objects on and around the Earth.

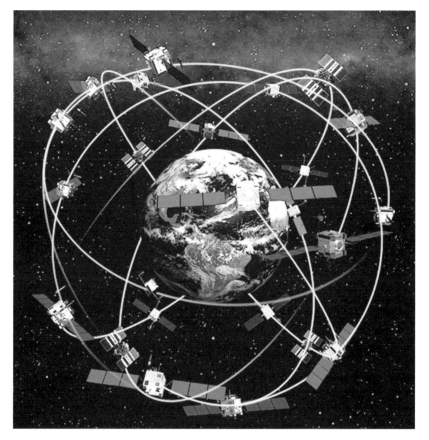

Figure 16.7 GPS Constellation. Courtesy of NOAA.

The accuracy of the navigational solutions for global navigation satellite systems (GNSS) depends critically on the precision of the clocks whose timing signals are broadcast by the satellites. The GPS satellites use multiple caesium or rubidium clocks, which are used to produce an internal timescale called GPS System Time or simply GPS Time. For the clocks on board, the satellites broadcast derived corrections that GPS receivers use to determine a geodetic position. The US Naval Observatory (USNO) monitors the satellite clocks to compare with its master clock, and this information is used to steer GPS time to better than within one microsecond of UTC(USNO), modulo integral seconds. (GPS Time does not insert leap seconds to maintain a close relationship to UT1. Consequently the difference between UTC and GPS Time continues to grow.)

For precision navigational systems operating in space on artificial satellites, three major systematic relativistic effects also need to be considered with regard to using the broadcast timing signal. Time dilation refers to the effect on the clock's frequency due to its velocity. This effect is given by:

$$\Delta t' = \frac{\Delta t}{\sqrt{1 - {v^2}/{c^2}}}, \tag{16.2}$$

where Δt is the time interval between events for an observer in some inertial frame, $\Delta t'$ is the time interval between those same events, as measured by another observer, moving with velocity v with respect to the former observer, and c is the speed of light. GPS satellites are moving with a velocity of 3.874 km/s, causing the clocks to appear to run slow in comparison to clocks on the Earth's surface by 7 μs/day.

Because the clocks, at an altitude of 20,184 km, experience a gravity field less than clocks on Earth, the gravitational redshift causes them to run fast. This effect can be estimated roughly, assuming circular orbits for the satellites, by:

$$\Delta t' = -\Delta t \left(\frac{\Delta \Phi}{c^2} \right) = -\Delta t \; \frac{1}{c^2} \left[\frac{-GM_E}{r} - \left(\frac{-GM_E}{a_E} \right) \right], \tag{16.3}$$

where Δt is the time interval between events for an observer in some gravitational potential, $\Delta t'$ is the time interval between those same events, as measured by another observer, experiencing a difference in the gravitational potential of $\Delta \Phi$ with respect to the former observer, G is the gravitational constant, M_E is the mass of the Earth, r is the distance of the satellite from the Earth's center of mass, and a_E is the radius of the Earth's geoid. This amounts to 45 μs/day for GPS satellites. The net effect is that a GPS satellite clock runs fast by approximately 38 μs/day compared to a clock at rest on the Earth's geoid. Considering that the timing signal travels at a rate of approximately 300 m in one μs, this large effect is compensated in the satellite clocks by offsetting the frequency before launch, so that the satellite clock appears to run at the same rate as a clock on the ground. Since the satellite orbits are not truly circular, but have an orbital eccentricity of about 0.02, there is also a sinusoidal semi-diurnal variation in the clocks' timing with an amplitude of 46 ns. This correction must be calculated and taken into account in the user's receiver.

The third effect is the Sagnac delay, which is caused by the motion of the receiver on the surface of the Earth due to the Earth's rotation during the time when the signal is on its way from the satellite. This delay is computed in the GPS receiver and is given by:

$$\Delta t_{Sagnac} = \frac{2\omega A}{c^2}, \tag{16.4}$$

where ω is the rotational speed of the Earth and A is the area swept out by the position vector with respect to the center of the Earth projected on to the equatorial plane. For a stationary receiver on the geoid, the Sagnac correction can be as large as 133 ns.

16.7 Space Mission Times

When spacecraft orbit around the Earth, the coordinates and times can be provided in terms of geocentric nonrotating coordinates or in terms of terrestrial coordinates that are rotating. When the spacecraft travel to other planets, barycentric coordinates are the most convenient. The observations typically are timed using UTC, and appropriate transformations can then be made. The round trip times to and from the spacecraft can be used as measures of the spacecraft position. Also the frequency of the received signal will be changed by the Doppler Effect, so the velocity of the spacecraft can be observed.

16.7.1 Doppler Effect

The Doppler Effect, which was proposed by Christian Doppler in 1842, is the change in frequency, or wavelength, of a signal as detected by an observer moving relative to the source of the signal. The total Doppler Effect results from the motion of the source and the motion of the observer. The change in frequency, Δf, is:

$$\Delta f = \frac{fv}{c} = \frac{v}{\lambda},\qquad(16.5)$$

and the observed frequency is:

$$f' = f + \Delta f,\qquad(16.6)$$

where f is the transmitted frequency, v is the velocity of the transmitter relative to the receiver in meters per second (positive when moving toward one another, negative when moving away), c is the speed of the wave traveling in air or a vacuum, and λ is the wavelength of the transmitted wave.

16.8 Proper Times at Planets

Chapter 9 discusses the differences between terrestrial proper and coordinate time scales in comparison to Barycentric Coordinate Time TCB. Similar relationships will hold for all of the other solar system bodies. Until recently there has been no need to define such time systems for those bodies. However, the accuracy requirements for spacecraft at Mars are approaching the point where Mars Time may need to be defined and implemented in the near future.

Table 16.3 *Magnitude of terms for relativistic timescales for the Earth and Mars*

		Maximum Amplitude	
Secular Drift		Diurnal Term	Principal Periodic Term
TT-TCG	60.2 μs/day	2.1 μs	
MT-TCA	12.1 μs/day	0.9 μs	
TCB-TCG	1.28 ms/day		1.7 ms
TCB-TCA	0.84 ms/day		11.4 ms

In a manner similar to Terrestrial Time, it would be possible to define Mars Time (MT), and just as we have defined Geocentric Coordinate Time (TCG), it would be possible to define Areocentric Coordinate Time (TCA) based on the Mars gravitational potential and orbital parameters. Such timescales would be used for spacecraft orbiting or vehicles on the surface of Mars. Table 16.3 indicates the size of the terms in comparison with the analogous timescales related to the Earth (Nelson, 2006). If such a time system were needed, IAU commission A3 on Fundamental Standards would be likely to define appropriate Martian time scales officially. For the Moon we expect the secular drift rate between TT and time on the lunar surface to be 56.0 μs/d and the amplitude of the periodic effect to be 0.48 μs at the Moon's orbital period of 27.3 days (Nelson, 2006).

16.9 Pulsars: An Independent Source of Time

Currently the assumption is that time systems based on atomic physics and those based on the dynamics of solar system bodies do not differ in a nonlinear manner. So Terrestrial Time and atomic time are related by a linear expression. Interest continues in the possibility of finding an additional independent source of time. One suggested source of such a time system is pulsars.

Pulsars are strongly magnetic rotating neutron stars that emit radiation in a beam that is only observed on Earth when that beam is pointed in the right direction. The beam rotates with the star with periods ranging from the order of milliseconds to seconds. Timings of their rotation periods are analyzed to determine possible improvements in the ephemerides of solar system bodies, and to establish evidence of gravitational waves. In addition, pulsar timings, particularly those of pulsars with periods of revolution of milliseconds, may identify irregularities in atomic time standards and contribute to improving the long-term accuracy of timescales. Analyses of these observations require very accurate ephemerides and physical models in order to obtain precise measures of the intervals. The ephemerides of solar system bodies and relativistic effects must be taken into account precisely. Other sources of error include receiver noise, modeling the gravity wave background, and propagation effects.

The utility of pulsar timings for constructing timescales is based on the knowledge of the total number of rotations of the millisecond pulsars. That depends on the accuracy with which the periods of their rotation can be modeled. Generally pulsars can be expected to have periods that change linearly. However, it is possible to observe apparent "glitches" in the periods, so operational monitoring of pulsars is critical.

R. N. Manchester (2008) pointed out the fundamental differences between atomic timescales and a possible pulsar-based timescale. A pulsar timescale is based on the physics of massive rotating bodies and is totally isolated from the solar system or the Earth. An atomic clock has a finite lifetime on the order of a decade, but pulsars are expected to continue rotating for billions of years. The precision of pulsar observations is on the order of 100 nanoseconds. Their contributions are based on the stability of observationally determined information regarding their periods and the rate of change of their periods, and not on physical principles.

A pulsar-based timescale based on stellar mass rather than the quantum processes of atomic clocks that would continue far longer than any physical clock could provide an independent check on terrestrial timescales. Such a timescale has been developed using observations at the Parkes Pulsar Timing Array (PPTA). The Ensemble Pulsar Scale (EPS) has a precision comparable to the uncertainties in international atomic timescales, and can be used to detect fluctuations in atomic timescales and provide a realization of terrestrial time, TT (PPTA11). Marginally significant differences between TT(PPTA11) and TT (BIPM11), an atomic timescale, are seen (Hobbs et al., 2012; Manchester & Hobbs, 2012).

16.10 White Dwarfs: An Independent Source of Time

White dwarf stars have also been proposed to provide an alternate source of uniform time. About 95% of all stars in our galaxy evolve from main-sequence masses to white dwarf stars. They are compact with high internal temperatures, and some also exhibit pulsations. White dwarf G117-B15A, for example, shows pulsation periods of around 215 seconds and its long-term stability is being investigated. A timescale long into the future might be a possibility. See Chapter 8 (Isern et al., 2012; Kepler et al., 2000; Kepler, 2012).

References

Comptes Rendus de la 15ᵉ CGPM (1975), 1976, 104, available at www1.bipm.org/en/convention/cgpm/resolutions.html.

Currie, D. (2014). A Lunar Laser Ranging Retroreflector Array for the 21st Century: Review of the History, Science, Status and Future. *ESPC Abstracts*, 9, ESPC2014-632-2, 2014.

Danjon, A. (1959) *Astronomie Generale*, Paris.

Explanatory Supplement to the Astronomical Almanac (1992). P. Kenneth Seidelmann, editor. Mill Valley, CA: University Science Books.

Explanatory Supplement to the Astronomical Almanac (2012). Sean E. Urban and P. Kenneth Seidelmann, editors. Mill Valley, CA: University Science Books.

Hobbs, G., Coles, W., Manchester, R. N., Keith, M. J., Shannon, R. M., Chen, D., Bailes, M., Bhat, D. R., Burke-Spolaor, S., Champion, D., Chaudhary, A., Hotan, A., Khoo, J., Kocz, J. Levin, Y., Oslowski, S., Preisig, B., Ravi, V., Reynolds, J. E., Sarkissian, J., van Straten, W., Verbiest, J. P. W., Yardley, D., & You, X. P. (2012). Development of a pulsar-based timescale. *Mon. Not. R. Astron. Soc.* 427, 2780–2787.

Isern, J., Althaus, L., Catalan, S., et al. (2012). White dwarfs as physics laboratories: The case of axions. 8th Patras Workshop on Axions, WIMPS and WISPS. Chicago and Fermilab. Online at http://axion-wimp2012.desy.de/, id.37.

Kepler, S. O. (2012). White Dwarf Stars: Pulsations and Magnetism. In H. Shibahashi, M. Takata, & A. E. Lynas-Gray, eds., *61st Fujihara Seminar: Progress in Solar/Stellar Physics with Helio- and Asteroseismology, ASP Conference Series*, 462. San Francisco, CA: Astronomical Society of the Pacific, pp. 322–325.

Kepler, S. O., Mukadam, A., Winget, D. E., Nather, R. E., Metcalfe, T. S., Reed, M. D., Kawaler, S. D., & Bradley, P. A. (2000). Evolutionary Timescale of the Pulsating White Dwarf G117-b15a: The Most Stable Optical Clock Known. *Astron. J.*, 534, L185–L188.

Laskar, J. (1986). Secular Terms of Classical Planetary Theories Using the Results of General Relativity. *Astron. Astrophys.*, 157, 59–70.

Manchester, R. N. (2008). The Parkes Pulsar Timing Array Project. In *40 YEARS OF PULSARS: Millisecond Pulsars, Magnetars and More*. AIP Conference Proceedings, Volume 983, 584–592.

Manchester, R. N. & Hobbs, G. (2012). Pulsar Timing and a Pulsar-Based Timescale. *Proceedings of the Journees 2011 "Systemes de reference spatio-temporels (JSR2011: Earth Rotation, Reference Systems and Celestial Mechanics: Synergies of Geodesy and Astronomy."* H. Schuh, S. Boehm, T. Nilsson, and N. Capitaine, editors. Vienna: Vienna University of Technology.

Meuus, J. & Savoie, D. (1992). The History of the Tropical Year. *J. of British Astronomical Association*, 102, 40–42.

Murphy, T. W. (2013). Lunar Laser Ranging: The Millimeter Challenge. *Reports on Progress in Physics*, 76, 076901.

Nelson, R. A. (2006). Relativistic Transformations for Time Synchronization and Dissemination in the Solar System. *International Astronomical Union XXVIth General Assembly.*

Stephenson, F. R., Morrison, L. V., & Hohenkerk, C. Y. (2016). Measurement of the Earth's Rotation: 720 BC to AD 2015. *Proceedings of the Royal Society A*, 472, 2016.0404.

Viswanathan, V., Fienga, A., Manche, H., et al. (2016). Impact of Infrared Laser Ranging on Lunar Dynamics. *AAS DPS Meeting* 48, id#109.02.

Williams, J. G. & Boggs, D. H. (2016). Secular Tidal Changes in Lunar Orbit and Earth Rotation. *Celest. Mech. & Dyn. Astron.* 126, 89–129.

17

Time and Frequency Transfer

17.1 Historical Transfer Techniques

Time and frequency transfer refers to the techniques and models used to compare clocks and/or frequency devices. Usually such comparisons are made between a conventionally accepted standard and the user's device. The available techniques can range from a low-precision visual comparison to sophisticated systems capable of the highest precision. Historically, astronomical observations of star transits were used to adjust clock timing, and clocks driving bell towers served as local time standards adequate for most users' needs. However, as the requirements for precise time and frequency grew, the needs for improved time and frequency transfer processes grew. In the 19th century, when navigators were some of the most demanding users of time, a time ball was often used in seaports to distribute accurate time. A ball was dropped on a highly visible pole at a prearranged time, allowing those who could see the ball to synchronize their clocks. With the development of the railroad and the consequent need for schedule coordination, time signal distribution via telegraph became critical (Bartky, 2000). Later, in the 20th century, wireless radio time signals became the principal means of time and frequency transfer.

Today a number of time and frequency transfer techniques are available, depending on users' needs. The International Telecommunications Union – Radiocommunications Sector (ITU-R) provides a handbook, *Selection and Use of Precise Frequency and Time Systems* (1997), with detailed information on modern time dissemination systems. Table 17.1 outlines techniques commonly used in current practice. Each is described in the following sections. Precise time and frequency transfer using any technique, however, depends not only on the precision physically possible with the various means of sending and receiving the signals, but also on thorough attention to modeling the path of the comparison signals, and careful calibration of the equipment involved.

Table 17.1 *Techniques used currently in time and frequency transfer*

Type	Precision		Coverage
	Time	Frequency	
Coaxial Cable	1–10 ns	10^{-14}–10^{-15}	Local
Telephone	1–10 ms	10^{-6}–10^{-8}	Regional
Optical Fiber	10–50 ps	10^{-15}–10^{-17}	Local
	100 ns	10^{-13}–10^{-14}	Regional
Microwave Link	1–10 ns	10^{-14}–10^{-15}	Local
Television Broadcast	10 ns	10^{-12}–10^{-13}	Local
INTERNET	1–10 ms	10^{-6}–10^{-8}	Global
High-Frequency Broadcast	1–10 ms	10^{-6}–10^{-8}	Global
Low-Frequency Broadcast	1 ms	10^{-10}–10^{-11}	Regional
Low-Frequency Navigation	1 μs	10^{-12}	Regional
Navigation Satellite Broadcast	10–500 ns	10^{-9}–10^{-13}	Global
Navigation Satellite Carrier Phase	0.5–1 ns	10^{-14}–10^{-15}	Global
Communication Satellite Two-Way	0.5–1 ns	10^{-14}–10^{-15}	Global

17.2 Time and Frequency Dissemination Modeling

Time and frequency transfer systems can use either one-way or two-way methods of exchanging information. In the former, it is necessary to estimate the delay in the propagation of the time signal. Errors due to the modeling of these estimates can range from a few nanoseconds to more than a few milliseconds depending on the technique used. Two-way techniques use the nearly simultaneous exchange of time signals along the same path. In this case, various effects, including the propagation delay, can be practically eliminated in the data processing. If timing better than ±1 μs is required, relativistic effects should also be considered.

17.2.1 Propagation Effects

The nature of the medium through which the information is sent has a significant influence on the accuracy of the comparisons. The attenuation of the signals as well as delays and possible phase shifts differ among the various techniques, but experimental testing of cables and physical modeling of tropospheric and iono-spheric delays generally permit users to estimate the quantitative nature of the effects, and make the appropriate corrections.

Electromagnetic signals travel with the speed of light, c, in a vacuum. In other media, the speed is reduced, causing delays and possible shifts in the phase of the signals. In free space, we also know that the signal strength falls off following a $1/r^2$ law, where r is the distance from the transmitter to the receiver. In media such as coaxial cables or optical fibers, however, the loss of signal strength can vary considerably depending on the materials used, and must be determined experimentally.

Propagation of electromagnetic waves is usually considered to follow a line-of-sight path, but the presence of the ionosphere, with its variable electron content and changing height, complicates estimates of delays in free space. Lower frequencies, in particular, propagate via a ground wave signal that can be affected by changes in ground conductivity, solar flares, geomagnetic storms, and variations in the tilts of the layers in the ionosphere (Middleton, 2001).

17.2.2 Calibration

A critical element in accurate time comparison is the calibration of signal delays through the equipment in the chain of the time/frequency transfer. Laboratory calibration of transmitter, receiver, and cable delays can provide accuracies on the order of single nanoseconds. However, environmental changes, particularly in temperature, can produce significant systematic errors, and careful attention to time variations in the calibrated delays is often necessary for applications requiring the highest accuracy.

17.2.3 Relativistic Effects

In the language of general relativity, two clocks are said to be synchronized in a coordinate system, when their readings are equal at common dates in the coordinate time t of the system. A "coordinate clock comparison" is the difference of clock readings at the same date in t. For terrestrial applications up to the orbit of geostationary satellites, two coordinate systems are often used. Both are geocentric. One rotates with the Earth, and is called the Geocentric Terrestrial Reference System (GTRS). It is the system originally defined by the IAU in 1991 and improved in subsequent years. The second, called the Geocentric Celestial Reference System (GCRS), is nonrotating. The rotating system is obtained from the latter by a transformation of space coordinates that does not modify the coordinate time, so using either system should lead to the same result. The following formulae assume that Terrestrial Time (TT) (practically realized by TT = TAI + 32.184 s) is the coordinate time t. However, using either TT or Geocentric Coordinate Time (TCG) leads to the same results, if properly handled.

Comparison of clock readings requires the evaluation of the accumulated coordinate time during the time transfer, either by the physical transportation of a clock or the transmission of a signal. A key quantity for a clock, A, generating proper time, τ_A, is the ratio, $d\tau_A/dt$, which is, in general, variable along the world line of the clock. Discussions of relativistic time comparisons appear in *Explanatory Supplement* (1992, 2012) and in Guinot (1997). Relativistic theory for time and frequency transfer by satellites has been developed. For fibers, the signal propagation is not geodesic, and the state of the fiber given by its position, velocity, and refractive index must be considered. Time and frequency transfer by optical fiber seeking accuracies at the 10^{-18} level requires evaluating relativistic effects for the fiber moving with a velocity due to the Earth's surface motions, such as rotation and tides, and being exposed to the Earth's gravitational field. For a 1,000 km fiber with a refractive index of 1.5, the Newtonian term (c^{-1}) is 5 ms. For a fiber at the equator, the Sagnac correction (c^{-2}) term is ±5 ns. The c^{-3} term is 3 ps at the surface of the Earth (Gersl et al., 2015).

17.2.3.1 Clock Transport in a Rotating Reference Frame

When time is transferred between two points, P and Q, using a portable clock, it is convenient to use the rotating reference frame with the potential $W(\mathbf{r})$, which includes the potential of the Earth's rotational motion. The coordinate time accumulated during transport is:

$$\Delta t = \int_P^Q \left[1 - \frac{\Delta W(\mathbf{r})}{c^2} + \frac{v^2}{2c^2} \right] d\tau + \frac{2\omega}{c^2} A_E, \tag{17.1}$$

where c is the speed of light, ω is the angular velocity of the Earth's rotation, v is the speed of the clock with respect to the ground, $\mathbf{r}$ is a vector from the center of the Earth to the clock, which is moving from P to Q, and A_E is the equatorial projection of the area swept out during the time transfer by the vector $\mathbf{r}$ as the clock moves from P to Q. $\Delta W(\mathbf{r})$ is the potential difference between the location of the clock at $\mathbf{r}$ and the geoid in an Earth-fixed coordinate system with the convention that $\Delta W(\mathbf{r})$ is positive when the clock is above the geoid. $d\tau$ is the increment of proper time accumulated on the portable clock as measured in the rest frame of the clock, i.e., the reference frame traveling with the clock. A_E is measured in an Earth-fixed coordinate system. As the area A_E is swept out, it is taken as positive when the projection of the clock's path on the equatorial plane is eastward. For a clock above the geoid by a height, h, of 20 km, the approximation of $\Delta W(\mathbf{r})$ by gh, where g is the total acceleration due to gravity (including the rotational acceleration of the Earth) evaluated at the geoid, leads to an error that may reach about 1×10^{-14} in relative frequency. For a better approximation, the potential difference $\Delta W(\mathbf{r})$ can be calculated to greater accuracy by:

$$\Delta W(\mathbf{r}) = -GM_e \left(\frac{1}{r} - \frac{1}{a_e} \right) - \frac{1}{2} \omega^2 \left(r^2 \sin^2\theta - a_e^2 \right)$$
$$+ \frac{J_2 GM_e}{2a_e} \left[1 + \left(\frac{a_e}{r} \right)^3 [3\cos^2\theta - 1] \right], \tag{17.2}$$

where a_e is the equatorial radius of the Earth, r is the magnitude of the vector $\mathbf{r}$, θ is the colatitude, GM_e is the product of the Earth's mass and the gravitation constant, and J_2 is the quadrupole moment coefficient of the Earth ($J_2 = +1.083 \times 10^{-3}$). See also Klioner (1992) for an evaluation of $\Delta W(\mathbf{r})/c^2$ as a function of the altitude.

17.2.3.2 Nonrotating Local Inertial Reference Frame

When time is transferred between points P and Q by means of a clock, the coordinate time elapsed during the motion of the clock is:

$$\Delta t = \int_P^Q \left[1 - \frac{V(\mathbf{r}) - U_g}{c^2} + \frac{v^2}{2c^2} \right] d\tau, \tag{17.3}$$

where $V(\mathbf{r})$ is the potential at the location of the clock and v is the velocity of the clock, both as viewed – in contrast to the expression for clock transport in a rotating reference frame given in Equation (17.1) – from a geocentric nonrotating reference frame. U_g is the potential at the geoid, including the effect of the potential of the Earth's rotational motion. Note that $V(\mathbf{r})$ does not include the effect of the Earth's rotation and should include the tide-generating potential of external bodies. This equation also applies to clocks in geostationary orbits, but should not be used beyond a distance of about 50,000 km from the center of the Earth.

17.2.3.3 Electromagnetic Signals Transfer in a Rotating Reference Frame

In a geocentric, Earth-fixed, rotating frame, the coordinate time elapsed between emission and reception of an electromagnetic signal in vacuum is:

$$\Delta t = \frac{1}{c} \int_P^Q \left[1 - \frac{\Delta V(\mathbf{r})}{c^2} \right] d\sigma + \frac{2\omega}{c^2} A_E, \tag{17.4}$$

where $d\sigma$ is the increment of standard length, or proper length, along the transmission path, and $\Delta V(\mathbf{r})$ is the difference between the potential at the point $\mathbf{r}$ and that at the geoid, as viewed from an Earth-fixed coordinate system. A_E is the area circumscribed by the equatorial projection of the triangle, whose vertices are at the center of the Earth, at the point of transmission of the signal P, and at the point of reception of the signal Q. The area A_E is positive when the signal path has an eastward component.

The second term in the integral amounts to about a nanosecond for a round-trip trajectory from the Earth to a geostationary satellite. In the third term, $2\omega/c^2 = 1.6227 \times 10^{-6}$ ns km^{-2}; this term can contribute hundreds of nanoseconds for practical values of A_E. The increment of proper length $d\sigma$ can be taken as the length measured using standard rigid rods at rest in the rotating system. This is equivalent to measurement of length by taking $c/2$ times the time (normalized to vacuum) of a two-way electromagnetic signal sent from P to Q along its transmission path.

17.2.3.4 Electromagnetic Signals Transfer in a Nonrotating
Local Inertial Reference Frame

In a geocentric nonrotating, local inertial frame, the coordinate time elapsed between emission and reception of an electromagnetic signal is:

$$\Delta t = \frac{1}{c} \int_P^Q \left[1 - \frac{V(\mathbf{r}) - U_g}{c^2} \right] d\sigma, \qquad (17.5)$$

where $V(\mathbf{r})$ and U_g are defined as earlier, and $d\sigma$ is the increment of standard length, or proper length, along the transmission path.

17.3 Time and Frequency Dissemination Systems

Table 17.1 shows some of the systems used more commonly for time and frequency transfer. Their capabilities vary significantly, and details regarding each of them are given in what follows.

17.3.1 Coaxial Cable

Direct time and frequency transfer using coaxial cable connections is commonly used in laboratory settings where distances between clocks are less than a few hundred meters. The physical parameters of the cable can vary with environmental parameters, so careful attention should be given to the environmental temperature and its stability as well as the type and length of the cable when making comparisons. Extreme accuracy may require frequent calibration. Speeds of signal transmission in such cables vary, but typical values can range from 65% to 85% of the speed of light in a vacuum.

17.3.2 Telephone

Time transfer can be provided by regional telephone services. These range from a simple voice recording of the current time to coded information for use with automated equipment. Without compensation for path delays, errors can be of the order of 0.1 to 1 second. With compensation, these errors can be of the order of 1 to 10 ms.

17.3.3 Optical Fiber

Optical fiber connections between clocks can provide very high-accuracy capabilities for time and frequency transfer. Sub-nanosecond time transfer can be possible over relatively short distances, but careful attention to the environmental stability of the fiber is essential. It is also important that such systems be calibrated frequently to make operational use of the capability of optical fiber.

Optical fibers offer excellent accuracy to meet future requirements for time comparisons. However, the normal use of optical fibers involves sharing of the fibers, so users do not have dedicated time on the fiber and the ability to obtain accurate time transfers.

Frequency comb technology is applied to the terahertz (THz) domain for microwave-to-THz synthesizers. A THz frequency reference transfer is based on a THz-to-optical synthesizer and an optical-to THz synthesizer connected by optical fiber with carrier-phase noise cancellation. One optical carrier synthesized from a THz standard is sent to the remote end as an intermediary over a phase-noise-cancelled fiber. At the remote end, the THz standard is retrieved from the carrier without loss of phase coherence. This allows dissemination of an accurate THz reference with 4×10^{-18} fractional frequency accuracy at 0.3 THz (Nagano et al., 2017).

A dark fiber network is a network that makes use of dark (unlit) optical fibers leased from a network service provider. The Dense Wavelength Division Multiplexing (DWDM) technology that puts data from different sources together on an optical fiber, with each signal carried at the same time on its own separate light wavelength, has been developed. With this technique, one channel provides an ultra-stable signal, leaving the rest of the spectrum for data transmission. RENATER, the French national telecommunications network for technology education and research, offers dark fiber infrastructure with high bandwidth availability for research projects. Two strontium optical clocks with uncertainties of 5×10^{-17} were compared via a phase-coherent frequency link connecting Paris and Braunschweig using 1,415 km of telecom fiber with fractional precision of 3×10^{-17} after 1,000 s averaging time (Lisdat et al., 2016). The REFIMEVE+ project (or MEFINEV+ project [Metrological Fiber Network with European Vocation +]) has used the RENATER dark network to transfer an ultra-stable optical frequency on the Internet over long distances without any traffic disruption (Lopez et al., 2015). A four-span cascaded link of 1,100 km has achieved relative frequency stability of 4×10^{-16} with 1 s integration time and 5×10^{-20} at 60,000 s integration time. Extending the distance to 1,480 km degrades the robustness of the link, but both links offer the possibility of comparing future optical clocks at 10^{-19} (Chiodo et al., 2015).

17.3.4 Microwave Links

Within local areas, microwave links can be used to distribute time and frequency with high accuracy. Two-way connections are necessary to provide the highest accuracy. Capabilities depend on atmospheric conditions as well as possible problems with signal reflections caused by objects along the distance traveled (multipath).

17.3.5 Television Broadcast

It is possible to take advantage of the timing of local television signals for time and frequency transfer. Common-view monitoring of a television signal can be used to provide time transfer accuracy of the order of ±10 ns. Each site involved in the transfer must monitor the same synchronization pulse in the television signal. The difference of the measurements then provides the time comparison. Specialized equipment is required at each site to use this technique.

17.3.6 Internet

A number of stratum 1 servers on the Internet can also be used to provide accurate time. Network Time Protocol (NTP), the most commonly used time protocol, provides the best performance. User software runs continuously as a background task that automatically updates the computer clock using servers at many different locations around the world.

17.3.7 High-Frequency Radio Signals

High-frequency (short-wave) radio signals continue to be an easy way of receiving timing information. Table 17.2 shows the high-frequency time signals available. The transmitted time is generally UTC, with a code or voice transmission of DUT1, so that an estimate of UT1 accurate to the nearest 10th of a second can be obtained using the relation UT1 = UTC + DUT1. ITU Recommendation ITU-R TF.768 contains a complete listing of HF services, including details of the content and format of the broadcasts. A comprehensive listing of stations and codes that are broadcast is also available in National Geospatial-Intelligence Agency Publication 117.

Broadcasts of voice time announcements provide accuracy of the order of a few 10ths of a second, but with special equipment the accuracy of the received time signal can be on the order of a few milliseconds. It is limited by unmodeled variations in the travel time of the radio signals, particularly at higher frequencies. Reception is usually better for frequencies less than 10 MHz during nighttime hours and for the higher frequencies during daytime hours. The quality of the reception depends on tropospheric and ionospheric conditions. Uncertainty in the number of reflections of the signal off of the ionospheric reflecting layers complicates the calculation of propagation delays. Generally a single reflection can be assumed for distances less than 1,600 km.

Table 17.2 *High-frequency time signals*

Frequency	Call Sign	Location	Broadcast Times
2.5 MHz	BPM	Xian, China	Various
	WWV	Fort Collins, Colorado, USA	Continuous
	WWVH	Kekaha, Hawaii, USA	Continuous
3.33 MHz	CHU	Ottawa, Canada	Continuous
3.81 MHz	HD2IOA	Guayaquil, Ecuador	0500–1700 UTC
4.996 MHz	RWM	Moscow, Russia	Continuous
4.998 MHz	EBC	Cadiz-San Fernando, Spain	1000–1100 UTC Monday–Friday
5 MHz	BPM	Xian, China	Various
	HLA	Taejon, Korea	Continuous
	LOL 1	Buenos Aires, Argentina	Various
	WWV	Fort Collins, Colorado, USA	Continuous
	WWVH	Kekaha, Hawaii, USA	Continuous
	YVTO	Caracas, Venezuela	Continuous
7.85 MHz	CHU	Ottawa, Canada	Continuous
9.996 MHz	RWM	Moscow, Russia	Continuous
10 MHz	BPM	Xian, China	Various
	LOL	Buenos Aires, Argentina	Various
	PPE	Rio de Janeiro, Brazil	Continuous
	WWV	Fort Collins, Colorado USA	Continuous
	WWVH	Kekaha, Hawaii USA	Continuous
14.67 MHz	CHU	Ottawa, Canada	Continuous
14.996 MHz	RWM	Moscow, Russia	Continuous
15 MHz	BPM	Xian, China	Various
	WWV	Fort Collins, Colorado, USA	Continuous
	WWVH	Kekaha, Hawaii, USA	Continuous
15.006 MHz	EBC	Cadiz-San Fernando, Spain	1000–1100 UTC Monday–Friday
20 MHz	WWV	Fort Collins, Colorado, USA	Continuous
	WWVH	Kekaha, Hawaii, USA	Continuous
25 MHz	WWV	Fort Collins, Colorado, USA	Continuous
	MIKES	Espoo, Finland	Continuous

17.3.8 Low-Frequency Broadcast Radio Signals

Low-frequency (long-wave) signals are also used because they are not as seriously affected by ground features or buildings. They can cover larger areas than high-frequency signals and pass through many types of building walls. Signal coverage can range from a few hundred to a few thousand kilometers. Signal stability is

Table 17.3 *Low-frequency time signals*

Station Call Name	Frequency	Power	Location
DCF77	77.5 kHz	50 kW	Mainflingen, Germany
HBG	75 kHz	20 kW	Prangins, Switzerland
JJY	40 kHz	50 kW	Fukushima Prefecture, Japan
	60 kHz	50 kW	Saga Prefecture, Japan
MSF	60 kHz	25 kW	Rugby, United Kingdom
RBU	66.66 kHz	10 kW	Moscow, Russia
RTZ	50 kHz	10 kW	Irkutsk, Russia
WWVB	60 kHz	50 kW	Fort Collins, Colorado, USA

affected by ionospheric variations, and for large distances between the transmitter and the user, it is important to account for the effect of sunrise and sunset along the signal path. Time signals with accuracy of the order of a millisecond are possible, but it is also possible to use these signals to calibrate the frequency of local oscillators by continuously monitoring the difference in phase between the signal and the local oscillator. After accounting for possible cycle slips, phase measurements with an accuracy of the order of a few tens of microseconds are possible. Table 17.3 shows providers of low-frequency timing signals.

17.3.9 Low-Frequency Navigation Signals

LORAN-C was used in the past as a means to transfer time. However, the US Coast Guard terminated the transmission of all US Loran-C signals in February 2010 and began dismantling stations. Most European systems have also been terminated. The system relied on synchronized broadcasts of low-frequency signals from a chain of stations separated by distances of kilometers to determine the location of the receiver. The stations emitted coded signals with 100 kHz bursts with varying strengths. The signals did not contain complete timing information and were not a source of UTC. However, if a user's clock was initially set to UTC by another means, LORAN-C could be used to keep the clock to within a few microseconds. Although three stations were required for navigation, only one was necessary for time and frequency transfer. Specially designed timing receivers were used to track the third sub-pulse of the received bursts. Each LORAN-C station had caesium standards that were used to synchronize the signals. Coverage was variable, depending on radiated power and the surface conductivity along the path between transmitter and receiver. The accuracy could be of the order of 50 ns to 300 ns. The limiting source of error was modeling the variations in path delay due mainly to weather-related events. Enhanced versions of LORAN are being planned or implemented in South Korea and Russia.

17.3.10 Navigation Satellite Broadcast Signals

Navigation satellite systems including global systems such as the US Global Positioning System (GPS), the Russian Globalnaya Navigazionnaya Sputnikovaya Sistema (GLONASS), and the European GALILEO system, as well as regional systems such as the Chinese BeiDou system, the Indian Regional Navigational Satellite System (IRNSS) – renamed Navigation Indian Constellation (NavIC) – and the Japanese Quasi-Zenith Satellite System (QZSS) use atomic standards on the satellites to provide global navigation solutions. This timing capability also provides the possibility of cheap, global, highly accurate time and frequency transfer. Reception of signals from four satellites is required to determine the three-dimensional position of the user's receiver and the user's time. Timing information can be obtained directly from a single satellite, if the user's position is already known. Sources of error for this one-way approach include errors in the geometrical delay, errors in the estimates of the effects of the ionosphere and troposphere on the propagation time, spurious reflections of the transmitted signals, and hardware delays. Uncertainties can be better than ±100 ns in a few minutes, and about ±10 ns with a 24-hour average.

Another means of time and frequency transfer is to use a common-view technique in which two standards can be compared by recording timing information from the same satellite at the same time. This technique eliminates errors in the satellite clock and minimizes ephemerides errors in the transmitted data and errors that affect the computation of the signal paths. It does require the receivers at both clocks to observe according to a schedule. A variation of this technique, called the "all-in-view" technique, eliminates the need for scheduling by requiring the receivers to observe all of the satellites in view at participating stations. Typical uncertainties are of the order of single nanoseconds.

A time-transfer technique based on Precise Point Positioning (PPP) has allowed clock comparison with 100 picosecond-level precision (Defraigne et al., 2008). PPP (Kouba & Héroux, 2001) is based on a consistent modeling and analysis of GPS and/or possibly GLONASS dual-frequency code and carrier-phase measurements. These data coming from all the satellites are processed together in a filter that solves for receiver coordinates, receiver clocks, zenith tropospheric delays, and phase ambiguities. Since the carrier frequency is about 1,000 times higher than the frequency of the timing code, the carrier-phase methods have much higher resolution. Clocks at two separated sites observe the same satellites at the same time recording the measured phase difference between the carrier and the local frequency reference. These data can be analyzed using post-processed satellite ephemerides, and ionospheric and tropospheric models to provide estimates of the time and frequency differences between the two sites. While time and frequency transfer based on the timing codes provides unambiguous measures of the time delay between the satellite and the receiver, the carrier phase measures contain unknown multiples of 2π radians that correspond to the integral wavelengths of the carrier signal between the satellite and receiver. These are estimated in the analyses to provide high-precision timing data (Ray & Senior, 2003; Delporte et al., 2008).

17.3.10.1 Global Positioning System (GPS)

One of the most common means to obtain accurate timing information is through GPS (www.gps.gov). The current GPS constellation consists of at least 27 satellites, each carrying multiple caesium and/or rubidium atomic clocks. They orbit at an altitude of approximately 20,200 km completing two orbits per sidereal day. Their signals are broadcast in L band with frequencies ranging from 1176.45 to 1575.42 MHz. Time transfer accuracy on the order of tens of nanoseconds is possible. That figure can be improved further by common-view or all-in-view techniques.

   GPS provides accurate UTC, but the system makes use of an internal timescale called GPS System Time, sometimes just called GPS Time. GPS Time is maintained close to UTC, or TAI, modulo one second, the offset being less than one microsecond and generally of order of tens of nanoseconds. The clocks of the GPS system, located in monitoring sites and in the satellites, are used to realize this internal scale. Leap seconds are not inserted in this timescale, and its starting epoch is midnight of January 5/6, 1980, so that TAI is ahead of GPS Time by 19 s, a nearly constant value. User receivers, however, generally provide UTC, making use of information that the satellites broadcast containing the time offset between UTC and GPS Time.

17.3.10.2 GLONASS

The Russian GLONASS provides services similar to GPS. However, it uses Moscow Time (UTC + 3 h) as its time reference instead of an internal timescale. Thus, this system is directly affected by leap second insertions. General information is provided at www.glonass-center.ru.

17.3.10.3 GALILEO

The European system of satellite positioning (GALILEO) will provide time information similar to that of GPS. The fully deployed GALILEO system will consist of 24 operational satellites, plus six in-orbit spares, positioned in three circular Medium Earth Orbit (MEO) planes at 23,222 km altitude above the Earth, and at an inclination of the orbital planes of 56 degrees to the equator. The system is expected to be complete in 2020. Good coverage is expected even at latitudes up to 75°N, the most northerly tip of Europe. The number of satellites with the optimized constellation design, and the three active spare satellites per orbital plane, is expected to ensure that the loss of one satellite should have no discernible effect on the user.

17.3.10.4 BeiDou/Compass

BeiDou-1 was an experimental regional navigation system, with three working satellites and one backup satellite in geostationary orbits decommissioned at the end of 2012. Since 2000 it provided navigational information for locations from longitude 70°E to 140°E and from latitude 5°N to 55°N. The broadcast frequency of the system was 2,491.75 MHz.

The first satellite of BeiDou-2 (formerly known as Compass) was launched in 2007. That was followed by nine more from 2009 to 2011, achieving functional regional coverage. In December 2011, the system went into operation on a trial basis, offering positioning data between longitude 55°E and 180°E and from latitude 55°S to 55°N. A total of 16 satellites were launched during this phase. In 2015, the system began transitioning toward global coverage with the first launch of a new satellite. By 2016, 23 satellites had been launched. The global navigation system should be finished by 2020 with a constellation of 35 satellites, offering complete coverage of the globe. Methods for improving the orbit determination and clock errors by use of TWSTFT, SLR, and solutions for selected variables for the BeiDou system have been developed (Tang et al., 2016).

17.3.11 Two-Way Satellite Time and Frequency Transfer (TWSTFT)

The two-way satellite time and frequency transfer technique is used to provide highly accurate time and frequency transfer independent of navigational satellite techniques. TWSTFT makes use of geostationary communications satellites to transmit simultaneously spread-spectrum timing codes using specialized modems between two users desiring to compare their clocks (see Figure 17.1). The process does not depend on knowing the satellite's position with high precision and takes advantage of the fact that the most significant propagation errors cancel. Accuracy at the sub-nanosecond level has been achieved when the locations of the antennas are well known and the electronic equipment is carefully calibrated. The process requires the exchange of data after the measurements have been made. Asymmetric delays in the communication satellite can be a source of error and relativistic corrections should be applied to achieve the highest precision (Kirchner, 1999; Schaefer et al., 2000).

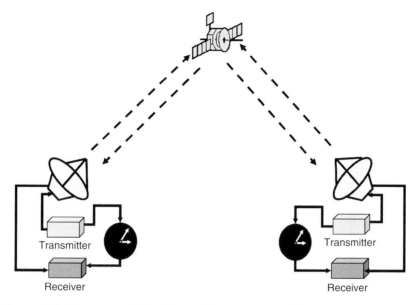

Figure 17.1 Two-way satellite time and frequency transfer

Frequency can be transferred using two-way carrier Doppler measurements. This method requires each station to observe both its and the other station's transponder signals as transferred by satellite. This carrier Doppler TWSTFT (CD-TWSTFT) is better than TWSTFT in short-term (1–100 s) frequency transfer and about the same in long-term (> 100 s) frequency transfer (Liu et al., 2014).

17.3.12 Optical Two-way Time and Frequency Transfer (TWTFT)

Optical TWTFT uses an exchange of pulses to synchronize the sites at start and stop time of the time intervals. Continuous observations are not necessary. At each site a coherent pulse train (comb) is generated by phase locking a femtosecond fiber laser to an optical oscillator with repetition rates differing by Δf between sites. Line of sight between the mismatched optical pulse trains gives time and phase information about the optical waveform over broad optical bandwidths (1 THz) as a train of interferograms (cross-correlations). These interferograms repeat at $\tau_0 = 1/\Delta f$, so that the 0th and nth interferograms are separated by exactly $\tau = n\,\tau_0$. The two exchanges are introduced to deal with variations in the optical path-length due to atmospheric turbulence, platform motion, and changes in air temperature and pressure. Instability below 1×10^{-18} at 1,000 s and systematic offsets below 4×10^{-19} over 2 km links have been experienced (Giorgetta et al., 2013).

17.4 Atomic Clock Ensemble in Space (ACES)

ACES is a European Space Agency (ESA) mission operating atomic clocks in the microgravity environment of the International Space Station (ISS) along with a network of stable clocks on Earth. On ISS two atomic clocks will be located in the Columbus module. One, PHARAO (see Chapter 11), is a frequency standard based on laser-cooled caesium atoms and the other is an active hydrogen maser (SHM). SHM provides short-term stability and PHARAO provides long-term stability and accuracy. Frequency transfer with time deviation better than 0.3 ps at 300 s, 7 ps at 1 day, and 23 ps at 10 days of integration time is expected. This performance would be one to two orders of magnitude better than TWSTFT and GPS, and would enable comparisons of ground clocks with 10^{-17} frequency resolution after a few days of integration time. ACES was scheduled to be launched to ISS in 2016, but has been delayed until 2018 (Cacciapuoti & Salomon, 2009; Delva et al., 2012).

References

Bartky, I. R. (2000). *Selling the True Time, Nineteenth-Century Timekeeping in America.* Stanford, CA: Stanford University Press.

Cacciapuoti, L. & Salomon, C. (2009). Space Clocks and Fundamental Tests: The ACES Experiment. *Eur. Phys. J. Special Topics*, **172**, 57–68.

Chiodo, N., Quintin, N., Stefani, F., Wiotte, F., Camisard, E., Chardonnet, C., Santarelli, G., Amy-Klein, A., Pottie, P.-E., & Lopez, O. (2015). Cascaded Optical Fiber Link Using the Internet Network for Remote Clocks Comparison. *Optics Express*, **23**, 33927–33937.

Defraigne, P., Guyennon, N., & Bruyninx, C. (2008). GPS Time and Frequency Transfer: PPP and Phase Only Analysis. *Int. J. of Nav. and Obs.*, **2008**, Article ID 175468, 7 pages. doi:10.1155/2008/175468.

Delporte, J., Mercier, F., Laurichesse, D., & Galy, O. (2008). GPS Carrier-Phase Time Transfer Using Single-Difference Integer Ambiguity Resolution. *International Journal of Navigation and Observation*, **2008**, Article ID 273785.

Delva, P., Le Poncin-Lafitte, C., Laurent, P., Meynadler, F., & Wolf, P. (2012). Time and Frequency Transfer with the ESA/CNES ACES-PHARAO Mission. *Highlights of Astronomy*, **16**, 211–212.

Explanatory Supplement to the Astronomical Almanac (1992). P. K. Seidelmann, ed., Mill Valley, CA: University Science Books.

Explanatory Supplement to the Astronomical Almanac (2012). S. E. Urban & P. K. Seidelmann, eds., Mill Valley, CA: University Science Books.

Gersl, J., Delva, P., & Wolf, P. (2015). Relativistic Corrections for Time and Frequency Transfer in Optical Fibres. *Metrologia*, **52**, 552–564.

Giorgetta, F. R., Swann, W. C., Sinclair, L. C., Baumann, E., Coddington, I., & Newbury, N. R. (2013). Optical Two-Way Time and Frequency Transfer over Free Space. *Nature Photonics Letters*. doi:10.1038/nphoton.2013.69.

Guinot, B. (1997). International Report: Application of General Relativity to Metrology. *Metrologia*, **34**, 261–290.

ITU Recommendation ITU-R TF.768. Available at www.itu.int/rec/R-REC-TF.768–6-200305-I/en.

Kirchner, D. (1999). Two-Way Satellite Time and Frequency Transfer (TWSTFT): Principle, Implementation and Current Performance. In W. R. Stone, ed., *Review of Radio Science 1996–1999.* Oxford: Oxford University Press.

Klioner, S. A. (1992). The Problem of Clock Synchronization: A Relativistic Approach. *Celest. Mech. & Dyn. Astr.*, **53**, 81–109.

Kouba, J. & Héroux, P. (2001). GPS Precise Point Positioning Using GPS Orbit Products. *GPS Solutions*, **5**, 12–28.

Lisdat, C., Grosche, G., Quinton, N., Shi, C., Raupach, S. M. F., Grebing, C., Nicolodi, D., Stefani, F., Al-Masoudi, A., Dörscher, S., Häfner, S., Robyr, J.-L., Chiodo, N., Bilicko, S., Bookjans, E., Koczwara, A., Koke, S., Kuhl, A., Wiotte, F., Meynadier, F., Camisard, E., Abgrall, M., Lours, M., Legero, T., Schnatz, H., Sterr, U., Denker, H., Chardonnet, C., Le Coq, Y., Santarelli, G., Amy-Klein, A., Le Targat, R., Lodewyck, J., Lopez, O., & Pottie, P.-E. (2016). A Clock Network for Geodesy and Fundamental Science. *Nat. Commun.* 7:12443 doi:10.1038/ncomms12443.

Liu, Z., Gong, H., Yang, W., Zhu, X., & Ou, G. (2014). Carrier-Doppler-Based Real-Time Two Way Satellite Frequency Transfer and Its Application in BeiDou System. *Advances in Space Research*, **54**, 896 900.

Middleton, W. M. (ed.) (2001). *Reference Data for Engineers Radio, Electronics, Computer & Communications (Reference Data for Engineers) (9th edn.)*. Newnes.

Nagano, S., Kumagai, M., Ito, H., Kajita, M., & Hanado, Y. (2017). Phase-Coherent Transfer and Retrieval of Terahertz Frequency Standard over 20 km Optical Fiber with 4×10–18 Accuracy. *Applied Physics Express*, **10**, 012502.

National Geospatial-Intelligence Agency Publication *117, Radio Navigational Aids* (2005). Available at www.nga.mil/portal/site/maritime/.

Ray, J. & Senior, K. (2003). IGS/BIPM Pilot Project: GPS Carrier Phase for Time/Frequency Transfer and Timescale Formation. *Metrologia*, **40**, S270–S288.

Schaefer, W., Pawlitzki, A., & Kuhn, T. (2000). New Trends in Two-Way Time and Frequency Transfer via Satellite. In *Proceedings of the 31st Annual Precise Time and Time Interval (PTTI) Systems and Applications Meeting: 1999*. Dana Point, CA, pp. 505–514.

Selection and Use of Precise Frequency and Time Systems (1997). Available at www.itu.int/publ/R-HDB-31–1997/en.

Tang, C., Hu, X., Zhou, S., Guo, R., He, F., Liu, L., Zhu, L., Li, X., Wu, S., Zhao, G., Yu, Y., & Cao, Y. (2016). Improvements of Orbit Determination Accuracy for BeiDou Navigation Satellite System with Two-way Satellite Time Frequency Transfer. *Advances in Space Research*, **58**, 1390–1400.

18

Modern Earth Orientation

18.1 Terrestrial to Celestial Reference Systems

The need for celestial and terrestrial reference systems is well established, and the mathematical procedures by which they can be related have become standardized. However, the different motions affecting these relationships still cannot be predicted with the accuracy needed for the most demanding applications. The details of the physical causes for the motions are not all well known. Consequently, observations are required to augment the models in order to provide the highest level of accuracy possible. These are implemented in established procedures for the transformations between the different reference systems.

Chapter 5 describes the concepts of the procedures used to transform from a terrestrial reference system (TRS) to the celestial reference system (CRS) at any epoch t. There we saw that:

$$[\text{CRS}(t)] = \mathbf{Q}(t)\ \mathbf{R}(t)\ \mathbf{W}(t)\ [\text{TRS}(t)], \tag{18.1}$$

where $\mathbf{Q}(t)$, $\mathbf{R}(t)$, and $\mathbf{W}(t)$ are the transformation matrices describing precession/nutation, the rotation of the Earth, and polar motion, respectively (Petit & Luzum, 2010), and t is defined by:

$$t = [TT - 2000\ \text{January}\ 1, 12h\ TT\ \text{in days}]/36525. \tag{18.2}$$

Note that 2000 January 1.5 TT = Julian Date 2451545.0 TT.

The precession/nutation rotation is given by:

$$\mathbf{Q}(t) = \begin{bmatrix} 1 - aX^2 & -aXY & X \\ -aXY & 1 - aY^2 & Y \\ -X & -Y & 1 - a(X^2 + Y^2) \end{bmatrix} \bullet \begin{bmatrix} \cos s & \sin s & 0 \\ -\sin s & \cos s & 0 \\ 0 & 0 & 1 \end{bmatrix}, \tag{18.3}$$

$$a = \tfrac{1}{2} + \tfrac{1}{8}(X^2 + Y^2), \tag{18.4}$$

where X and Y are the angular "coordinates" of the Conventional Intermediate Pole (CIP) in the CRS, provided in part by the conventional mathematical models, and s is given by Equations (5.3–5.5). For the highest precision, it is necessary to account for the differences between the observed value and the theoretical model. Mathematically:

$$\begin{aligned} X &= X_{\text{MODEL}} + \delta X \\ Y &= Y_{\text{MODEL}} + \delta Y, \end{aligned} \tag{18.5}$$

where X_{MODEL} and Y_{MODEL} are the values provided by the models and δX and δY are the celestial pole along the meridian of 90° east observations.

The Earth's rotation angle is handled by the matrix:

$$\mathbf{R}(t) = \begin{bmatrix} \cos\theta & -\sin\theta & 0 \\ \sin\theta & \cos\theta & 0 \\ 0 & 0 & 1 \end{bmatrix}, \tag{18.6}$$

θ being the Earth Rotation Angle given by:

$$\theta(T_u) = 2\pi \left(\begin{array}{l} \text{UT1 Julian Days elapsed since } 2451545.0 + 0.7790572732640 \\ + 1.00273781191135448\ T_u \end{array} \right). \tag{18.7}$$

where T_u = (Julian UT1 date − 2451545.0), and $UT1 = UTC + (UT1-UTC)$.

Finally the polar motion rotation is given by:

$$\mathbf{W}(t) = \begin{bmatrix} \cos s' & -\sin s' & 0 \\ \sin s' & \cos s' & 0 \\ 0 & 0 & 1 \end{bmatrix} \bullet \begin{bmatrix} 1 & 0 & 0 \\ 0 & \cos y & \sin y \\ 0 & -\sin y & \cos y \end{bmatrix} \bullet \begin{bmatrix} \cos x & 0 & -\sin x \\ 0 & 1 & 0 \\ \sin x & 0 & \cos x \end{bmatrix}. \tag{18.8}$$

where x and y are the angular "coordinates" of the Conventional Intermediate Pole in the TRS and s' in microarcseconds (µas) can be approximated for the 21st century by:

$$s' = -47\ \mu\text{as}\ t. \tag{18.9}$$

where t is given in Equation (18.2) and the angles X, Y, UT1−UTC, x, and y collectively are known as the Earth orientation parameters, which must be determined by observation.

18.2 Determination of Earth Orientation Parameters

Chapter 5 outlines some of the techniques used in the past to provide the Earth orientation data. These have included visual and photographic telescopic

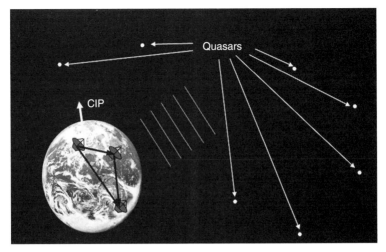

Figure 18.1 The concept of VLBI used to monitor the Earth orientation parameters

observations, Doppler observations of artificial satellites, laser ranging to the Moon and artificial satellites, very long baseline interferometry, and analysis of the orbits of navigational satellites. A solution of Earth orientation parameters based on optical astrometry demonstrated the 28-year "Markowitz wobble" and an apparent 78-year period in polar motion (Vondrak et al., 2009). Not all of those techniques continue to be used. A variety of high-precision techniques are used currently to relate the CRS and the TRS and to make predictions of future Earth orientation. These are discussed in what follows.

18.2.1 *Very Long Baseline Interferometry (VLBI)*

The VLBI technique makes use of multiple radio telescopes observing very distant radio sources to determine the Earth's orientation with respect to the celestial reference frame realized by the conventionally adopted radio source positions. The baselines defined by the vectors between the positions of the telescopes, rigidly attached to the Earth's surface, are defined in the terrestrial frame (Altamimi et al., 2016). The radio sources are quasi-stellar radio sources, quasars, which are powerful radio emitters located at such great distances from the Earth that they can be considered to show minimal space motion. Monitoring of the changing aspect of the baseline vectors with respect to these sources provides the Earth orientation parameters required to implement the mathematical relationship shown in Equation (18.1) (see Johnston, 1979). The concept is shown in Figure 18.1.

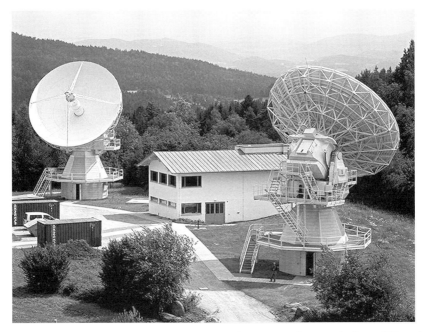

Figure 18.2 Twin VLBI telescopes at Wetzel, Germany by permission of Wetzel Observatory

Quasars are thought to be the centers of some distant galaxies, probably with massive black holes at the center. Those that appear to be the most "point-like" with minimal observed space motion and variations in observed signal intensity form a group that defines the International Celestial Reference Frame (ICRF) (Fey et al., 2009). These objects broadcast strong radio signals composed essentially of random noise.

The radio telescopes used to monitor the signals have diameters on the order of tens of meters and are located at continental distances to provide a better representation of the global Earth orientation and better resolution of the radio sources themselves. Their positions are well determined and provide part of the definition of the International Terrestrial Reference Frame. Figure 18.2 shows the telescope located at Wetzel, Germany. Figure 18.3 shows the worldwide distribution of the antennas currently in use in geodetic VLBI operations and those sites used in the past for such operations. The operations are organized through the International VLBI Service for Geodesy and Astronomy (IVS). (See Chapter 19.)

Observations are made by multiple telescopes observing quasars widely distributed over the sky in order to determine all of the Earth orientation parameters with the best precision. Because the telescopes are located at different distances from the radio sources, differences occur in the times of arrival of the signals at each

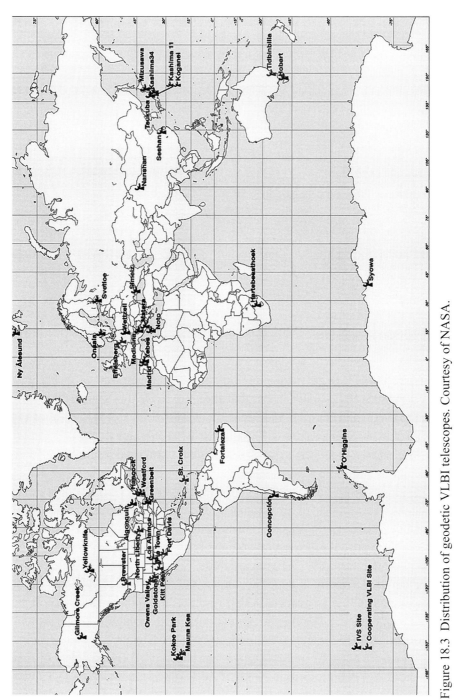

Figure 18.3 Distribution of geodetic VLBI telescopes. Courtesy of NASA.

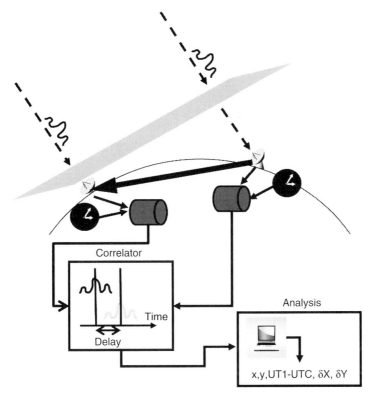

Figure 18.4 VLBI process

telescope. Each site uses a hydrogen maser that serves as a clock to measure precisely the difference in the arrival times. The digitized signals and the times are then sent to a correlator, where the signals are analyzed to determine the delays and the rate of change of the delays among the telescopes involved in the observation. The development of high-speed data transfer networks permits near real-time data transfer, without the necessity of recording the data on electronic media that have to be shipped to the correlator (Carter & Robertson, 1986; Carter et al., 1989).

At the correlator, the estimated delays can be calculated using the known positions of the telescopes and quasars and a priori estimates of the Earth orientation parameters. This information is used to determine the actual delays and delay rates, which can then be analyzed to determine the corrections to the a priori estimates of the Earth's orientation. Figure 18.4 outlines the process graphically. Observations are typically made at frequencies of S band (2.3 GHz) and X band (8.4 GHz) in order to take advantage of the dispersive nature of the ionosphere and calculate the delays due to the total electron content (TEC) in the ionosphere. This delay, $\tau_{\text{ionosphere}}$, depends on the frequency of the signal, f, and is proportional to the TEC and inversely proportional to f^2.

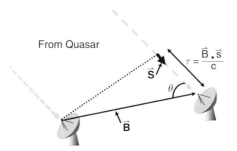

Figure 18.5 Mathematical representation of VLBI delay

$$\tau_{\text{ionosphere}} \propto \frac{TEC}{f^2}. \tag{18.10}$$

Two general types of delay measurements are possible using VLBI: group delay and phase delay. If φ is the measured difference in phase (in radians) between the signals at the two telescopes, the phase delay is given by:

$$\tau_p = -\frac{\varphi(\omega)}{\omega}, \tag{18.11}$$

and the group delay is given by:

$$\tau_g = -\frac{d\varphi(\omega)}{d\omega}. \tag{18.12}$$

Phase delays can be measured with precisions of a picosecond or better, but with an unknown integer number of phase cycles, which are 451 ps at S band and 122 ps at X band. Group delays can be measured with precisions of a few picoseconds without cycle ambiguities and these are the measures typically used in geodetic VLBI observations.

Mathematically, referring to Figure 18.5, where $\vec{B}$ and $\vec{s}$ are the baseline vector and unit vector in the direction of the source, respectively, the delay can be expressed as:

$$\tau = \frac{\vec{B} \bullet \vec{s}}{c} = \frac{|\mathbf{B}||\mathbf{s}| \cos \theta}{c}, \tag{18.13}$$

where c is the speed of light and θ is the angle between $\vec{B}$ and $\vec{s}$. The station coordinates used to create $\vec{B}$ must be corrected for tectonic plate motion, the tides, and the loading of the site by the atmosphere and possibly nearby oceans. $\vec{s}$ is computed using the conventionally adopted precession/nutation and the a priori Earth orientation estimates. Observational estimates of τ must be corrected for atmospheric and ionospheric delays as well as the delays caused by the gravitational attraction of solar system bodies along the path of the signal. The observations themselves are used to calculate the atmospheric and ionospheric effects.

Observations can be affected by instrumental errors, propagation modeling errors, and the fact that the sources may exhibit some extended structure. The instrumental errors include possible deformation of the structure due to temperature and wind loading and possible clock errors. The delay caused by the neutral atmosphere depends on temperature, pressure, and humidity along the path of the signal, and can vary significantly with direction and time at the observing site. Sources are chosen to be used in the observations with concern for any evidence of motion or extended structure. Only the most "point-like" are used to mitigate issues regarding source structure. The strategy for weighting observations has been investigated and improved solutions can be achieved, but automated VLBI analysis is difficult and user attention appears desirable (Wielgosz et al., 2016).

VLBI precision allows analysts to determine the relative distances between the telescopes to a few millimeters and the positions of the radio sources to fractions of a milliarcsecond. Because the telescopes are fixed rigidly in the terrestrial reference frame, the variations in the observed delays provide the information necessary to determine the orientation of the baseline in the celestial frame. From these data corrections to the a priori estimates of x, y, $UT1-UTC$, X, and Y can be derived. The VLBI operations and analyses are coordinated internationally through the IVS (see Chapter 19) (Schlüter & Behrend, 2007) (http://ivscc.gsfc.nasa.gov/).

The CONT11 observing campaign, carried out between September 15 and 29, 2011, with 14 globally distributed VLBI stations, produced estimates of Earth orientation parameters (EOP) with rms differences of about ±31 µas in polar motion and ±7 µs in length of day (Nillsson et al., 2014), compared with Global Navigation Satellite Systems (see Section 18.2.2) data.

The Very Long Baseline Array (VLBA) is a network of 25-meter-diameter radio telescopes in North America, Hawaii, and Saint Croix. In 1997, Research and Development with VLBA (RDV) sessions were begun. These are a joint effort between NASA Goddard Spaceflight Center, the US Naval Observatory (USNO), and the National Radio Astronomy Observatory (NRAO) using the 10 VLBA antennas and 4–10 geodesy network antennas. Since January 2012, daily USNO/ NRAO UT1 40-minute sessions using Pietown and VLBA antennas have been conducted. The UT1 precision averages ~29 µs with a median of ~21 µs. The daily IVS UT1 sessions achieve 13 µs mean and 11 µs median. The lower precision of the prior sessions is due to the shorter baselines, but they provide rapid and independent UT1 measurements (Gordon, 2016). VLBI data can be used to determine the sub-daily terms in the tidal model at the level of 2–4 µas for polar motion and 0.2–0.4 µs for UT1 (Panafidina et al., 2012).

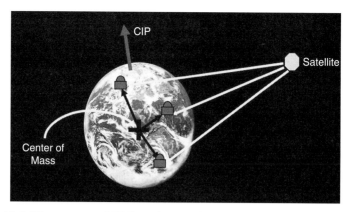

Figure 18.6 The concept of GPS used to monitor the Earth orientation parameters

The VLBI 2010 technology involves upgrading to next-generation VLBI stations around the world with fast slewing antennas, broadband observing systems, and a software correlator. The goals are 1 mm position accuracy on global scales, continuous measurements of time series of station positions and Earth orientation parameters, and turnaround times of less than 24 hours (Petrachenko et al., 2013).

18.2.2 Global Positioning System (GPS)

Although VLBI is the only technique that provides all five of the Earth orientation parameters, GPS now provides the most accurate information regarding the polar motion portion (x, y) of the Earth orientation parameters. Monitor stations, with locations well known with respect to the terrestrial reference system, receive signals from the satellites. This information is used to determine the precise orbits of the satellite in an inertial system. As part of the process, the polar motion parameters can be derived by relating the satellite positions with the locations of the monitor stations. (See Figure 18.6.)

The satellites orbit the Earth in planes that rotate in space. The motion of the planes cannot be separated strictly from variations in the Earth's rotation. Consequently, GPS cannot be used to provide accurate UT1–UTC without some form of a priori knowledge of that part of the rotation of the planes that is independent of the Earth's rotation. The daily rate of UT1–UTC is equivalent to the excess length of day (LOD), and this can be estimated. By integrating the LOD time series for relatively short periods of time it is possible to determine a UT1–UTC time series if an accurate integration constant can be determined using VLBI observations.

The determination of the satellite orbits requires accurate knowledge of the Earth's gravitational field, including any time variations. While the orbits are

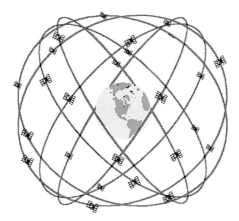

Figure 18.7 The GPS constellation of satellites. University of Colorado Boulder.

sensitive to the Earth's center of mass, they are relatively insensitive to elements of the celestial reference system. As a result, the GPS orbits can provide no information regarding the celestial pole offsets, X and Y.

The constellation of GPS satellites is shown in Figure 18.7. It is composed nominally of a set of six orbital planes with inclinations of 55 degrees, each being populated with four satellites. The satellites orbit at an altitude of 20,200 km. In reality, more than 24 satellites are in orbit at any one time in order to ensure continuous operation of the system. GPS is one of a set of operational, or proposed, navigational satellite systems, but it is currently the only one that has been used operationally to provide reliable Earth orientation information.

The observations that are analyzed for Earth orientation information are coordinated by the International Global Navigation Satellite Services (GNSS) Service (IGS). (See Chapter 19.) The observing network consists of more than 500 sites around the world.

The accuracy of the polar motion observations is made possible by tracking the phase of the signals broadcast from the satellites. Currently, the GPS satellites make use of a number of L-band frequencies, but all GPS satellites broadcast signals at 1.227 and 1.575 GHz. These are modulated by a pseudo-random noise code. Most of the ionospheric delay in the reception of the signals can be eliminated by using the two frequencies, just as in the case of the analysis of the VLBI observations. Tracking the phase of the carrier signal provides no information regarding the time of transmission. That information is only provided by the code. Carrier-phase measurements are differences in carrier phase cycles and fractions of cycles over time. The carrier phase data and code information gathered from the observing sites shown in Figure 18.8 are used by the analysis centers of the

IGS to derive ionospheric maps, orbits of the satellites, and Earth orientation information, as well as to maintain a precise terrestrial reference frame.

The accuracy of the results depends on the accuracy of various physical models used in the orbital analyses. These include models of the tropospheric delays, geopotential, solid Earth tides, ocean tides, ocean and atmospheric loading at the observing sites, solar pressure, atmospheric drag, relativistic effects, and Earth albedo (Rothacher, 1999; Kouba et al., 2000). GPS observations can be particularly useful for investigating sub-daily changes in Earth orientation parameters (Panafidina et al., 2012).

18.2.3 Satellite Laser Ranging (SLR)

Satellite laser ranging is a technique that uses a network of special-purpose telescopes to measure the travel times of very short laser light pulses to and from a set of artificial Earth satellites equipped with laser retro reflectors. Just as in the case of the GPS analyses, these data are used to relate the orbits of the satellites determined in inertial space to a terrestrial reference frame defined by the accurate locations of the telescopes that contribute their observations. The concept is shown in Figure 18.8. These observations, like the GPS observations, are only used to provide polar motion and length of day observations. They also are important for the definition of the Earth's gravitational field model.

Typically, pulses of light with wavelengths of 532 nanometers (nm) (green) and pulse lengths of 10–100 picoseconds (ps) are sent to these satellites and the length of time for the signal to be returned to the telescope is measured with nanosecond precision. Figure 18.9 shows the telescope used to carry out these observations at the Goddard Spaceflight Center and Figure 18.10 shows the geographic distribution of sites where observations have been made. SLR stations have been added in Brasilia, La Plata, India, and Hartebeethoek, providing better coverage from the Southern Hemisphere (Kehm et al., 2016). Not all of those sites have been used for

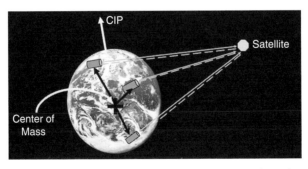

Figure 18.8 The concept of SLR used to monitor the Earth orientation parameters

Figure 18.9 Goddard Spaceflight Center SLR telescope. Courtesy of NASA.

Earth orientation determinations. Many have been used in various geodetic research efforts and to establish an accurate terrestrial reference frame (Schutz et al., 1989). However, the scale of the International Terrestrial Reference System (ITRF) determined from the SLR technique has been about 0.7 parts per billion too small due to systematic errors in the measurements, or treatment, of range measurements (Appleby et al., 2016).

The satellites primarily used for Earth orientation measurements are the Laser Geodynamics Satellite (LAGEOS) satellites, which are essentially spherical satellites covered with laser retro reflectors designed to improve the signal-to-noise ratio of the observations. (See Figure 18.11.) LAGEOS-I was built by NASA and launched in 1976 into a nearly polar circular orbit with an altitude of 6,000 km. A second satellite of the same type (LAGEOS-II) was built by the Agenzia Spaziale Italiana of Italy and launched in 1992 into a similar orbit, but with an inclination of 51 degrees. Both are approximately 60 cm in diameter, weigh 411 kilograms, and are covered by 426 retro reflectors. Other satellites that have been used include Starlette (1,000 km) and Stella (800 km) developed and launched by France, Etalon-I and -2 (19,000 km) developed and launched by the former USSR, and Ajisai (1,500 km) developed and launched by Japan.

As was the case with GPS, the Earth orientation information is derived in the process of determining the satellite orbits using the ranges from the network of observing locations. However, in the case of SLR, ranges measured using visual light are used instead of ranges measured using radio transmissions. As a result, somewhat different physical effects must be accounted for in the SLR analyses. While both techniques are sensitive to the geopotential, solid Earth tides, ocean

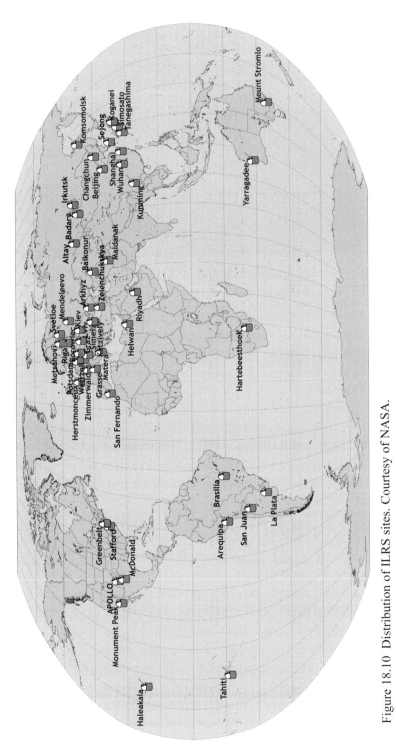

Figure 18.10 Distribution of ILRS sites. Courtesy of NASA.

Figure 18.11 LAGEOS satellite. Courtesy of NASA.

tides, ocean and atmospheric loading at the observing sites, solar pressure, atmospheric drag, relativistic effects, and Earth albedo, the SLR technique does not have to contend with the ionospheric problems. On the other hand, it does require clear skies to receive returns.

SLR has been used for ranging to GPS and GLONASS satellites to improve the orbit determinations. Two GPS satellites and all GLONASS satellites are equipped with retro reflectors for this purpose. Single-photon SLR stations achieve residuals with precision less than 1 mm. Using the center for orbit determination in Europe (CODE) new orbit model, the SLR mean residual is in the range of 0.1–1.8 mm observing GLONASS-M satellites with uncoated corner cubes (Sosnica et al., 2015).

The International Laser Ranging Service (ILRS) was created in 1998 to coordinate the observations and analyses (Gurtner et al., 2005). (See Chapter 19.) It provides data regarding the Earth orientation and terrestrial reference frame to the International Earth Rotation and Reference System Service (IERS) routinely. These data have also been used to produce a long wavelength gravity field reference model that supports all precision orbit determination and provides the basis for studying temporal gravitational variations due to mass redistribution and accurate determinations of tectonic plate motions.

In the past, Earth orientation information was also provided by ranging to targets placed on the Moon's surface by the *Apollo* astronauts. While ranging to lunar targets continues, operational Earth orientation information is no longer contributed from the lunar ranges.

18.2.4 *Doppler Orbit Determination and Radio Positioning Integrated on Satellite (DORIS)*

A fourth technique, Doppler Orbit Determination and Radio Positioning Integrated on Satellite, also contributes data to the IERS, primarily for the extension of the terrestrial reference frame. While it can also be used to determine polar motion, the precision of the derived *x, y* coordinates is not adequate for operational application in the area of Earth orientation parameters.

DORIS is a French system developed by the Centre National d'Etudes Spatiales (CNES) in conjunction with the Institut Géographique National (IGN) and the Groupe de Recherche de Géodésie Spatiale (GRGS). It uses specialized satellites to monitor signals from beacons placed at a number of locations worldwide. As the satellite approaches the beacon on the ground the Doppler effect shifts the frequency seen on the satellite to a higher frequency than that actually broadcast. When the satellite moves away from the beacon, the observed frequency is lower, again as predicted by the Doppler effect. This information, when combined with other data from a network of beacons (see Figure 18.12), allows analysts to determine the location of the beacons with high precision as well as determine the satellite orbits. The system has been placed on a number of satellites intended to carry out geodetic and geophysical research, including Jason-1 and ENVISAT altimetry satellites and the remote-sensing satellites SPOT-2, SPOT-4, and SPOT-5. It also flew with SPOT-3 and TOPEX/POSEIDON. As with the other techniques, a service organization called the International DORIS Service (IDS) has been created to coordinate the observations and data analyses (see Chapter 19) (Tavarnier et al., 2005).

The IDS provided the IERS with 1,140 weekly solutions of station coordinates and Earth orientation parameters from 1993.0 to 2015.0 for the ITRF 2014. The data were from 11 DORIS satellites and six analysis centers. The internal position consistency of 10 mm or better was achieved (Moreaux et al., 2016). Details concerning the accuracies and problems with the data are given by Blossfeld et al. (2016).

18.2.5 *Geophysical Modeling*

Modern astronomical, meteorological, and geophysical observations have made clear that the atmosphere and oceans do affect the Earth's orientation. This relationship has been explained by analyses of the conservation of angular momentum in the solid Earth, ocean, and atmosphere systems. If, for example, a component of atmospheric angular momentum (AAM) increases, the analogous angular momentum component of the solid Earth would decrease to conserve the system angular

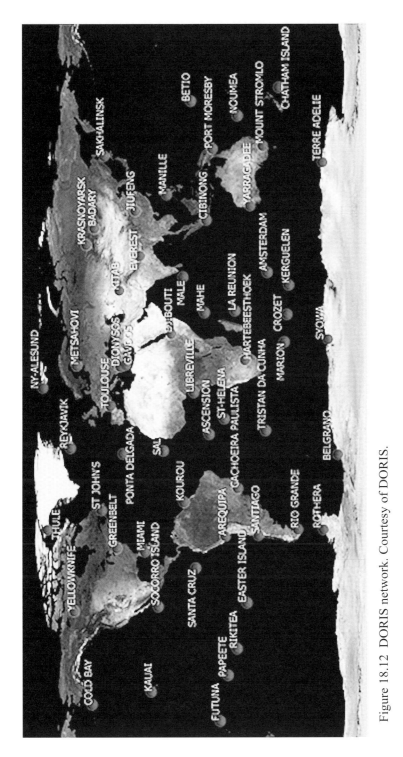

Figure 18.12 DORIS network. Courtesy of DORIS.

momentum. This does appear to be the case and the increasing number of accurate environmental observations has led to timely accurate measurements of the angular momentum variations in the atmosphere and ocean that can be compared with the astronomical observations of the Earth's orientation (Koot et al., 2006).

It is possible to take advantage of this relationship to improve predictions of the Earth's orientation. Near-term predictions of expected variations in the atmospheric and ocean angular momentum (OAM) can be used to assist in making forecasts of the expected variations in the Earth's orientation. The predictions of the angular momentum of the atmosphere and the ocean are made possible by a network of global environmental observatories and sophisticated mathematical forecast models of these systems. So, while geophysical modeling might not be considered strictly a means of observing the Earth's changing orientation, it does make significant contributions in the short-term prediction of polar motion and the Earth's rotation angle.

The relationship of Earth orientation data and angular momentum data is facilitated through the use of the dimensionless "effective" angular momentum functions (Barnes et al., 1983):

χ_1 along the meridian of 0° longitude,

χ_2 along the meridian of 90° east longitude,

χ_3 axial component.

$$\chi_1 = x(t) + \frac{1}{\sigma}\frac{d}{dt}y(t), \quad \chi_2 = -y(t) + \frac{1}{\sigma}\frac{d}{dt}x(t), \quad \chi_3 = LOD(t)\Big/_T, \quad (18.14)$$

where σ is the frequency of the free Chandler wobble, $2\pi/435 \text{ day}^{-1}$, $LOD(t)$ is the excess length of day, and T is the length of day, 86,400 s.

The relationship of χ_3 and LOD is well documented (Hide et al., 1980; Gross, 2012; Dill et al., 2013), and is shown in Figure 18.13.

For polar motion, the relationship is expressed in terms of the excitation functions where the geodetic excitation can be computed from the astronomical observations.

$$\chi_1^{\text{geodetic}}(t) = x(t) + \frac{2\pi}{T_C}\dot{y}(t), \quad \chi_2^{\text{geodetic}}(t) = -y(t) + \frac{2\pi}{T_C}\dot{x}(t). \quad (18.15)$$

The glacial isostatic adjustment caused by rapid ice melting is now seen to affect the motion of the pole (Chen et al., 2013). Seasonal station displacements can affect the celestial reference frame and Earth orientation parameters. From VLBI observational data no systematic seasonal signal is found in the orientation of the celestial reference frame, but positional changes appear in radio sources observed

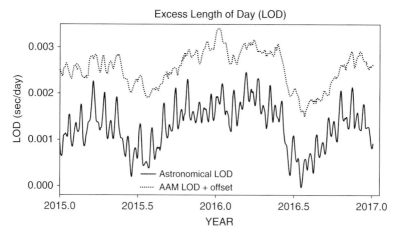

Figure 18.13 Comparison of astronomical observations of excess length of day with effects derived from analyses of atmospheric angular momentum (AAM).

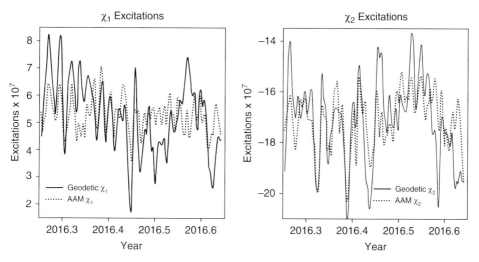

Figure 18.14 shows the relationship between the geodetic and AAM excitation functions

unevenly over the year. The omitted annual and semi-annual seasonal harmonic signal in the horizontal station coordinates affects the Earth orientation parameters, causing differences of several tens of microarcseconds (Krasna et al., 2015).

The core–mantle boundary (CMB) topography is bumpy (the mountains and valleys represent height differences of a kilometer) and causes pressures and torques that may enhance length of day changes and nutations at selected frequencies. These enhancements are not due to the choice of topography, but are a general fact for some frequencies close to internal wave frequencies. These variations are due to resonances with inertial waves in the incremental core flow. For frequencies far from internal wave frequencies, nutations might still be due to the larger topography variations (Dickman, 2003; Puica et al., 2015).

18.2.6 Geomagnetic Field

The geomagnetic field is generated in the Earth's liquid metallic outer core. The Earth's rotation and the topography of the coupling surfaces strongly affect core motions and consequently the geomagnetic field (Lambeck, 1980; Hide & Dickey, 1991; Hide et al., 2000; Miyagoshi & Hamano, 2013). The connection between the decadal geomagnetic field and LOD variations has been investigated (Yoshida & Hamano, 1995; Dumberry & Bloxham, 2006). Geomagnetic jerks (GMJ) are abrupt changes in the secular acceleration (second derivative) of motion of the Earth's magnetic field. They are thought to be due to changes in the fluid flow at the surface of the Earth's core, and they are not always observed at all stations and not always simultaneously (Gokhberg et al., 2016). Jerks occurred in 1910, 1969, 1972, 1978, 1982, 1991, 1999, 2003, 2007, 2011, and 2014 (Bloxham et al., 2002; Chulliat & Maus, 2014; Kotze, 2017). There appears to be a correlation between GMJs and changes in all five Earth orientation parameters (Vondrak & Ron, 2015). There is also evidence of free core nutation (FCN) changes correlated with GMJs (Malkin, 2013, 2016; Vondrak & Ron, 2014). Variations in the LOD within a period of about 10 years might be associated with GMJs (Gokhberg et al., 2016). The shorter period components of LOD variations correlate with GMJs for LOD oscillations contemporaneous with jerks in 1969, 1972, 1978, and 1982 (Holme & de Viron, 2013).

18.3 Earth Orientation Data

The IERS (see Chapter 19) routinely provides Earth orientation data produced from the observations available from the sources outlined earlier. Historical, current, and forecast values of the parameters are made available to users from two product centers, the Earth Orientation Center and the Rapid Service/Prediction Service. The Earth Orientation Center Product Center publishes long-term Earth orientation parameters with the monthly publication of IERS Bulletin B that lists the most recent values of the Earth's orientation in the IERS Reference System. The IERS Rapid Service/Prediction Center provides Earth orientation parameters on a rapid turn-around basis, primarily for real-time users and others needing the highest-quality EOP information sooner than that available in the final series published by the IERS Earth Orientation Center. Table 18.1 lists the uncertainties of the IERS products.

Table 18.1 *Uncertainties of IERS products*

		Precision			
Technique	Sampling Time	Polar Motion	UT1–UTC	LOD	Celestial Pole Offsets
VLBI	1 day	0.000 08″	0.000 005 s	0.000 007 s/day	0.000 085″
VLBI	1 hr	–	0.000 020 s	0.000 028 s/day	–
SLR	1 day	0.000 15″	–	–	–
GPS	1 day	0.000 02″	–	0.000 020 s/day	–
Combination		**0.000 004″**	**0.000 006 s**	**0.000 009 s/day**	**0.000 1″**
Five-day Prediction		**0.000 09″**	**0.000 02 s**	**0.000 03 s/day**	**0.000 1″**

References

Altamimi, Z., Rebischung, P., Métivier, L., & Collilieux, X. (2016). ITRF2014: A New Release of the International Terrestrial Reference Frame Modeling Nonlinear Station Motions, *J. Geophys. Res. Solid Earth*, **121**. doi:10.1002/2016JB013098

Appleby, G., Rodriguez, J., Altamimi, Z. (2016). Assessment of the Accuracy of Global Geodetic Satellite Laser Ranging Observations and Estimated Impact on ITRF Scale: Estimation of Systematic Errors in LAGEOS Observations 1993–2014, *J. Geodesy*, **90**, 1371–1388.

Barnes, R. T. H., Hide, R., White, A. A., & Wilson, C. A. (1983). Atmospheric Angular Momentum Fluctuations, Length-of-Day Changes and Polar Motion. *Proceedings of the Royal Society of London Series A*, **387**, 31–73.

Blossfeld, M., Seitz, M., Angermann, D., & Moreaux, G. (2016). Quality Assessment of IDS Contribution to ITRF2014 Performed by DGFI-TUM. *Advances in Space Research*, **58**, 2505–2519.

Bloxham, J., Zatman, S., & Dumberry, M. (2002). The Origin of Geomagnetic Jerks. *Nature*, **420**, 65–68.

Carter, W. E. & Robertson, D. S. (1986). Studying the Earth by Very-Long-Baseline Interferometry, *Sci. Am.*, **255**, 44–52.

Carter, W. E., Robertson, D. S., & Fallon, F. W. (1989). Polar Motion and UT1 Time Series Derived from VLBI Observations. *IERS Tech. Notes*, No. 2, 35–39.

Chen, J. L., Wilson, C. R., Ries, J. C., & Tapley, B. D. (2013). Rapid Ice Melting Drives Earth's Pole to the East. *Geophys. Res. Let.*, **40**, 2625–2630. doi:10.1002/grl.50552

Chulliat, A. & Maus, S. (2013). Geomagnetic Secular Acceleration, Jerks, and a Localized Standing Wave at the Core Surface from 2000 to 2010. *J. of Geophysical Research: Solid Earth*, 10.1002/2013 JB010604.

Dickman, S. R.(2003). Evaluation of "Effective Angular Momentum Function" Formulations with Respect to Core–Mantle Coupling. *J. Geophys. Res.*, **108**(B3), 2150. doi:10.1029/2001JB001603

Dill, R., Dobslaw, H., & Thomas, M. (2013). Combination of Modeled Short-Term Angular Momentum Function Forecasts from Atmosphere, Ocean, and Hydrology with 90-Day EOP Predictions. *Journal of Geodesy*, **87**, 567–577.

Dumberry, M. & Bloxham, J. (2006). Azimuthal Flows in the Earth's Core and Changes in Length of Day at Millennial Timescales. *Geophys. J. Int*. **165**, 32–46. doi:10.1111/j.1365-246X.2006.02903.x

Fey, A., Gordon, D., & Jacobs, C., eds. (2009). The Second Realization of the International Celestial Reference Frame by Very Long Baseline Interferometry, Presented on Behalf of the IERS/IVS Working Group. In *(IERS Technical Note 35)*. Frankfurt am Main: Verlag des Bundesamts für Kartographie und Geodäsie.

Gokhberg, M. B., Olshanskaya, E. V., Chkhetiani, O. G., Shalimov, S. L., & Barsukov, O. M. (2016). Correlation between Large Scale Motions in the Liquid Core of the Earth and Geomagnetic Jerks, Earthquakes, and Variations in the Earth's Length of Day. *Doklady Earth Sciences*, **467**, 280–283.

Gordon, D. (2016). Impact of the VLBA on Reference Frames and Earth Orientation Studies. *J. Geodesy*. doi:10.1007/s00190-016–0955–0

Gross, R. S. (2012). Improving UT1 Predictions Using Short-Term Forecasts of Atmospheric, Oceanic, and Hydrologic Angular Momentum. In H. Schuh, S. Boehm, T. Nilsson, & N. Capitaine, eds., *Journées Systèmes de Référence Spatio-temporels 2011*. Vienna: Vienna University of Technology, pp. 117–120.

Gurtner, W., Noomen, R., & Pearlman, M. R. (2005). The International Laser Ranging Service: Current Status and Future Developments. *Advances in Space Research*, **36**, 327–332.

Hide, R., Birch, N. T., Morrison, L. V., Shea, D. J., & White, A. A. (1980). Atmospheric Angular Momentum Fluctuations and Changes in the Length of the Day. *Nature*, **286**, 114–117.

Hide, R., Boggs, D. H., & Dickey, J. O. (2000). Angular Momentum Fluctuations within the Earth's Liquid Core and Torsional Oscillations of the Core–Mantle System. *Geophys. J. Int*. **143**, 777–786. doi:10.1046/j.0956-540X.2000.01283.x

Hide, R. & Dickey, J. O. (1991). Earth's Variable Rotation. *Science* **253**, 629–637. doi:10.1126/science.253.5020.629

Holme, R. T. & de Viron, O. (2013). Probing Geomagnetic Jerks Combining Geomagnetic and Earth Rotation Observations. *American Geophysical Union, Fall Meeting*, #GP52A–01.

Johnston, K. J. (1979). The Application of Radio Interferometric Techniques to the Determination of Earth Rotation. In D. D. McCarthy & J. D. Pilkington, eds., *Time and the Earth's Rotation*. Dordrecht: Reidel, pp. 183–190.

Kehm, A., Blossfeld, M., & Pavlis, E. C. (2016). Future Global SLR Network Evolution and Its Impact on the Terrestrial Reference Frame. *Geophysical Research Abstracts*, **18**, EGU2016–5848.

Kotze, P. B. (2017). The 2014 Geomagnetic Jerk as Observed by Southern African Magnetic Observatories. *Earth, Planets, and Space*, 69(17). doi:10.1186/s40623-017–0605–7.

Koot, L., De Viron, O., & Dehant, V. (2006). Atmospheric Angular Momentum Time-Series: Characterization of Their Internal Noise and Creation of a Combined Series. *J. Geodesy*, **79**, 663–674.

Kouba, J., Beutler, G., & Rothacher, M. (2000). IGS Combined and Contributed Earth Rotation Parameter Solutions. In S. Dick, D. McCarthy, & B. Luzum, eds., *Polar Motion: Historical and Scientific Problems*. ASP Conference Series, Vol. 208, also IAU Colloquium #178. San Francisco, CA: ASP, pp. 277.

Krasna, H., Malkin, Z., & Bohm, J. (2015). Non-Linear VLBI Station Motions and Their Impact on the Celestial Reference Frame and Earth Orientation Parameters. *J. Geodesy*, **89**, 1019–1033.

Lambeck, K. (1980). *The Earth's Variable Rotation*. Cambridge: Cambridge University Press.

Malkin, Z. (2013). Free Core Nutation and Geomagnetic Jerks. *J. Geodynamics*, **72**, 53–58.

Malkin, Z. (2016). Free Core Nutation: New Large Disturbance and Connection Evidence with Geomagnetic Jerks. arXiv:1603.03176v1.

Miyagoshi, T. & Hamano, Y. (2013). Magnetic Field Variation Caused by Rotational Speed Change in a Magnetohydrodynamic Dynamo. *Phys. Rev. Lett.* **111**, 124501. doi:10.1103/Phys Rev Lett.111.124501

Moreaux, G., Lemoine, F. G., Capdeville, H., Kuzin, S., Otten, M., Štěpánek, P., Willis, P., & Ferrage, P. (2016). The International DORIS Service Contribution to the 2014 Realization of the International Terrestrial Reference Frame. *Advances in Space Research*, **58**, 2479–2504.

Nilsson, T., Heinkelmann, R., Karbon, M. R., Raposo-Pulido, V., Soja, B., & Schuh, H. (2014). Earth Orientation Parameters Estimated from VLBI during the CONT11 Campaign. *J. Geodesy*, **88**, 491–502.

Panafidina, N., Kurdubov, S., & Rothacher, M. (2012). Empirical Model of Subdaily Variations in the Earth Rotation from GPS and Its Stability. In H. Schuh, S. Boehm, T. Nilsson, & N. Capitaine, eds., *Journées Systèmes de Référence Spatio-temporels 2011*. Vienna: Vienna University of Technology, pp. 148–151.

Petit, G. P. & Luzum, B. J., eds. (2010). *IERS Conventions (2010)* (IERS Technical Note 36). Frankfurt am Main: Verlag des Bundesamts für Kartographie und Geodäsie.

Petrachenko, W., Behrend, D., Hase, H, Ma, C., Niell, A., Schuh, H., & Whitney, A. (2013). The VLBI2010 Global Observing System (VGOS). *Geophysical Research Abstracts*, 15, EGU2013–12867.

Puica, M., Dehant, V., Folgueira, M., Trinh, A., & van Hoolst, T. (2015). Topographic Coupling at Core–Mantle Boundary in Rotation and Orientation Changes of the Earth. *Geophysical Research Abstracts*, 17, EGU2015-13930–1.

Rothacher, M. (1999). The Contribution of GPS Measurements to Earth Rotation Studies. In *Journées 1998: Systèmes de référence spatio-temporels: Conceptual, Conventional and Practical Studies Related to Earth Rotation*. Paris: Observatoire de Paris, France. Département d'Astronomie Fondamentale, pp. 239–247.

Schlüter, W. & Behrend, D. (2007). The International VLBI Service for Geodesy and Astrometry (IVS): Current Capabilities and Future Prospects. *J. Geodesy*, **81**, 379–387.

Schutz, B. E., Tapley, B. D., Eanes, R. J., & Watkins, M. M. (1989). Earth Rotation from LAGEOS Laser Ranging. *IERS Tech. Notes*, No. 2, 53–57.

Sosnica, K., Thaller, D., Dach, R., Steigenberger, P., Beutler, G., Arnold. D., & Jäggi, A. (2015). Satellite Laser Ranging to GPS and GLONASS. *J. Geodesy*, **89**, 725–743.

Tavernier, G., Fagard, H., Feissel-Vernier, M., Lemoine, F., ... Willis, P. (2005). The International DORIS Service. *Advances in Space Research*, **36**, 333–341.

Vondrak, J. & Ron, C. (2014). Geophysical Excitation of Nutation and Geomagnetic Jerks. *Geophysical Research Abstracts*, **16**, EGU2014–5691.

Vondrak, J. & Ron, C. (2015). Earth Orientation and Its Excitations by Atmosphere, Oceans, and Geomagnetic Jerks. *Serb. Astron. J.*, **191**, 59–66.

Vondrak, J., Ron, C., & Stefka, V. (2009). New Solution of Earth Orientation Parameters in 20th Century. *Highlights of Astronomy*, **15**, XXVIIth IAU General Assembly, I. F. Corbett, editor.

Wielgosz, A., Tercjak, M., & Brzezinski, A. (2016). Testing Impact of the Strategy of VLBI Data Analysis on the Estimation of Earth Orientation Parameters and Station Coordinates. *Reports on Geodesy and Geoinformatics*, **101**, 1–15.

Yoshida, S. & Hamano, Y. (1995). Geomagnetic Decadal Variations Caused by Length-of-Day Variation. *Phys. Earth Planet. Int.* **91**, 117–129. doi:10.1016/0031–9201(95)03038-X

19

International Activities

19.1 Time and International Activities

Because time has become an international standard, a variety of international organizations have evolved to deal with various aspects of time and time-keeping. These range from the political and commercial agencies concerned with issues related to international standards to scientific organizations dealing with subtle aspects of the technical definitions of timescales and organizations that promote the development of even more accurate devices and means for time dissemination.

19.2 Treaty of the Meter

Although the subject of time was not covered originally in the Treaty of the Meter signed on May 20, 1875 (Bureau International des Poids et Mesures, 2006), it has since become part of the mission of the organizational structure put in place by the Treaty. The Treaty of the Meter is also known as the Meter Convention or in French as the Convention du Mètre. Written in the French language, it was signed by 17 countries at the International Metric Convention that was called to organize formally the means to maintain the metric standards. The number of signatories increased to 21 in 1900, 32 in 1950, 44 by 1975, 48 by 1997, and 49 by 2001. As of August 2016, 58 signatories and 41 associate states and economies had signed the Treaty. It was revised in 1921, and the system of units it established was renamed the Système international d'unités (International System of Units) (SI) in 1960. To carry out the intentions of the treaty, three organizations were created: the Conférence Générale des Poids et Mesures (CGPM), the Comité International des Poids et Mesures (CIPM), and the Bureau International des Poids et Mesures (BIPM). The CGPM took responsibility for an international standard time in 1985 (Guinot, 2000).

19.2.1 General Conference on Weights and Measures (CGPM)

Delegates from each of the signatories along with observers from each of the associates comprise the CGPM. It meets every four years at the BIPM, where it receives the official report of the CIPM, discusses possible improvements in the SI units, and endorses new metrological results and international scientific recommendations regarding the fundamental units. It also makes decisions regarding the future direction of the BIPM. For example, the SI was established in 1960 by the 11th CGPM and is modified by the CGPM as required to reflect the latest advances.

19.2.2 International Committee on Weights and Measures (CIPM)

Eighteen individuals, each from a different member state, comprise the CIPM. Its mission is to promote uniformity in the international measurement units principally by submitting draft resolutions to the CGPM for its approval. It discusses the work of the BIPM and issues an annual report on the operations of the BIPM to the governments of the member states. Its members discuss and coordinate current metrological activities and prepare other reports, including the SI Brochure.

The CIPM has created a number of Consultative Committees (in French: Comités Consultatifs) to provide technical information on a wide range of metrological activities. Each committee is composed of technical experts from national metrology institutes, and the chair of each committee usually serves on the CIPM. These committees discuss scientific and technical advances related to metrology and formulate recommendations for the CIPM. They also advise the CIPM on the work of the BIPM. The committees with titles current as of 2017 are:

Consultative Committee for Acoustics, Ultrasound and Vibration (CCAUV),
Consultative Committee for Electricity and Magnetism (CCEM),
Consultative Committee for Length (CCL),
Consultative Committee for Mass and Related Quantities (CCM),
Consultative Committee for Photometry and Radiometry (CCPR),
Consultative Committee for Amount of Substance – Metrology in Chemistry (CCQM),
Consultative Committee for Ionizing Radiation (CCRI),
Consultative Committee for Thermometry (CCT),
Consultative Committee for Time and Frequency (CCTF), and the
Consultative Committee for Units (CCU).

The CIPM meets annually at the BIPM to discuss the reports of the Consultative Committees. The CCU assists in the preparation of the SI Brochure. Suggested modifications of the SI are submitted to the CGPM by the CIPM for formal

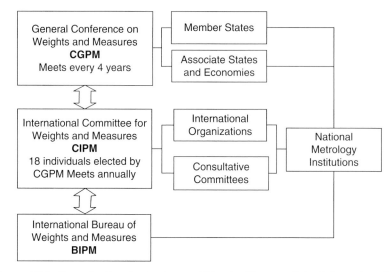

Figure 19.1 Organizational structure within the Treaty of the Meter that deals with time

adoption. On matters relating to interpretation or usage of the SI, the CIPM may also adopt its own resolutions and recommendations.

19.2.3 Bureau International des Poids et Mesures (BIPM)

The third organization created by the Meter Convention is the BIPM, located in Sèvres, a suburb of Paris. Its status is that of an intergovernmental organization financed by the member states of the Meter Convention. Its operations fall under the supervision of the CIPM, and its staff carries out its mission to ensure international unification of physical measurements. It provides the basis for a single, coherent system of measurements traceable to the SI.

The BIPM is currently organized in five technical departments: Physical Metrology, Time, Ionizing Radiation, Chemistry, and Information Technology Services (see Figure 19.1). These carry out a variety of tasks, including maintaining the kilogram, coordinating international measurement standards, and in the case of time, providing the actual SI unit, the second.

19.3 Scientific Unions

International scientific organizations have contributed to accurate timekeeping by promoting investigations of the associated scientific and technical problems. They have made recommendations that have affected modern timekeeping significantly in the past, and it is expected that they will continue to contribute to future developments. Two scientific unions concerned with precise time, the International Astronomical Union (IAU) and the International Union of Geodesy and Geophysics (IUGG), are members of the International Council for Science (ICSU), a nongovernmental organization of 31 international scientific unions and 22 national scientific bodies representing 142 countries. In pursuit of its mission to strengthen international science it seeks to address major scientific issues and facilitate interaction across disciplines and countries. It maintains close working relationships with a number of intergovernmental and nongovernmental organizations, especially the United Nations Educational, Scientific and Cultural Organization (UNESCO) and the Third World Academy of Sciences (TWAS). ICSU holds a General Assembly every three years to set general direction, policies, and priorities. Its funding is mainly through national members and scientific unions along with grants from UNESCO, the United States, and France.

19.3.1 International Astronomical Union (IAU)

The International Astronomical Union was founded in 1919 to promote the science of astronomy through international cooperation. It is made up of national and individual members. National members are organizations that represent national professional astronomical communities within their countries, and individual members are professional scientists whose research relates to astronomy. Individual members are elected by the Union's Executive Committee following the recommendation of a National Member.

The IAU is currently organized into nine divisions. Each division is broken down further into commissions that deal with specific specialized topics. The number of commissions now totals 35. The organization also allows for any number of working groups that can report either to divisions or to commissions. As of 2017, there are 79 national members and more than 12,740 individual members. The Executive Committee sets and implements the overall policy, and the operations are overseen by a set of elected officers. The center for its business activities is the IAU Secretariat, which is hosted by the Institut d'Astrophysique de Paris in France.

In addition to sponsoring a number of symposia each year, the IAU holds a General Assembly every three years. The IAU defines fundamental astronomical and physical constants and astronomical nomenclature. It also promotes educational activities in astronomy and discusses future developments dealing with the science of astronomy. Matters related to the subject of time are discussed in Division A, Fundamental Astronomy, which operates through three commissions within the division: Commission A1, Astrometry; Commission A2, Rotation of the Earth; Commission A3, Fundamental Standards; one Inter-Division Commission

A4: Celestial Mechanics and Dynamical Astronomy; and one Cross-Division Commission: Solar System Ephemerides. The latter two are shared with Division F, Planetary Systems and Bioastronomy.

To undertake well-defined tasks, the IAU has also established Working Groups within Division A. These include Working Groups on (1) the Theory of Earth Rotation and Validation (joint with the International Association of Geodesy), (2) Multi-Waveband Realizations of International Celestial Reference System, (3) Numerical Standards in Fundamental Astronomy (NSFA), (4) Standards of Fundamental Astronomy (SOFA), (5) Third Realization of International Celestial Reference Frame, and (6) Time Metrology Standards. In addition, a joint Working Group with Division F has been established dealing with Cartographic Coordinates & Rotational Elements. The commissions are composed of technical experts dealing with detailed aspects of the commission's tasks. Commissions A2 and A3 are of particular interest for those dealing with time.

The objectives of Commission A2, Rotation of the Earth, are to:

1. Encourage and develop cooperation and collaboration in observation and theoretical studies of Earth orientation variations.
2. Serve the astronomical community by linking it to the official organizations that provide the International Terrestrial and Celestial Reference Systems/Frames and Earth orientation parameters (EOP), namely International Association of Geodesy (IAG), International Earth rotation and Reference systems Service (IERS), International VLBI Service for Geodesy and Astrometry (IVS), International GNSS Service (IGS), International Laser Ranging Service (ILRS), and International DORIS Service (IDS).
3. Develop methods to improve the accuracy and understanding of Earth orientation variations and related reference systems/frames.
4. Ensure agreement and continuity of the reference frames used for studying Earth orientation variations with other astronomical reference frames and their densification.
5. Provide means of comparing observational and analysis methods and results to ensure accuracy of data and models and encourage the development of new observation techniques.

The activities of Commission A3, Fundamental Standards, include facilitating advances in astronomy and other fields in science and engineering, by developing, implementing, and communicating fundamental IAU-endorsed standards for fundamental astronomy. Such standards include, but are not limited to:

1. celestial and terrestrial reference systems/frames and the transformations among them;

2. timescales;
3. precession/nutation models;
4. Earth rotation and polar motion, including physical models (e.g., Earth's gravity field, lunar gravity field, Earth interior model, solid Earth–tide modeling);
5. star catalogs;
6. ephemerides of solar system bodies; and
7. special and general relativistic models for time and space.

The Working Group Time Metrology Standards has the objective of enhancing the interaction between astronomers and the time and frequency metrology community. The time metrology community provides the reference atomic timescale, which is the basis of coordinate times used for space-time referencing in astronomy. The dynamical timescale based on precision timing of pulsars has the potential of improving the long-term standard of time. Potential future developments will require common actions of both astronomers and metrologists. Examples include the possible new definition of the second, a consequence of the advent of optical clocks, and the contribution to the studies on the adoption of a uniform international reference timescale.

19.3.2 International Union of Geodesy and Geophysics (IUGG)

The International Union of Geodesy and Geophysics is a nongovernmental, scientific organization, established in 1919 to promote international coordination of scientific studies of the Earth and its environment in space. These studies include the shape of the Earth, its gravitational and magnetic fields, the dynamics of the Earth as a whole and of its component parts, the Earth's internal structure, composition, and tectonics, the generation of magmas, volcanism, and rock formation, the hydrological cycle including snow and ice, all aspects of the oceans, the atmosphere, ionosphere, magnetosphere and solar–terrestrial relations, and analogous problems associated with the Moon and other planets. It is made up of eight semiautonomous associations, each responsible for specific topics within the Union activities:

• International Association of Cryospheric Sciences (IACS)
• International Association of Geodesy (IAG)
• International Association of Geomagnetism and Aeronomy (IAGA)
• International Association of Hydrological Sciences (IAHS)
• International Association of Meteorology and Atmospheric Sciences (IAMAS)
• International Association for the Physical Sciences of the Ocean (IAPSO)
• International Association of Seismology and Physics of the Earth's Interior (IASPEI)

• International Association of Volcanology and Chemistry of the Earth's Interior (IAVCEI)

These associations can organize individual assemblies in the interim between the IUGG General Assemblies that are held every four years. The IUGG has 59 regular member countries and 10 associate members, most of whom participate in the Union through their national academy or other adhering body.

Most of the activity dealing with timekeeping is carried out through the IAG, which was originally organized as the Mitteleuropäische Gradmessung (Central European Arc Measurement) in 1862 as the first significant international scientific organization. It became the Europäische Gradmessung (European Arc Measurement) in 1867 and in 1886 the Internationale Erdmessung. At the first IUGG General Assembly in 1922, it became one of the sections, and it took its present name at the IUGG General Assembly of 1930.

The official mission of the IAG is the advancement of geodesy. It is concerned with establishment of reference systems, monitoring the gravity field, Earth rotation, and deformation of the Earth's surface, including oceans and ice. It holds symposia and workshops to promote international cooperation and knowledge. Its components include:

• Commission 1: Reference Frames,
• Commission 2: Gravity Field,
• Commission 3: Geodynamics and Earth Rotation,
• Commission 4: Positioning and Applications,
• Inter-Commission Committee on Theory,
• Inter-Commission Committee on Theory,
• Global Geodetic Observing System (GGOS),
• 14 International Scientific Services.

The Global Geodetic Observing System is the observing system of the IAG. It works with the IAG components providing the geodetic infrastructure necessary for monitoring the Earth system and for global change research. To maintain a stable, accurate, and global reference frame, GGOS provides observations of the Earth's shape, gravity field, and rotational motion. It does this by integrating different geodetic techniques, models, and approaches in order to ensure long-term, precise monitoring of geodetic observables.

19.3.3 International Telecommunications Union (ITU)

The International Telecommunications Union is a United Nations organization that deals with information and communications technologies. Based in Geneva,

Switzerland, the ITU has 193 member states and almost 800 sector members and associates. Sector members are recognized operating agencies, scientific or industrial organizations, financial or development institutions, and organizations of an international character representing them. The ITU is comprised of three sectors:

• Radiocommunication (ITU-R)
• Telecommunications Standardization (ITU-T)
• Telecommunication Development (ITU-D)

The ITU-R manages the international radio frequency spectrum and satellite orbit resources and is the primary sector of the ITU dealing with issues of time and frequency. The ITU-T deals with setting technical specifications so that elements of communications systems can interoperate seamlessly, and ITU-D creates policies and regulations and provides training programs and financial strategies in developing countries. The ITU also organizes TELECOM events that bring together leading elements of the information and communication technologies (ICT) as well as ministers and regulators for exhibitions, and high-level forums.

The ITU dates back to the days following the establishment of telegraph networks. To facilitate international communications, countries gradually established regional agreements, and in 1865 the International Telegraph Convention was signed, resulting in the formation of the International Telegraph Union. With the development of the telephone and wireless telegraphy, it was necessary to establish international agreements regarding radiotelegraphy. The first International Radiotelegraph Conference was held in 1906 in Berlin, resulting in the first International Radiotelegraph Convention and a set of regulations. These regulations, which have since been expanded and revised by the following radio conferences, are now known as the Radio Regulations.

Within the ITU, the International Telephone Consultative Committee (CCIF) was established in 1924, followed by the International Telegraph Consultative Committee (CCIT) in 1925, and the International Radio Consultative Committee (CCIR) in 1927. These organizations coordinated the technical studies, tests, and measurements and drew up international standards to ensure international communications. The 1927 International Radiotelegraph Conference was the first to allocate frequency bands to existing radio services, including fixed, maritime and aeronautical mobile, broadcasting, amateur, and experimental.

In 1932, the International Telecommunication Convention was formed by combining the International Telegraph Convention and the International Radiotelegraph Convention. At the same time, the name of the Union was changed to International Telecommunication Union to reflect its expanding scope. In 1947, it became a United Nations specialized agency and in 1956, the CCIT and the CCIF were merged to form the International Telephone and Telegraph Consultative

Committee (CCITT). In 1992, a plenipotentiary conference revised the structure of the ITU into the three sectors that integrated the functions carried out by the CCIR and the CCITT.

Some documents of the ITU have the status of international treaties. These are (1) the Constitution and Convention of the International Telecommunication Union originally signed in 1992 and subsequently amended at following plenipotentiary conferences and (2) the Administrative Regulations, which include the Radio Regulations (www.itu.int/publ/R-REG-RR/en) and the International Telecommunication Regulations (www.itu.int/ITU-T/itr/), which complement the Constitution and the Convention. The first version of what came to be called the Radio Regulations appeared in 1906 and they have been revised frequently since then. Previously adopted Telephone and Telegraph Regulations were merged in 1998 to create the International Telecommunication Regulations, and these were revised in 2012. The Radio Regulations incorporate the decisions of the World Radiocommunication Conferences, including all appendices, resolutions, recommendations, and ITU-R recommendations incorporated by reference. World radiocommunication conferences (WRCs) are held every two to three years to review and, if required, revise the Radio Regulations. The general program of world radiocommunication conferences is established four to six years in advance and the final agenda set by the ITU Council two years before the conference, with the concurrence of a majority of member states.

The Plenipotentiary Conference is held every four years to set the Union's general policies, adopt plans for the future, and elect the management team. At this conference ITU member states decide on the future of the organization and sector members can attend as observers. The ITU Council, in the interval between Plenipotentiary Conferences, deals with broad telecommunication policies and prepares a report on the policy and strategic planning of the ITU. It is responsible for ensuring the smooth operation of the Union and facilitates the implementation of the provisions of the ITU Constitution, the ITU Convention, and the Administrative Regulations.

Within the ITU, matters relating to precise time and its dissemination fall within the tasks of the ITU-R, and within the ITU-R they are part of the agenda of Study Group 7. Study Group 7 is part of the structure of the Study Groups, which includes:

- Study Group 1 (SG 1) – Spectrum management
- Study Group 3 (SG 3) – Radiowave propagation
- Study Group 4 (SG 4) – Satellite services
- Study Group 5 (SG 5) – Terrestrial services
- Study Group 6 (SG 6) – Broadcasting services
- Study Group 7 (SG 7) – Science services

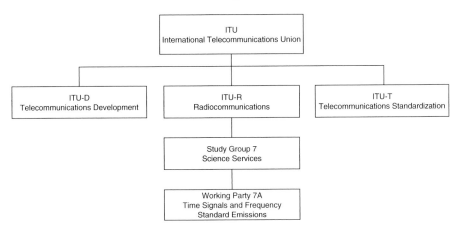

Figure 19.2 Organizational structure within the ITU dealing with time

Within Study Group 7, issues related to precise time fall within the purview of Working Party 7A. Study Group 7 is structured as follows:

- Working Party 7A (WP 7A) – Time signals and frequency standard emissions
- Working Party 7B (WP 7B) – Space radiocommunication applications
- Working Party 7C (WP 7C) – Remote sensing systems
- Working Party 7D (WP 7D) – Radio astronomy

Issues related to precise time that might be expected to be included in the Radio Regulations would then be expected to be brought up first with Working Party 7A for discussion. They then would go to Study Group 7, then to ITU-R, before being accepted at a World Radio Conference (see Figure 19.2).

19.4 Service Organizations

Another set of time-related organizations deals with the coordination and analyses of observations relating to the Earth's rotation. Administratively these organizations fall within the World Data System (ICSU-WDS), an interdisciplinary body of the ICSU created in 2008. Its mission is to promote long-term stewardship of, and access to, quality-assured scientific data and data services, products, and information across disciplines in the natural and social sciences, and the humanities. It aims to facilitate scientific research by coordinating and supporting trusted scientific data services for the provision, use, and preservation of relevant data, and is meant to combine the efforts of two previous organizations, World Data Centers and the Federation of Astronomical and Geophysical Data Analysis Services. The ICSU-WDS has 105 member organizations. Of these, 68 are regular members that are data stewards and/or data analysis services, 11 are network members representing groups of data stewardship organizations and/or data analysis services, 8 are partner members that contribute support or funding, and 18 are associate members organizations that are interested in the endeavor, but do not contribute direct funding or other material support.

19.4.1 *International Earth rotation and Reference systems Service (IERS)*

The IERS was established as the International Earth Rotation Service in 1987 by the International Astronomical Union and the International Union of Geodesy and Geophysics, and it began operation on January 1, 1988. In 2003, it was renamed the International Earth Rotation and Reference Systems Service, but retained the acronym IERS. It is a regular member of the ICSU-WDS whose objectives are to provide the International Celestial Reference System (ICRS) and its realization, the International Celestial Reference Frame (ICRF); the International Terrestrial Reference System (ITRS) and its realization, the International Terrestrial Reference Frame (ITRF); Earth orientation parameters required to transform between the ICRF and the ITRF; geophysical data to interpret and model variations in the ICRF, ITRF, or Earth orientation parameters; and the standards, constants, and models (i.e., conventions) necessary to use the products. Its products include the International Celestial Reference Frame, the International Terrestrial Reference Frame, monthly Earth orientation data, daily rapid service estimates of near real-time Earth orientation data and their predictions, announcements of the differences between astronomical and civil time for time distribution by radio stations, leap second announcements, products related to global geophysical fluids such as mass and angular momentum distribution, an annual report and technical notes on conventions and other topics, and long-term Earth orientation information.

It carries out its objectives with an organization made up of Technique Centers, Product Centers, Combination Centers, an Analysis Coordinator, and a Central Bureau. The Technique Centers are independent service organizations that have made commitments to the IERS to contribute observational material regarding various aspects of the Earth's rotation. They control the organization of their observations, the analyses and archiving of data, and the development of possible improvements either in the technique or in the analyses of their data. The data are delivered without interruption and with minimal delay. Currently, the four Technique Centers are the International GNSS Service (IGS), the International Laser Ranging Service (ILRS), the International VLBI Service (IVS), and the International DORIS Service (IDS). The organization is shown in Figure 19.3 and the details of its individual products are outlined later in this chapter (Dick & Richter, 2004).

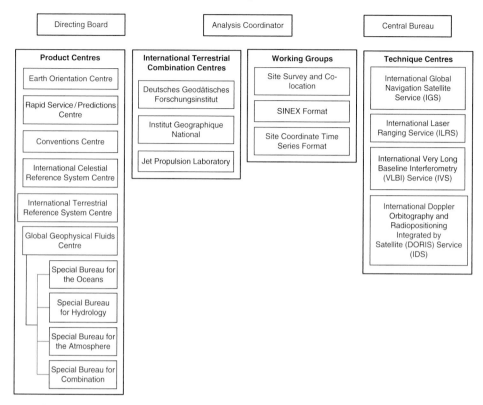

Figure 19.3 The organizational structure of the IERS

The Product Centers provide the actual products of the IERS. These are various self-supported organizations that have committed to provide these products operationally. The Product Centers include the Earth Orientation Center, the Rapid Service/Prediction Center, the Conventions Center, the International Celestial Reference System Center, the International Terrestrial Reference System Center, and the Global Geophysical Fluids Center. Within the latter Product Center, a number of sub-bureaus exist to handle particular aspects of the work. These include the Special Bureau for the Atmosphere, the Special Bureau for the Oceans, the Special Bureau for Tides, the Special Bureau for Hydrology, the Special Bureau for the Mantle, the Special Bureau for the Core, the Special Bureau for Gravity/ Geocenter, and the Special Bureau for Loading.

The Earth Orientation Center is situated at the Bundesamt für Kartographie und Geodäsie in Frankfurt am Main, Germany, and is responsible for monitoring the long-term aspects of the variations of Earth orientation parameters. It publishes monthly bulletins containing Earth orientation data as well as announcements regarding leap seconds and the difference between UT1 and UTC. The IERS Rapid Service/Prediction Center is provided by the US Naval Observatory, and it

Table 19.1 *Special Bureaus of the Global Geophysical Fluids Product Center*

Special Bureau	Location	Mission
Atmosphere	Atmospheric and Environmental Research, Inc.	relevant atmospheric data
Oceans	Jet Propulsion Laboratory	data relating to nontidal changes in oceanic processes
Hydrology	Center for Space Research, University of Texas at Austin	datasets and numerical models related to the changing distribution of water over the planet
Combination	Faculté des Sciences, de la Technologie et de la Communication	combination of geophysical fluid data

provides Earth orientation parameters on a rapid turnaround basis, primarily for real-time users and others needing the highest-quality EOP information sooner than that available in the final series published by the IERS Earth Orientation Center. It also provides forecasts of future variations in the Earth orientation parameters.

The Conventions Center is a joint operation of the Bureau International des Poids et Mesures and the US Naval Observatory. It is responsible for the maintenance of the IERS conventional models, constants, and standards used in the definition and realization of the reference systems. The ICRS Center is responsible for the definition of the ICRS and its realization, the ICRF. This effort is carried out jointly by the Observatoire de Paris and the US Naval Observatory. Similarly, the ITRS Center is responsible for the definition of the ITRS and its realization, the ITRF. This work involves network coordination including collocation, local ties, and site quality, and is carried out by the Institut Géographique National (IGN) in France.

The Global Geophysical Fluids Center, through its four Special Bureaus, supports research in areas related to the variations in Earth rotation, gravitational field, and geocenter that are caused by mass transport in the geophysical fluids. It is housed at the University of Luxembourg. The sub-bureaus and their locations are shown in Table 19.1.

IERS Combination Research Centers develop methods to combine data or products provided by different techniques. The ITRS Combination Centers provide ITRF products to the ITRF Product Center after combining inputs from the Technique Centers. The Analysis Coordinator is responsible for the long-term internal consistency of the IERS products, and the Central Bureau provides the general administration of the IERS consistent with the Directing Board policies. It is the executive arm of the Directing Board, and facilitates communications among the components of the organization.

19.4.2 International VLBI Service for Geodesy and Astrometry (IVS)

The International VLBI Service for Geodesy and Astrometry, one of the services of GGOS and the ICSU-WDS, is an international collaboration of organizations that provides VLBI data to the IERS and other astrogeodetic users (Schlüter & Behrend, 2007; Schuh & Behrend, 2012; Behrend, 2013). The objectives of the IVS are to enhance the individual VLBI programs of the member organizations by creating a joint service that coordinates their activities, to promote research and development of the technique, and to serve as an interface with the users of VLBI data. To meet its objectives, the IVS coordinates observing programs, sets standards for observing stations, establishes VLBI data formats, recommends analysis software, and sets up data delivery processes. It carries out its tasks through an organization that includes a Coordinating Center, network stations, operation centers, correlators, data centers, analysis centers, and technology development centers.

The Coordinating Center coordinates both the day-to-day and the long-term activities of the IVS following the directions established by the Directing Board. The network stations are the global VLBI observing sites with geodetic capability that comply with the IVS performance standards for data quality. The operation centers coordinate routine operations of one or more networks. These activities include planning observing programs, establishing operating plans and procedures for the stations in the network, generating the observing schedules, and posting the observing schedule to an IVS Data Center for distribution and to the Coordinating Center for archiving.

19.4.3 International Laser Ranging Service (ILRS)

The International Laser Ranging Service (Pearlman et al., 2002; Pearlman et al., 2005) is an international collaboration of organizations providing data obtained by laser ranging to artificial Earth satellites or to the Moon. It is one of the services of GGOS and the ICSU-WDS. The ILRS collects, archives, and distributes observational data and provides products, including polar motion and excess length of day, coordinates and velocities of the ILRS observing sites, time-variable coordinates of the geocenter, static and time-variable models of the Earth's gravity field, satellite ephemerides, lunar ephemerides and librations, and fundamental physical constants. An important user of this information is the IERS, where the data are of particular importance for the maintenance of the ITRF.

The ILRS observational data are obtained by its member tracking stations that range to a constellation of approved satellites and the Moon, using the most advanced laser-tracking equipment. They transmit the ranging data at least daily to one or more operations and/or data centers. These collect and merge the data from other stations in sub-networks, perform data quality checks, and reformat the data as necessary. Global data centers are the primary interfaces to the analysis centers and the outside users and they archive and provide online access to tracking data received from the operational/regional data centers. Analysis centers process tracking data from one or more data centers to produce the ILRS products operationally. At a minimum, every analysis center must process the global LAGEOS-1 and LAGEOS-2 datasets and provide Earth orientation parameters on a weekly or sub-weekly basis, as well as other products, such as station coordinates, as required by the IERS. Lunar Analysis Centers process normal point data from the Lunar Laser Ranging (LLR) stations and generate a variety of scientific products, including precise lunar ephemerides, librations, and orientation parameters, which provide insights into the composition and internal makeup of the Moon, its interaction with the Earth, tests of general relativity, and solar system ties to the ICRF.

A central bureau coordinates the activities on a daily basis, facilitating communications and information transfer within the organization and with outside users. It maintains the list of satellites approved for tracking support and their priorities, and carries out the directions of an international governing board composed of representatives of the member organizations.

19.4.4 International GNSS (Global Navigational Satellite Service) Service (IGS)

The International GNSS Service (Kouba et al., 1998; Beutler et al., 1999; Dow et al., 2005; Dow et al., 2009) is an international consortium of organizations that provides GNSS tracking data, orbits, and other data products in near real time. The IGS currently offers these products for two GNSS, the Global Positioning System (GPS) and GLONASS, and expects to provide similar data in the future for other systems as they become available. It is affiliated with the International Association of Geodesy, GGOS, and the ICSU-WDS.

IGS products are critical for the improvement and extension of the ITRF, monitoring of solid Earth deformations, and monitoring of polar motion. The accuracy of the monitoring station horizontal positions is ±3 mm, and station motion accuracy is ±2 mm/year. IGS data are also used to evaluate Earth rotation and variations in sea level, ionospheric monitoring, and measuring precipitable water vapor in the atmosphere. The accuracy of the rapid polar motion data is ±0.4 mas, and excess length of day information is available with an accuracy of ±0.01 ms/day.

The IGS accomplishes its objectives with an international network of more than 500 continuously operating dual-frequency GPS monitoring stations, more than 20 regional and operational data centers, four global data centers, 12 analysis centers, and a number of associate or regional analysis centers. The Central Bureau for the service maintains an Information System that provides access to IGS products. An international Governing Board oversees the services.

The monitoring stations track the satellites with high-accuracy geodetic receivers and send the tracking data to the data centers. They validate and archive the data, making it available to the other elements of the IGS and external users. The analysis centers process the data from the data centers and provide ephemerides, Earth orientation parameters, station coordinates, and clock information. An Analysis Coordinator develops analysis standards and monitors the activities of the analysis centers, providing quality control and performance evaluation. The Analysis Coordinator is also responsible for combining the output of the individual analysis centers into the single set of "official" IGS products that is sent to the global data centers for distribution. The Central Bureau coordinates the activities and operations of the service consistent with the policies set by an international directing board.

19.4.5 International DORIS Service (IDS)

Like the IVS, the ILRS, and the IGS, the International DORIS Service (Tavernier et al., 2006; Willis et al., 2010; Moreaux et al., 2016) provides data from an observational system related to the determination of Earth orientation parameters. DORIS is an acronym for "Doppler Orbitography and Radio-positioning Integrated by Satellite." It is a microwave tracking system used to determine the precise location of the satellites on which it is installed. DORIS systems have been installed on 14 artificial Earth satellites. It measures the Doppler frequency shift of a radio signal transmitted from ground stations and received by the satellite. Within the IERS, the DORIS data are used primarily for the maintenance and extension of the ITRF.

The IDS, which is also a service affiliated with the IAG, the ICSU-WDS, and GGOS, collects, archives, and distributes DORIS observations that are then used to determine coordinates and velocities of the ground stations, the geocenter and scale of the Terrestrial Reference Frame, ionospheric information, high-accuracy ephemerides of DORIS satellites, and Earth orientation parameters. Its organization is much like those of the previously described organizations. The IDS coordinates an observational network made up of more than 50 ground stations along with two data centers and six analysis centers.

Web Sites of International Organizations

Organization	Web Site
BIPM	www.bipm.org
IAG	www.iag-aig.org
IAU	www.iau.org
ICSU	www.icsu.org
IDS	ids-doris.org
IERS	www.iers.org
IGS	www.igs.org
ILRS	ilrs.cddis.eosdis.nasa.gov
ITU	www.itu.int
IUGG	www.iugg.org
IVS	ivscc.gsfc.nasa.gov

References

Behrend, D. (2013). Data Handling within the International VLBI Service. *Data Science Journal*, **12**, WDS81–WDS84, ISSN 1683–1470. doi:10.2481/dsj.WDS–011

Beutler, G., Rothacher, M., Schaer, S., Springer, T. A., Kouba, J., & Neilan, R. E. (1999). The International GPS Service (IGS): An Interdisciplinary Service in Support of Earth Sciences. *Adv. Space Res.* **23**, 631–635.

Bureau International des Poids et Mesures (2006). *The International System of Units (SI)* (8th edn.). Paris: Bureau International des Poids et Mesures.

Dick, W. R. & Richter, B. (2004). The International Earth Rotation and Reference Systems Service (IERS). *Organizations and Strategies in Astronomy*, **5**. André Heck, ed., Strasbourg Astronomical Observatory, France. *Astrophysics and Space Science Library* **310**, Dordrecht: Kluwer Academic Publishers, 159–168.

Dow, J. M., Neilan, R. E., & Gendt, G. (2005). The International GPS Service (IGS): Celebrating the 10th Anniversary and Looking to the Next Decade. *Adv. Space Res.* **36**, 320–326.

Dow, J. M., Neilan, R. E., & Rizos, C. (2009). The International GNSS Service in a Changing Landscape of Global Navigation Satellite Systems. *Journal of Geodesy*, **83**, 191–198. doi:10.1007/s00190-008–0300-3

Guinot, B. (2000). History of the Bureau International de l'Heure. In S. Dick, D. McCarthy, & B. Luzum, eds., *Polar Motion: Historical and Scientific Problems*. ASP Conference Series, **208**. San Francisco, CA: Astronomical Society of the Pacific, pp. 175–184.

Kouba, J., Mireault, Y., Beutler, G., Springer, T., & Gendt, G. (1998). A Discussion of IGS Solutions and Their Impact on Geodetic and Geophysical Applications. *GPS Solutions*, **2**, 3–15.

Moreaux, G., Lemoine, F. G., & Capdeville, H. (2016). The International DORIS Service Contribution to the 2014 Realization of the International Terrestrial Reference Frame. *Advances in Space Research*, **58**(12), 2479–2504.

Pearlman, M. R., Degnan, J. J., & Bosworth, J. M. (2002). The International Laser Ranging Service. *Advances in Space Research*, **30**, 135–143.

Pearlman, M., Noll, C., Gurtner, W., & Noomen, R. (2005). The International Laser Ranging Service and Its Support for IGGOS. In P. Tregoning & C. Rizos, eds., *Dynamic Planet: Monitoring and Understanding a Dynamic Planet with Geodetic and Oceanographic Tools*. Berlin: Springer, p. 741.

Schlüter, W. & Behrend, D. (2007). The International VLBI Service for Geodesy and Astrometry (IVS): Current Capabilities and Future Prospects. *Journal of Geodesy*, **81** (6–8), 379–387.

Schuh, H. & Behrend, D. (2012). VLBI: A Fascinating Technique for Geodesy and Astrometry. *Journal of Geodynamics*, **61**, 68–80. doi:10.1016/j.jog.2012.07.007.

Tavernier, G., Fagard, H., Feissel-Vernier, M., Le Bail, K., Lemoine, F., Noll, C., Noomen, R., Ries, J. C., Soudarin, L., Valette, J. J., & Willis, P. (2006). The International DORIS Service: Genesis and Early Achievements. In DORIS Special Issue, P. Willis ed. *Journal of Geodesy* **80**, 403–417.

Willis, P., Fagard, H., Ferrage, P., Lemoine, F. G. . . . Valette, J. J. (2010). The International DORIS Service (IDS): Toward Maturity. *Advances in Space Research*, **45**(12), 1408–1420.

20

Time Applications

20.1 Time Enables the Infrastructure

The infrastructure of modern society depends critically on time and frequency services. We have grown to expect the universal convenience of time and frequency just as we expect the accessibility of electrical power and water services. However, the users of these services are often not aware of the fact that they play such important roles in their lives. Requirements for time and frequency with widely varying precision and accuracy exist in the areas of utility services, banking and finance, emergency services, communications, navigation, inventory control, environmental services, transportation management, surveying, agriculture, and recreation, as well as in scientific and technical applications.

20.2 Positioning and Navigation Services

The advent of Global Navigation Satellite Systems (GNSS), such as the Global Positioning System (GPS), GLONASS, and GALILEO, has not only provided users an easily accessible source of positioning, navigation, and timing information on a global scale, but it has also imposed new requirements on the providers of time and frequency. Synchronized timing signals are critical to the operation of these systems. (See Chapter 18.) The fact that a timing error of one nanosecond is roughly equivalent to an error of 30 cm, or one foot, in position means that accurate positioning and navigation requires very precise timing information.

The requirements for timing are, in part, driven by the positioning and navigation applications of the GNSS signals. In addition to the military applications for which GPS was originally intended, these applications now include operation of emergency responders, location of transportation services such as trucks and trains, ship navigation, air traffic control, geodetic land surveys, earthquake monitoring, agricultural and fishing applications, and identification and location of inventory items, as well as a host of recreational uses. Land surveying applications and optimization of agricultural operations are further examples of specific applications of improved positioning made possible through precise timing, and some intelligent transportation systems require navigation accuracies of a few centimeters. See Section 20.3.

Navigation with systems other than GNSS that depend on the timing of electronic signals also requires a level of precision in timing synchronization that is compatible with the system's expected positional accuracy. As an example, LORAN (**LO**ng **R**ange **A**id to **N**avigation) is a hyperbolic navigation system that provides multiple synchronized low-frequency transmitters so that the user equipment can determine the difference in the time of arrival from pairs of transmitters to determine a position. e-LORAN (enhanced LORAN) can be expected to need timing at the level of ±5 to a few tens of nanoseconds to meet their navigational objectives. The status of e-LORAN, as a backup to GPS, is under consideration in a number of countries.

20.3 Time Domain Astronomy (TDA)

Time Domain Astronomy is the study of astronomical objects and phenomena that change with time, such as asteroids, comets, variable stars, quasi-stellar objects, eclipses, planetary transits, gravitational lensing, and such. Advances in observing, data storage, and processing technologies have made possible astronomical surveys of sky regions, e.g., MACHO (Alcock et al., 2000), EROS (Rahal et al., 2009), OGLE (Udalski, Kubiak, & Szymanski, 1997), Pan-STARRS (Kaiser et al., 2002), and CRTS (Larson et al., 2003). New telescopes are being planned with the Large Synoptic Survey Telescope (LSST) (Ivezic et al., 2011) expected to generate a 150 Petabyte (150×2^{50} bytes) database, and a 40 Petabyte catalog of 50 billion astronomical objects during 10 years. LSST could issue 2 million alerts of transient events nightly. These events are supernovae, planetary occultations, near-Earth objects, gamma-ray bursts, stellar flares, comets, gravitational microlensing events, and supermassive black holes flares, for example (Tyson & Borne, 2012).

This flow of data will require real-time mining of approximately 2 terabytes per hour, real-time classification of 50 billion objects, and analysis, evaluation, and knowledge extraction of 2 million events per night (Borne, 2013). This will require accurate timing of events and processing the observations with appropriate timescales.

20.4 Intelligent Transportation Systems

Intelligent transportation systems include self-driving cars, connected vehicles, driving assistance and warning systems, car navigation, emergency vehicle notification systems, automatic road enforcement, variable speed limits, collision avoidance systems, dynamic traffic light sequence, traffic signal control systems, container management systems, variable message signs, automatic number plate recognition, speed cameras, parking guidance and information systems, weather information, bridge de-icing systems, and others. These systems require time, frequency, and navigation resources. It is likely that such systems will impose requirements for navigational errors less than a few centimeters, implying timing precision needs better than a 10th of a nanosecond. History of intelligent transportation systems is given at www.its.dot.gov/index.htm. For example, self-driving cars are being developed by Waymo (Google), Apple, Tesla, Volvo, Ford, General Motors, Fiat, Toyota, and others.

20.5 Communications

Precise timing is perhaps most critical in communications applications. Common clocks are used, for example, in switching voice and data traffic through the telephone networks. Precise timing is also needed to time stamp information packets and to allow several transmitters to send information simultaneously over a single communication channel. This process, called "multiplexing," can be accomplished by means of various techniques. All of them require some form of precise timing or frequency. In this application, it is often the case that, while precision may be critical, accuracy may not be required. This is because a communications network may be operated with internal standards for time and frequency, and accuracy is only needed if different networks are required to interoperate with others. At that point, they may require an external standard for time and frequency.

Time-division-multiplexing (TDM) is a process in which multiple bit streams of information are sent as sub-channels on one communication channel at what appears to be at the same time. In reality, the streams are taking turns in time. Fixed time intervals are created so that, for example, a block of data from sub-channel 1 is sent during time interval 1, a block from sub-channel 2 is sent during time interval 2, etc. The process continues until a frame, which consists of one time interval for each sub-channel, is completed. At that point, a new cycle begins with the next block of information from each sub-channel being transmitted. TDM, in the case where several stations are connected to the same physical medium such as the same frequency channel, is called time division multiple access (TDMA). This process is often used in cellular phone networks, for example. Clearly the time slots must be synchronized at each end of the transmit–receive network to enable the process to work.

Frequency division multiple access (FDMA) divides the radio spectrum into a series of individual frequency bands allowing each user access to an allocated frequency band without interfering with other users. Code division multiple access (CDMA) uses spread-spectrum technology and a coding scheme to allow a number of users to share a frequency band. Spectrum management is critical and that capability depends on accurate frequency standards being available along the communications networks. Each transmitter has an assigned code that is known to the user. The spread-spectrum technique spreads the signal in a particular bandwidth over a frequency domain. To reconstruct the signal at the receiver, the transmitter and receiver must use the same coding scheme to spread the signal over the frequency domain.

The American National Standards Institute (ANSI) describes telecommunications timing requirements by means of four "stratum" levels. According to the ANSI standard "Synchronization Interface Standards for Digital Networks" (ANSI/T1.101–1987), a Stratum 1 timing source is an autonomous source of timing requiring no other input than possibly a yearly calibration. It must perform with a maximum drift rate, defined in terms of fractional frequency $\Delta f/f$, better than $\pm 1 \times 10^{-11}$ over a year. A properly calibrated Stratum 1 source, then, is capable of providing bit-stream timing that will not slip relative to a perfect standard more than once every four to five months. When a frame slip does occur in network voice equipment, it generally reacquires frame synchronization quickly, resulting in an audible pop or click, but data circuits lose a number of bits depending on the data rate being transmitted, and on whether or not an error correction protocol is being used. A typical Stratum 1 timing source is often an atomic standard (caesium beam or hydrogen maser) or an oven-controlled quartz crystal oscillator, even though atomic standards are capable of better accuracy than these specifications. A properly calibrated clock system controlled by means of a GPS timing signal may also be considered for use as a Stratum 1 timing source. In telecommunications networks, Stratum 1 sources are considered the primary sources of time and frequency for networks.

Stratum 2 sources must be capable of being adjusted to match the frequency of the Stratum 1 source within a range of $\pm 1.6 \times 10^{-8}$ in one year and drift no more than $\pm 1 \times 10^{-10}$ in a day. This will provide a frame slip rate of about one slip in seven days when the Stratum 2 source is not being updated by a Stratum 1 source. Typical Stratum 2 sources are rubidium standards and oven-controlled quartz oscillators. Table 20.1 compares all of the stratum level definitions.

Stratum 3E was developed after Stratum 3 was standardized in order to meet SONET (synchronous optical networking) equipment requirements. SONET is a protocol that is used to transfer multiple digital bit streams

Table 20.1 *Standards for stratum levels*

Stratum	Adjustment Accuracy	Maximum Drift Rate	Time between Frame Slips
1	1×10^{-11}	—	72 days
2	1.6×10^{-8}	1×10^{-10}/day	7 days
3E	4.6×10^{-6}	1×10^{-8}/day	3.5 hours
3	4.6×10^{-6}	3.7×10^{-7}/day	6 minutes
4	32×10^{-6}	32×10^{-6}/day	—

over an optical fiber using lasers or light-emitting diodes (LEDs). Stratum 4 sources are designed to track Stratum 2 or 3 sources and have no holdover capability. They are not recommended as a timing source for any other system outside the network.

In Network Time Protocol (NTP) applications, the term "stratum level" is used differently. For these purposes, stratum level refers to the distance of a network server from a reference clock. Thus, a Stratum 0 source is assumed to be an accurate source of time and frequency with minimal delays that cannot be used on the network. They can be connected to computers and these act as Stratum 1 servers. A Stratum 1 time server must be directly linked to a reliable source of UTC time, not over a network path, and acts as a primary network time standard. A Stratum 2 server gets its time over a network link, using NTP, from a Stratum 1 server, and a Stratum 3 server gets its time over a network link, using NTP, from a Stratum 2 server, etc.

20.6 Power Grid

Precise timing becomes a concern for the distribution of electrical power in more than one specific application. The efficiency of transmitting power throughout grids depends on precise matching of the phases of the alternating electrical current. This requires an accurate frequency reference across the network of power grids. Typically, this phase information is obtained by individual power companies by maintaining local sources of time that are referred to a standard timing signal. GPS supplies a cheap, easy-to-use timing signal that provides time with an accuracy orders of magnitude better than that necessary for phase matching. For example, a ±one-degree phase difference in a 60 Hz oscillation corresponds to a ±46 μs time difference, so maintaining phase matching at the level of ±0.1° would require timing at the level of ±5 μs. In general, adequate capability can be maintained using rubidium standards and quartz oscillators that are steered to a GNSS timing signal. Temporary loss of that signal could be mitigated by the holdover capacity of the time and frequency standards and their backups.

A second application of precise timing in electrical power applications is in locating faults or breaks in the grid. A ± 1 μs timing capability is adequate to meet these applications. Other applications of timing information deal with billing and secondary aspects of power distribution. These requirements lie in the range of ± 0.1 s to ± 1 ms.

20.7 Banking and Finance

Timing information with relatively low accuracy is used by banking and finance institutions to time stamp financial transactions. Generally, an accuracy of only three seconds is needed for time stamping transactions. For this purpose, a timing system with accuracy of ± 0.1 s is more than adequate and easily achievable by a number of means. Typically timing networks are used that involve rubidium or quartz clocks that are steered using NTP or GPS. These clocks would be capable of relatively long periods without being steered in the event of the loss of an outside source of timing information.

20.8 Emergency Services

Timing requirements for time stamping standard emergency service operations is about the same as that for the banking and finance sector. Emergency service centers generally rely on radio services and NTP as sources for accurate time, with backup systems of rubidium or quartz clocks. Enhanced emergency services locate callers through the GNSS services. This capability delivers fast responses and good geolocation of required services. It is an application of existing GNSS positioning, and, as such, the timing requirements are equivalent to those discussed in Section 20.2.

20.9 Water Flow

Precise timing information is even used for time stamping by water utilities to control flow in and out of local systems through remotely controlled valves and to control the chemical treatment of water and sewage. As with banking and finance applications, the accuracy required is relatively low and easily achievable using widely available sources of time and low precision clocks.

20.10 Scientific

There are a wide range of scientific applications of time and frequency in many fields. In astronomy, time is critical for observations and theories of the motion of objects. Pulsar observations and analysis requires precise time and ephemerides. Determination of Earth rotation, kinematics, and geodesy depends on precisely timed observations. Planetary sciences and ephemerides computations require the family of relativistic dynamical timescales. Measurements in physics depend on accurate time and the relativistic relations between timescales. Precise frequencies are required for many scientific purposes.

20.11 Religions

The religious uses of time do not require great accuracy and are basically tied to the motion of the Sun and Moon. Jewish Sabbath and holidays begin at sunset, which is location dependent and the times are affected by the weather and terrain. Islamic prayer times are determined by the altitude of the Sun and the lunar months are determined by the observations of the crescent Moon after new moon, which can be predicted computationally. Both are location dependent.

20.12 General Public

The general public obtains time from many sources and at various accuracies. Many people get the time from their smartphones or from their computers. Radio signals can automatically maintain the time on clocks and wristwatches. Communication systems, such as radio and television, provide the time.

20.13 Summary

Table 20.2 provides an outline of the applications of precise time and frequency along with representative one-sigma accuracy and precision needs. Some need only precision, but others require accuracy or the ability to be traceable to a conventional standard. The table shows that many users of precise time and frequency information do not require the ultimate in either quantity, and that there is a very large range in the needs of users. Consequently, the relative cost of the apparatus is likely to drive the choices of equipment used in the various applications.

Table 20.2 *Representative estimates of time and frequency needs*

Community	Application	Purpose	Accuracy		Precision	
			Time	Frequency	Time	Frequency
Positioning and Navigation	Aviation	Fuel management, traffic spacing, course navigation	±3 ns	±3.5 × 10^{-14}		
	Space	Artificial satellite location, surveillance	±25 ns	±3.0 × 10^{-13}		
	Maritime	Fuel management, routing, cargo location	±25 ns	±3.0 × 10^{-13}		
	Transportation	Fuel management, real-time routing, cargo location, intelligent transportation system	±1 ns	±1.0 × 10^{-14}		
	Agriculture	Field management, fertilizer optimization, livestock tracking	±10 ns	±1.0 × 10^{-13}		
	Railroad	Asset location, real-time routing	±10 ns	±1.0 × 10^{-13}		
	Automotive	Intelligent highway system	±1 ns	±1.0 × 10^{-14}		
	Recreation	Small boat navigation, hiking, bicycle touring	±25 ns	±3.0 × 10^{-13}		
Communications	Voice	Cellular phones			±1.0 ns	±1.0 × 10^{-14}
	Data	Data transmission, secure communications			±1.0 ns	±1 × 10^{-11}
Survey and Mapping	Geographic information systems	Asset location	±10 ns	±1.0 × 10^{-13}		
	Datum management	Land management	±25 ns	±3.0 × 10^{-13}		

Table 20.2 (cont.)

Community	Application	Purpose	Accuracy		Precision	
			Time	Frequency	Time	Frequency
Energy	Power grid	Phase matching	±50 μs	±6.0 × 10^{-10}		
	Oil and gas location	Exploration	±25 ns	±3.0 × 10^{-13}		
	Power line management	Fault location			±10 ns	±1.0 × 10^{-13}
Banking and Finance	Asset management	Time stamping	±0.1 s	±1.0 × 10^{-6}		
Emergency Services	Fire, police, medical services	Search and rescue, vehicle location	±10 ns	±1.0 × 10^{-13}		
Water	System operation	Flow management			±0.1 s	±1.0 × 10^{-6}
Environmental	Resource management	Hazardous waste containment	±25 ns	±3.0 × 10^{-13}		
Scientific	Geodesy	Plate tectonics, ocean level	±1.0 ns	±1.0 × 10^{-14}		
	Astronomy	Pulsar investigations	±1.0 ns	±1.0 × 10^{-14}		
	Physics	Measurement precision	±0.01 ns	±1.0 × 10^{-16}		

References

Alcock, C., Allsman, R. A., Alves, D. R., Axelrod, T. S., Becker, A. C., Bennett, D. P., Cook, K. H., Dalal, N., Drake, A. J., Freeman, K. C., Geha, M., Griest, K., Lehner, M. J., Marshall, S. L., Minniti, D., Nelson, C. A., Peterson, B. A., Popowski, P., Pratt, M. R., Quinn, P. J., Stubbs, C. W., Sutherland, W., Tomaney, A. B., Vandehei, T., & Welch, D. (2000). The MACHO Project: Microlensing Results from 5.7 Years of LMC Observations. *Astrophys. Journal*, 542, 281–307.

Borne, K. (2013). Virtual Observatories, Data Mining and Astroinformatics. In T. Oswalt & H. Bond, eds., *Planets, Stars and Stellar Systems. Astronomical Techniques, Software, and Data, vol. 2*. Hoboken, NJ: Wiley, pp. 404–443.

Ivezic, Z. et al. (2011). LSST: From Science Drivers to Reference Design and Anticipated Data Products, *ArXiv e-prints*, 0805.2366, www.lsst.org/lsst/overview/.

Kaiser, N., Aussel, H., Burke, B. E., Boesgaard, H., Chambers, K., Chun, M. R., Heasley, J. N., Hodapp, K.-W., Hunt, B., Jedicke, R., Jewitt, D., Kudritzki, R., Luppino, G. A., Maberry, M., Magnier, E., Monet, D. G., Onaka, P. M., Pickles, A. J., Rhoads, P. H. H., Simon, T., Szalay, A., Szapudi, I., Tholen, D. J., Tonry, J. L., Waterson, M., & Wick, J. (2002). Pan-STARRS: A Large Synoptic Survey Telescope Array. *Society of Photo-Optical Instrumentation Engineers (SPIE) Conference Series*, 4836, 154–164.

Larson, S., et al. (2003). The CSS and SSS NEO Surveys. *AAS/Division for Planetary Sciences Meeting Abstracts #35, Bulletin of the AAS*, 982.

LSST Science Collaborations and LSST Project 2009 (2013). *LSST Science Book*, Version 2.0, arXiv:0912.0201, www.lst.org/lsst/scibook.

Rahal, Y. R., Afonso, C., Albert, J.-N., Andersen, J., Ansari, R., et al. (2009). The EROS-2 Search for Microlensing Events towards the Spiral Arms: The Complete Seven Season Results. *Astron. Astrophys.*, 500, 1027–1044.

"Synchronization Interface Standards for Digital Networks" (ANSI/T1.101–1987), American National Standards Institute, New York.

Tyson, J. & Borne, K. (2012). Future Sky Surveys, New Discovery Frontiers. In M. Way, J. D. Scargle, K. Ali, & A. Srivastava, eds., *Advances in Machine Learning and Data Mining for Astronomy*. Boca Raton, FL: CRC Press, pp. 161–181.

Udalski, A., Kubiak, M., & Szymanski, M. (1997). Optical Gravitational Lensing Experiment. OGLE-II: The Second Phase of the OGLE Project. *Acta Astronomica*, 47, 319–344.

York, D. G., et al. (2000). The Sloan Digital Sky Survey: Technical Summary. *Astron. J.*, 120, 1579.

21

Future of Timekeeping

21.1 Future Needs for Time

The demand for improvements in the precision and accuracy of time and frequency will drive the developments in the field for years to come. The applications on the horizon that would make use of this improved capability are as varied as the current applications.

Efficient means to improve the transportation methods of the future are just one of those applications. More accurate positioning of aircraft made possible through improved synchronization of future timing standards will enable more efficient use of air space. This might mean that, not only could aircraft be spaced more efficiently, but air traffic could make use of more direct flight paths, rather than being constrained. Methods of managing ground transportation are being developed. Self-driving cars and vehicle-to-vehicle communication systems are being developed and tested. This could mean that at some future time drivers could take advantage of a managed network of roads controlled through precise navigation of vehicles. The efficiency of such a system will depend on precise positioning, again made possible by synchronized timing signals.

Currently, the inability to navigate indoors, or in areas where Global Navigation Satellite System (GNSS) signals are too weak to be effective, limits the effectiveness of responses to emergency situations. It is likely that building planners of the future will consider indoor navigation capabilities, enabled by precise time signals within buildings. GNSS signals currently aid in locating emergency situations using existing GNSS timing signals. These types of services are likely to be improved in the future as timing capabilities improve.

The management of resources currently takes advantage of accurate location of inventory items. Identification of items and their whereabouts is an important tool available to managers. The usefulness of such identification and location systems depends directly on the resolution precision of those systems. This capability depends on precise positioning made possible by timing capabilities and can be expected to be expanded and improved by increased accuracies and signal availability.

The communications facilities of the future may be direct beneficiaries of improved time and frequency. Currently systems are operating at the edge of our capability to provide sufficient bandwidth. More precise spectrum allocation, made possible by improved frequency standards, will enable a greater throughput of data in the future.

The scientific community will also benefit significantly with enhanced time and frequency capability. Measurements depending on accurate time or precise frequency resolution will be instrumental in future developments. This should benefit the fields of geodesy, physics, astronomy, space sciences, Earth dynamics, and optical measurements, to mention just a few fields affected by the accuracies of time and frequency.

21.2 Modeling the Earth's Rotation

Future improvements in Earth rotation modeling are likely to depend on both improved observations and improvements in understanding the physics of the Earth. In the area of astronomical observations of the Earth's orientation, we can expect observations to be made more accurately and with higher time resolution, so that observations of the Earth's rotation vector in space and in the terrestrial frame could be available routinely with sub-daily frequency. We can also expect that observations of the phenomena known to affect the rotation vector will also be made available with higher resolution in both time and space. For example, observations of winds and ocean currents are likely to be available from observing platforms in space and from a dense grid of terrestrial observing locations. This information will conceivably lead to estimates of atmospheric and ocean angular momentum with hourly frequency.

Improved modeling of lower frequency variations in the Earth's rotation will depend on improved understanding of the physical processes of the Earth's interior. Advances in the measurement of the gravity and magnetic fields could be helpful in this endeavor. The understanding of the variations in these fields, both in time and in space, is likely to improve knowledge of the response of the Earth's rotation vector to the physical phenomena occurring within the Earth. This knowledge may lead to understanding of the drivers of polar motion and the forces changing the Earth rotation.

Using these observations, together with improved capacity to analyze the data quickly, we might expect true real-time estimates of the Earth's rotation angle in space as well as nearly real-time values of the variations in the Earth's rotational speed. We may even get to the point that we could predict more reliably the Earth's rotation for months in advance, but that capability will continue to be a challenge for the future.

21.3 Clocks of the Future

Future clocks will be available to cover a wide range of users' needs. Chip-scale atomic clocks are becoming readily available now, and we can imagine a future when individuals interested in extremely precise time could be wearing watches that offer time based on atomic energy-level transitions. Assuming that such devices can be made sufficiently robust, these clocks are likely to be adapted to a number of applications depending on user requirements.

Optical clock technology is being developed and provides highly precise laboratory standards. Developments of spectroscopic techniques are making available extremely high-Q transitions for laboratory standards. Fountains using alkali atoms may gradually replace the caesium beam oscillators of today.

The necessity for development of improved devices will come from the need for improvements in positioning and navigation, as well as the growing requirements for the transfer of information. Expectations for improved communications capabilities continue to grow. These expected improvements include higher data rates and more efficient use of the radiofrequency spectrum in order to provide greater information throughput. All of these applications are likely to take advantage of improvements in frequency standards.

The use of pulsars to provide long-term frequency stability to current time-scales has led to an ensemble pulsar-based timescale that can be expected to improve in accuracy. The investigation of the physics of pulsars will lead to a better understanding of their rotational speeds and improved accuracies. The pulsar-based timescale will be an independent source of time, not subject to solar system gravitational effects and, thus, able to identify possible systematic variations in TAI.

Limited data are available for forming a timescale based on white dwarf rotations, but it may be possible to identify additional white dwarfs with accurate rotational rates that could form an ensemble white dwarf timescale (Kepler et al., 2000; Kepler, 2012).

21.4 Future Timescales

Timescales can be expected to keep pace with the development of the future time and frequency standards. This will not happen without development of statistical models to evaluate the performance of contributing clocks and the development of robust algorithms to combine the best features of the contributors to the timescale. The accuracy of clock comparisons will need to be developed to make this process viable, and for this purpose, improved calibration techniques will need to be developed.

The future of leap seconds, as a means to reconcile the use of atomic time with observations of the Earth's rotation, will not be resolved at least until 2023. The GNSS currently makes available a timescale free of unpredictable one-second adjustments. The future will decide if a leap-second-less timescale will be in the realm of the everyday user, or if the users of time and frequency will prefer two timescales: one for "everyday" use related to the Earth's rotation and another for use in precise time applications without leap seconds. By 2023, a more precise timescale, a redefinition of the second, and applications requiring more precision and accuracy may lead to a redefinition of timescales. In any event we can expect that the future timescales will continue to be based on atomic energy-level transitions.

A need for more accurate and rapid information that is readily available concerning the Earth's rotation, as given now by UT1, can be expected. Improved observations and distribution methods can be expected to make more accurate Earth rotation data available.

The SI second of today's timescales is defined by the hyperfine transition of the caesium atom. As new time and frequency standards are being developed that take advantage of higher-Q spectral lines, the redefinition of the second, its basis, and when it is introduced, will be considered. We would expect that the second will be redefined in some way so as to be compatible with the past and with the precision needed for future applications. Many scientific and technological standards today (e.g., the meter) are based on the current definition of the second, so it is difficult to imagine a revised definition of the second that would not be compatible with the past (Arias, 2009, 2017). In the future, we might expect observed timescales, such as TAI, pulsar, and maybe white dwarf-based timescales. We expect operationally standard timescales, such as the atomic timescale, UTC, or a more accurate version thereof, and the Earth rotation scale, UT1. We can also expect that the timescales implicit in the global navigation satellite systems of the future will play a large role in everyday usage. We hope that these scales might be part of the international standards.

The need for accurate, realizable timescales for use in space can be expected. We have defined such timescales, e.g., TT, TCB, TCG, and TDB, for use in dynamical applications in the solar system. With the expected improvements in accuracies, the family of dynamical timescales will need to have better definitions based on higher orders of relativity and the post-post-Newtonian parameters. While these scales are necessary for the development of ephemerides and reduction of planetary and lunar observations, observations made during space missions are made using UTC. We anticipate the need for relativistically correct timescales, easily comparable to clocks on the Earth, for future planetary expeditions. The timing accuracy required for navigation and communications for missions at Mars indicates the need for an accurate relation between UTC and a Martian proper time. Missions to other planets will require accurate models of transformations among the proper times for those planets.

21.5 Future Time Distribution

With all of the developments likely to occur as outlined earlier, questions remain about how we might expect to get time to the user of the future. There is no reason to expect that the future satellite navigation systems will not continue to provide precise time to the vast majority of precise time users. We might also expect that time provided by those systems will be limited only by the ability to calibrate the user's receiver. Those needing time and frequency with qualities similar to those provided by laboratory standards are likely to make use of optical fibers. This technology is currently being used, but requires dedicated fibers for precise time transfer. It is likely to be extended in the future, and we might imagine a future when the use of optical fiber for time and frequency applications is ubiquitous (Kodet et al., 2016).

An intriguing possibility that has been discussed is that of quantum entanglement. This is a theoretical phenomenon in which objects that are separated in space have quantum states that are linked in such a manner that, when one changes state, the other must also change. Should this be true, it would have obvious implications for instantaneous time synchronization. However, many have argued that information cannot be transferred faster than the speed of light, so that this technique would not be a viable means to transfer time (Coecke, 2003).

Another possibility more likely for development in the future is that of a distributed clock, in which the clock that is to be used as the conventional standard will actually be made up of a large number of interrelated clocks both on the Earth's surface and in space. The relationship between these clocks might be known through optical fiber connections or laser ranging connections. If the potential bias and rate of any clock in this distributed system is known nearly instantaneously, then that clock provides a nearly instantaneous traceability to the conventional standard. Such a system will depend critically on our future ability to calibrate the time comparison network of the future (Tanenbaum & van Steen, 2007).

References

Arias, E. F. (2009). Current and Future Realizations of Coordinate Time Scales. In A. Klioner, P. K. Seidelmann, & M. H. Soffel, eds., *Relativity in Fundamental Astronomy.* Proceedings of IAU Symposium 261. Paris: International Astronomical Union.

Arias, E. F. (2017). New Technologies and the Future of Time Keeping. In E. F. Arias, L. Combrink, P. Gabor, C. Hohenkerk, & P. K. Seidelmann, eds., *Science of Time 2016.* Springer.

Coecke, B. (2003). The Logic of Entanglement. *Research Report PRG-RR-03–12*, 2003. arXiv:quant-ph/0402014.

Kepler, S. O. (2012). White Dwarf Stars: Pulsations and Magnetism. In H. Shibahashi, M. Takata, & A. E. Lynas-Gray, eds., 61st Fujihara Seminar: Progress in Solar/Stellar Physics with Helio- and Asteroseismology. ASP Conference Series, 462. San Francisco, CA: Astronomical Society of the Pacific, pp. 322–325.

Kepler, S. O., Mukadam, A., Winget, D. E., & Bradley, P. A. (2000). Evolutionary Timescale of the Pulsating White Dwarf G117-b15a: The Most Stable Optical Clock Known. *Astron. J.*, **534**, L185–L188.

Kodet, J., Pánek, P., & Procházka, I. (2016). Accuracy of Two-Way Time Transfer via a Single Coaxial Cable. *Metrologia*, **50**, 18–26.

Tanenbaum, A. S. & van Steen, M. (2007). *Distributed Systems: Principles and Paradigms.* Upper Saddle River, NJ: Prentice-Hall.

Acronyms

See Table 14.1 for acronyms for institutions contributing to TAI

2MASS	2 Micron Astronomical Sky Survey
α	right ascension
A1	USNO atomic timescale
A3	BIH atomic timescale
AAM	atmospheric angular momentum
ACES	Atomic Clock Ensemble in Space
AD	Anno Domini(Common Era)
AGU	American Geophysical Union
AM	former BIH timescale
ANSI	American National Standards Institute
au	astronomical unit
BC	Before Christ (Before Common Era)
BCE	Before Common Era
BCRS	Barycentric Celestial Reference System
BDS	BeiDou Navigation Satellite System
BGI	Bureau Gravimétrique International
BIH	Bureau Internationale de l'Heure
BIPM	Bureau International des Poids et Mesures
c	speed of light
CCAUV	Consultative Committee for Acoustics, Ultrasound and Vibration
CCD	charge coupled device
CCDS	Comite Consultatif pour la Définition de la Seconde
CCEM	Consultative Committee for Electricity and Magnetism
CCIF	International Telephone Consultative Committee
CCIR	International Radio Consultative Committee
CCIT	International Telegraph Consultative Committee
CCITT	International Telephone and Telegraph Consultative Committee
CCL	Consultative Committee for Length

(cont.)

CCM	Consultative Committee for Mass and Related Quantities
CCPR	Consultative Committee for Photometry and Radiometry
CCQM	Consultative Committee for Amount of Substance – Metrology in Chemistry
CCRI	Consultative Committee for Ionizing Radiation
CCT	Consultative Committee for Thermometry
CCTF	Consultative Committee for Time and Frequency
CCU	Consultative Committee for Units
CDMA	code division multiple access
CDS	Centre des Données astronomiques de Strasbourg
CD-TWSTFT	Carrier Doppler TWSTFT
CE	Common Era (Anno Domini)
CEO	Celestial Ephemeris Origin
CEP	Celestial Ephemeris Pole
CGPM	Conférence Général des Poids et Mesures
CIF	Celestial Intermediate Frame
CIO	Celestial Intermediate Origin
CIP	Celestial Intermediate Pole
CIPM	Comite International des Poids et Mesures
CIRF	Celestial Intermediate (or True) Reference Frame
CIRS	Celestial Intermediate Reference System
CMB	cosmic microwave background
CMB	core–mantle boundary
CNES	Centre National d'Etudes Spatiales
CODE	Center for Orbit Determination in Europe
cpsd	cycles per sidereal day
CRS	Celestial Reference System
Cs	Caesium
CSNNPC	China Satellite Navigation Project Center
CTRF	Conventional Terrestrial Reference Frame
CTRS	Conventional Terrestrial Reference System
δ	declination
DE	Development Ephemeris
DGFI-TUM	Deutsches Geodatisches Forschungsinstitut der Technischen Universitat Munchen
DMA	Defense Mapping Agency (then NIMA, now NGA)
DORIS	Doppler Orbit Determination and Radiopositioning Integrated on Satellite
DoV	Deflection of the Vertical
DST	Daylight Savings Time
ΔT	TT – UT1

(*cont.*)

DUT1	low-precision prediction of UT1–UTC
DWDM	Dense Wavelength Division Multiplexing
EAL	Echelle Atomique Libre (Free Atomic Timescale)
EDT	Eastern Daylight Time
eLORAN	enhanced LORAN
EOP	Earth orientation parameters
EPS	Ensemble Pulsar Scale
ERA	Earth rotation angle
EROS	Earth Resources Observation Systems
ESA	European Space Agency
EST	Eastern Standard Time
ET	Ephemeris Time
ETR	Ephemeris Time Revised
FAGS	Federation of Astronomical and Geophysical Data Analysis Services
FCN	Free Core Nutation
FDMA	frequency division multiple access
FK3	Third Fundamental Katalog
FK4	Fourth Fundamental Katalog
FK5	Fifth Fundamental Katalog
FLRW	Friedman-Lemaître-Robertson-Walker
G	billion
GA	Greenwich atomic timescale
GAST	Greenwich Apparent Sidereal Angle
GCRF	Geocentric Celestial Reference Frame
GCRS	Geocentric Celestial Reference System
GCT	Greenwich Civil Time
GEM	Goddard Earth Model
GGOS	Global Geodetic Observing System
GHz	gigahertz
GHA	Greenwich Hour Angle
GIA	Glacial Isostatic Adjustment
GMAT	Greenwich Mean Astronomical Time
GMJ	geomagnetic jerks
GMST	Greenwich Mean Sidereal Time
GMT	Greenwich Mean Time
GNSS	Global Navigation Satellite Service
GPS	Global Positioning System
GR	general relativity
GRGS	Groupe de Recherche de Géodésie Spatiale
GRS	Geocentric Reference System
GRS	Geodetic Reference System
GSD	Greenwich Sidereal Date

(*cont.*)

GST	Greenwich Sidereal Time
GTRF	Geocentric True Reference Frame
HA	hour angle
HCRS	Hipparcos Catalog Reference System
HMNAO	Her Majesty's Nautical Almanac Office
HST	Hubble Space Telescope
IAA	Institute of Applied Astronomy
IACS	International Association of Cryospheric Sciences
IAG	International Association for Geodesy
IAGA	International Association of Geomagnetism and Aeronomy
IAHS	International Association of Hydrological Sciences
IAMAS	International Association of Meteorology and Atmospheric Sciences
IAPSO	International Association for the Physical Sciences of the Ocean
IASPEI	International Association of Seismology and Physics of the Earth's Interior
IAU	International Astronomical Union
IAVCEI	International Association of Volcanology and Chemistry of the Earth's Interior
ICET	International Center for Earth Tides
ICRF	International Celestial Reference Frame
ICRS	International Celestial Reference System
ICSU	International Council of Science (formerly the International Council of Scientific Unions)
IDS	International DORIS Service
IERS	International Earth Rotation and Reference System Service
IGN	Institut Géographique National
IGS	International GNSS Service
IHRF	International Hipparcos Reference Frame
ILE	Improved Lunar Ephemeris
ILRS	International Laser Ranging Service
ILS	International Latitude Service
IMCCE	Institut de Mécanique Céleste et de Calcul des Ephemerides
IPMS	International Polar Motion Service
ISES	International Space Environment Service
ISGI	International Service of Geomagnetic Indices
ISS	International Space Station
ITRF	International Terrestrial Reference Frame
ITRS	International Terrestrial Reference System
ITU-D	International Telecommunications Union – Development Sector
ITU-R	International Telecommunications Union – Radiocommunication Sector

(cont.)

ITU-T	International Telecommunications Union – Standardization Sector
IUGG	International Union of Geodesy and Geophysics
IVS	International VLBI Service for Geodesy and Astrometry
JD	Julian Date
JDN	Julian Day Number
JGM	Joint Gravity Model
K	degrees kelvin
kHz	kilohertz
LAGEOS	Laser Geodynamics Satellite
LAT	Local Apparent Solar Time
ΛCDM	Lambda–Cold Dark Matter
LCT	Local Civil Time
LED	light-emitting diode
LHA	local hour angle
LITS	Linear Ion Trap Frequency Standard
LLR	lunar laser ranging
LMT	Local Mean Time
LOD	length of day, commonly used to designate excess length of day
LORAN	Long Range Aid to Navigation
LSST	Large Synoptic Survey Telescope
LST	Local Sidereal Time
MACHO	MAssive Compact Halo Objects
mas	milliarcsecond
μas	microarcsecond
μs	microsecond
MCXO	microcomputer-controlled crystal oscillator
MEO	Medium Earth Orbit
MJD	Modified Julian Day
mm	millimeter
Mpc	Megaparsecs
ms	millisecond
MT	Mars Time
nas	nanoarcsecond
NASA	National Aeronautical and Space Agency
NBS	National Bureau of Standards (now NIST)
NGA	National Geospatial Intelligence Agency (Formerly NIMA and DMA)
NICT	National Institute of Information and Communications
NIM	National Institute of Metrology (Beijing, China)
NIMA	National Imaging and Mapping Agency (Now NGA)
NIST	National Institute of Standards and Technology
nm	nanometer
NPL	National Physical Laboratory

(cont.)

NRC	National Research Laboratory of Canada
ns	nanosecond
NSFA	Numerical Standards of Fundamental Astronomy
NTP	Network Time Protocol
OAM	ocean angular momentum
OCXO	oven-controlled crystal oscillator
OGLE	optical gravitational lensing experiment
ON	Observatoire de Neuchâtel
OP	Observatoire de Paris
OTS	optical-to-THz synthesizer
PanSTARRS	Panoramic Survey Telescope and Rapid Response System
PHARAO	Projet d'Horloge Atomique par Refroidissement d'Atomes en orbite
POSS	Palomar Optical Sky Survey
ppb	parts per billion
PPN	parameterized post-Newtonian
PPP	Precise Point Positioning
PPTA	Parkes Pulsar Timing Array
ps	picosecond
PSMSL	Permanent Service for Mean Sea Level
PTB	Physikalische Technische Bundesanstalt
PZT	photographic zenith tube
QBSA	Quarterly Bulletin of Solar Activity
Rb	Rubidium
RDV	Research and Development with the VLBA
RGO	Royal Greenwich Observatory
rms	root mean square
s	second
SAO	Smithsonian Astrophysical Observatory
SEC	Security and Exchange Commission
SHM	Satellite Hydrogen Maser
SI	Système International
SIDC	Solar Influences Data Analysis Center
SLR	Satellite Laser Ranging
SOFA	Standards of Fundamental Astronomy
SONET	synchronous optical networking
ST	sidereal time
STScI	Space Telescope Science Institute
TA(BIH)	BIH atomic timescale
TA(k)	atomic timescales from source k
TAI	International Atomic Time
TCA	Areocentric Coordinate Time

(*cont.*)

TCB	Barycentric Coordinate Time
TCG	Geocentric Coordinate Time
TCM	Mars Coordinate Time
TCXO	temperature-compensated crystal oscillator
TDA	Time Domain Astronomy
TDB	Barycentric Dynamical Time
TDM	time division multiplexing
TDMA	time division multiple access
TDT	Terrestrial Dynamical Time
TEC	total electron content
TEO	Terrestrial Ephemeris Origin
Teph	Barycentric Ephemeris Time
THz	terahertz
TIO	Terrestrial Intermediate Origin
TOS	THz-to-optical synthesizer
TT	Terrestrial Time
TVAR	time variation
TWAS	Third World Academy of Sciences
TWSFT	Two-way Satellite Time and Frequency Transfer
UCAC	USNO CCD Astrographic Catalog
UNESCO	United Nations Educational, Scientific and Cultural Organization
URSI	International Union of Radio Science
USNAO	US Nautical Almanac Office
USNO	US Naval Observatory
UT	Universal Time
UT0	Universal Time 0
UT1	Universal Time 1
UT2	Universal Time 2
UTC	Coordinated Universal Time
VGOS	VLBI2010 Global Observing System
VieVS	Vienna VLBI Software
VLBA	Very Long Baseline Array
VLBI	Very Long Baseline Interferometer
VLF	very low frequency
WDS	World Data System
WGD	World Geodetic Datum
WGD2000	World Geodetic Datum 2000
WGFS	Frequency Standards Working Group
WGMS	World Glacier Monitoring Service
WGS	World Geodetic System
WIPM	Wuhan Institute of Physics and Mathematics
WRC	World Radiocommunication Conference
Z	zenith distance

Glossary

ΔT: the difference between **Terrestrial Time** and **Universal Time**; specifically the difference between Terrestrial Time (TT) and UT1: $\Delta T = TT - UT1$.

aberration: the apparent angular displacement of the observed position of a celestial object from its **geometric position**, caused by the finite velocity of light in combination with the motions of the observer and of the observed object. Annual aberration is due to the motion of the Earth around the Sun, while diurnal aberration is due to the Earth's rotation.

accuracy: closeness of an estimated (e.g., measured or computed) value to a standard or accepted value of a particular quantity.

Allan variance: the square root of the sum of the squares of the differences between consecutive readings divided by twice the number of differences.

almanac, astronomical: an annual publication containing information on the locations of celestial bodies, together with the times and circumstances of various astronomical events such as sunset and sunrise, of particular use for navigation.

altitude: the angular distance of a celestial body above or below the horizon, measured along the great circle passing through the body and the zenith. Altitude is 90° minus zenith distance.

analemma: a curve showing the angular offset of a celestial body (usually the Sun) from its mean position on the celestial sphere as viewed from another celestial body (usually the Earth). In the case of the Sun as seen from the Earth, this is a curve resembling a figure of eight that is commonly printed on globes.

angular momentum: measure of the extent to which an object will continue to rotate about a reference point unless acted upon by an external torque. Angular momentum is related to the mass, velocity, and distance of an object from the reference point.

anomaly: angular measurement of a body in its orbit from its pericenter.

aphelion: the point in a planetary orbit that is at the greatest distance from the Sun.

apogee: the point at which a body in orbit around the Earth reaches its farthest distance from the Earth. Apogee is sometimes used with reference to the apparent orbit of the Sun around the Earth.

apparent place: the position at which the object would actually be seen from the center of the Earth, as described by apparent right ascension and declination.

apparent right ascension and declination: angular coordinates in the true equator and equinox of date reference system at a specified date. They are geocentric positions differing from the ICRS positions by annual parallax, gravitational light deflection due to the solar system bodies except the Earth, annual aberration, and the time-dependent rotation describing the transformation from the GCRS to the Celestial Intermediate Reference System (they are similar to intermediate positions in the CIO-based system but the apparent right ascension origin is at the equinox). Note that apparent declination is identical to intermediate declination.

apparent solar time: the measure of time based on the diurnal motion of the true Sun. The rate of diurnal motion undergoes seasonal variation because of the obliquity of the ecliptic and the eccentricity of the Earth's orbit. Additional small variations result from irregularities in the rotation of the Earth on its axis.

aspect: the apparent angular position of any of the planets or the Moon relative to the Sun, as seen from Earth.

asterism: a pattern of stars seen in the Earth's sky which is not an official constellation.

astrometric ephemeris: an ephemeris of a solar system body in which the tabulated positions are essentially comparable to catalog mean places of stars at a standard epoch. An astrometric position is obtained by adding to the geometric position, computed from gravitational theory, the correction for light-time.

astronomical coordinates: the longitude and latitude of a point on Earth relative to the geoid. These coordinates are influenced by local gravity anomalies.

astronomical horizon: locus of points with a zenith distance of 90°.

astronomical unit (au): 149 597 870 700 meters. Prior to 2012, the au was the distance from the center of the Sun at which a particle of negligible mass, in an unperturbed circular orbit, would have an orbital period of 365.2568983 days. This is slightly less than the semimajor axis of the Earth's orbit.

atomic second: see second, Système International.

axis of inertia: the axis of a principal moment of inertia. In the case of Earth, if it is considered symmetrical under rotation about a given axis, the symmetry axis is a principal axis. Also referred to as the axis of figure.

axis of rotation: the instantaneous axis about which the Earth's rotation is taking place. It moves slowly around the axis of figure and this motion is referred to as polar motion.

azimuth: the angular distance measured clockwise along the horizon from a specified reference point (usually north) to the intersection with the great circle drawn from the zenith through a body on the celestial sphere.

barycenter: the center of mass of a system of bodies, e.g., the center of mass of the solar system or the Earth-Moon system.

Barycentric Celestial Reference System (BCRS): a system of barycentric space-time coordinates for the solar system within the framework of general relativity with metric tensor specified by the IAU. Formally, the metric tensor of the BCRS does not fix the coordinates completely, leaving the final orientation of the spatial axes undefined. However, for all practical applications, the BCRS is assumed to be oriented according to the ICRS axes.

Barycentric Coordinate Time (TCB): the coordinate time of the BCRS; it is related to Geocentric Coordinate Time (TCG) by relativistic transformations that include secular terms.

Barycentric Dynamical Time (TDB): a timescale originally intended to serve as an independent time argument of barycentric ephemerides and equations of motion. In the

IAU 1976 resolutions, the difference between TDB and TDT was stipulated to consist of only periodic terms, a condition that cannot be satisfied rigorously. The IAU 1991 resolutions introducing barycentric coordinate time (TCB) noted that TDB is a linear function of TCB, but without explicitly fixing the rate ratio and zero point, leading to multiple realizations of TDB. In 2006, TDB was redefined through a linear transformation of TCB.

Barycentric Ephemeris Time (T_{eph}): the timescale of the JPL DE ephemeris that has been scaled to Terrestrial Time. It is equivalent to the redefined TDB.

Baryon: a composite subatomic particle made up of three quarks. Baryons and mesons belong to the hadron family of particles.

Baryonic dark matter: dark matter, which is undetectable from emitted radiation, but its presence is inferred from gravitational effects on visible matter. It is composed of baryons, which are protons and neutrons and combinations of these.

Big Bang theory: the favored cosmological model for the universe. This model explains the expansion of the universe from a high-density and high-temperature state, and a range of current observed phenomena, including abundance of light elements, cosmic microwave background, large-scale structure, and Hubble's law. The big bang is estimated to have occurred approximately 13.8 billion years ago, which is the current age of the universe.

Boltzmann constant: a physical constant relating energy at the individual particle level with temperature. Its dimension is energy divided by temperature. The SI value is $1.38064852(79) \times 10^{-23}$ J/K.

caesium fountain: an apparatus that realizes the SI definition of the second by vertically launching caesium atoms through a microwave cavity and allowing gravity to bring the atoms back down through the cavity.

calendar: a system of reckoning time in which days are enumerated according to their position in cyclic patterns.

catalog equinox: the intersection of the hour circle of zero right ascension of a star catalog with the celestial equator.

Celestial Ephemeris Origin (CEO): the original name for the Celestial Intermediate Origin (CIO) given in the IAU 2000 resolutions.

Celestial Ephemeris Pole (CEP): the reference pole for nutation and polar motion used from 1984 to 2003 with the IAU 1980 Theory of Nutation; the axis of figure for the mean surface of a model Earth in which the free motion has zero amplitude. This pole was originally defined as having no nearly diurnal nutation with respect to a space-fixed or Earth-fixed coordinate system and being realized by the IAU 1980 nutation. It is now replaced by the CIP.

celestial equator: the plane perpendicular to the Celestial Ephemeris Pole. Colloquially, the projection onto the celestial sphere of the Earth's equator.

Celestial Intermediate Origin (CIO): origin for right ascension on the intermediate equator in the Celestial Intermediate Reference System. It is the nonrotating origin in the GCRS that is recommended by the IAU. The CIO was originally set close to the GCRS meridian and throughout 1900–2100 stays within 0.1 arcseconds of this alignment.

Celestial Intermediate Pole (CIP): geocentric equatorial pole defined as being the intermediate pole, in the transformation from the GCRS to the ITRS, separating nutation from polar motion. It replaced the CEP on January 1, 2003. Its GCRS position results from (i) the part of precession/nutation with periods greater than two days, and (ii) the retrograde diurnal part of polar motion (including the free core nutation [FCN]) and (iii) the frame bias. Its ITRS position results from (i) the part of polar motion that is outside the retrograde diurnal band in the ITRS and (ii) the motion in the ITRS corresponding to nutations with periods less than two days. The motion of the CIP is realized by the IAU precession/nutation plus time-dependent corrections provided by the IERS.

Celestial Intermediate Reference System (CIRS): geocentric reference system related to the GCRS by a time-dependent rotation taking into account precession/nutation. It is defined by the intermediate equator (of the CIP) and CIO on a specific date. It is similar to the system based on the true equator and equinox of date, but the equatorial origin is at the CIO.

celestial pole: either of the two points projected onto the celestial sphere by the extension of the Earth's axis of rotation to infinity.

celestial pole offsets: time-dependent corrections to the precession/nutation model, determined by observations. The IERS provides the celestial pole offsets in the form of the differences, dX and dY, of the CIP coordinates in the GCRS with respect to the IAU 2000A precession/nutation model (i.e., the CIP is realized by the IAU 2000A precession/nutation plus these celestial pole offsets). In parallel the IERS also provides the offsets, $d\psi$ and $d\varepsilon$, in longitude and obliquity with respect to the IAU 1976/1980 precession/nutation model.

celestial pole offsets at J2000.0: offset of the direction of the mean pole at J2000.0, provided by the current model, with respect to the GCRS. These offsets are part of what is often called frame bias.

celestial sphere: an imaginary sphere of arbitrary radius upon which celestial bodies may be considered to be located. As circumstances require, the celestial sphere may be centered at the observer, at the Earth's center, or at any other location.

center of figure: that point so situated relative to the apparent two-dimensional figure of a body that any line drawn through it divides the figure into two parts having equal apparent volumes. If the body is oddly shaped, the center of figure may lie outside the figure itself.

Chandler wobble: the approximately 435-day periodic motion of the CIP in the ITRF corresponding to the free motion of the non-rigid Earth.

chronometer: high-precision, portable timekeeping device.

clepsydra: a device to measure time based on the uniform flow of water. From the Greek, literally it means a water thief.

clock: any device for indicating the time.

conjunction: the phenomenon in which two bodies have the same apparent celestial longitude or right ascension as viewed from a third body. Conjunctions are usually tabulated as geocentric phenomena. For Mercury and Venus, geocentric inferior conjunction occurs when the planet is between the Earth and Sun, and superior conjunction occurs when the Sun is between the planet and Earth.

constellation: a grouping of stars, usually with pictorial or mythical associations, that serves to identify an area of the celestial sphere. Also, one of the precisely defined areas of the celestial sphere, associated with a grouping of stars that the IAU has designated as a constellation.

Conventional International Origin: the international origin of polar motion adopted for use by the former International Latitude Service (ILS). It was defined in 1967 by an adopted set of astronomical latitudes of the five stations of the ILS. It approximately coincided with the mean pole of 1903.0 as determined by the ILS. To avoid ambiguity, this origin should be designated by its full name. This designation should be avoided for the current origin (the ITRF pole) of the polar motion, which no longer coincides with the conventional international origin.

coordinate time: the time coordinate, which, together with three spatial coordinates specify an event in a four-dimensional space-time reference system. An unambiguous way of dating and the time based on the theory of motion in a specific reference system.

Coordinated Universal Time (UTC): the timescale differing from TAI by an integral number of seconds; it is maintained within ± 0.90 second of UT1 by the introduction of one-second steps (leap seconds).

Coriolis effect: an apparent deflection of moving objects when they are viewed from a rotating frame. The Coriolis effect is caused by a fictitious Coriolis force, which appears in the equation of motion of an object in a rotating frame of reference.

Cosmic Microwave Background (CMB): the thermal radiation left from the hot early phase of the big bang. CMB is the oldest light in the Universe, which today is strongest in the microwave wavelengths.

cosmic time: proper time of co-moving observers in the universe.

cosmology: the study of the origin, evolution, and future of the universe and the physical laws involved.

Coulomb interaction: the static interaction between two charges, Q_1 and Q_2, separated by a distance r producing a force proportional to $Q_1 Q_2 / r^2$.

crustal motion: the motion of large surface areas of the Earth with respect to other surface areas, also referred to as plate tectonics and continental drift.

culmination: passage of a celestial object across the observer's meridian; also called "meridian passage." More precisely, culmination is the passage through the point of greatest altitude in the diurnal path. Upper culmination (also called "culmination above pole" for circumpolar stars and the Moon) or transit is the crossing closer to the observer's zenith. Lower culmination (also called "culmination below pole" for circumpolar stars and the Moon) is the crossing farther from the zenith.

dark energy: an unknown form of energy that is thought to exist and cause the acceleration of the expansion of the universe. Its primary characteristic is that its density does not decrease with cosmic expansion.

dark matter: the unidentified 27% of the mass and energy in the observable universe, not accounted for by dark energy, baryonic matter, radiation, and neutrinos. It does not emit or interact with electromagnetic radiation and is, hence, invisible in the electromagnetic spectrum.

day: an interval of 86 400 SI seconds, unless otherwise indicated.

decadal variations: quasi-periodic components of polar motion or observed variations of the Earth's rotation with periods between 2 and 40 years apparently due to interactions between the Earth's mantle and liquid core. Also decadal polar motion, decadal irregularities, decadal fluctuations.

decans: 36 groupings of stars used to reckon time in ancient Egypt.

declination: angular distance on the celestial sphere north or south of the celestial equator. It is measured along the hour circle passing through the celestial object.

deferent: In representing the motion of planets with epicycles, the circle centered around a point halfway between a point called the equant and the Earth on which an epicycle moves with uniform motion, not with respect to the center, but with respect to the equant.

deflection of light: the angle by which the apparent path of a photon is altered from a straight line by the gravitational field of a massive object situated along the path of the photon. In the case of the Sun, the path is deflected radially away from the Sun by up to

1″.75 at the Sun's limb. Correction for this effect, which is independent of wavelength, is included in the reduction from mean place to apparent place.

deflection of the vertical: the angle between the astronomical and the geodetic vertical.

delta T: See ΔT.

delta UT1: See $\Delta UT1$.

direct motion: for planetary movement in the solar system, motion that is counterclockwise in the orbit as seen from the north pole of the ecliptic; for an object observed on the celestial sphere, motion that is from west to east, resulting from the motion of the object relative to the Earth.

diurnal motion: the apparent daily motion, caused by the Earth's rotation, of celestial bodies across the sky from east to west.

Doppler cooling: a mechanism based on the Doppler effect used to trap and cool atoms, and sometimes used synonymously with laser cooling.

Doppler effect: the change in frequency of a received signal due to the relative motion of an observer with respect to the source of the signal.

DUT1: predicted value of UT1−UTC with an accuracy of ± 0.1s.

dynamical equinox: the ascending node of the Earth's mean orbit on the Earth's true equator; i.e., the intersection of the ecliptic with the celestial equator at which the Sun's declination changes from south to north.

dynamical time: the independent variable of the equations of motion of solar system bodies.

Earth's inner core: a solid sphere of nickel and iron at the center of the Earth about 1,220 km in radius.

Earth's mantle: a highly viscous layer directly under the crust and above the outer core, about 2,900 km thick.

Earth orientation: information specifying the relationship of terrestrial and celestial reference frames.

Earth's outer core: liquid layer, approximately 2,300 km thick, composed of iron and nickel above the solid inner core.

Earth Rotation Angle (ERA): angle measured along the intermediate equator of the Celestial Intermediate Pole (CIP) between the Terrestrial Intermediate Origin (TIO) and the Celestial Intermediate Origin (CIO), positively in the retrograde direction. It is related to UT1 by a conventionally adopted expression in which ERA is a linear function of UT1. Its time derivative is the Earth's angular velocity.

eccentric anomaly: in undisturbed elliptic motion, the angle measured at the center of the ellipse from pericenter to the point on the circumscribing auxiliary circle from which a perpendicular to the major axis would intersect the orbiting body.

eccentricity: a parameter that specifies the shape of a conic section; one of the standard elements used to describe an elliptic orbit.

eclipse: the obscuration of a celestial body caused by its passage through the shadow cast by another body.

eclipse, annular: a solar eclipse in which the solar disk is never completely covered but is seen as an annulus or ring at maximum eclipse. An annular eclipse occurs when the apparent disk of the Moon is smaller than that of the Sun.

eclipse, lunar: an eclipse in which the Moon passes through the shadow cast by the Earth. The eclipse may be total (the Moon passing completely through the Earth's umbra), partial (the Moon passing partially through the Earth's umbra at maximum eclipse), or penumbral (the Moon passing only through the Earth's penumbra).

eclipse, solar: an eclipse in which the Earth passes through the shadow cast by the Moon. It may be total (observer in the Moon's umbra), partial (observer in the Moon's penumbra), or annular.

ecliptic: the plane perpendicular to the mean heliocentric orbital angular momentum vector of the Earth-Moon barycenter in the Barycentric Celestial Reference System, commonly the mean plane of the Earth's orbit.

elements, Besselian: quantities tabulated for the calculation of accurate predictions of an eclipse or occultation for any point on or above the surface of the Earth.

elements, orbital: parameters that specify the position and motion of a body in orbit.

entropy: a measure of the number of microscopic configurations in a thermodynamic system specified by certain macroscopic variables. If each microscopic configuration is equally probable, the entropy of the system is the natural logarithm of that number of configurations multiplied by the Boltzmann constant.

epact: the number of days in the age of the moon on January 1 of any given year in the Gregorian calendar system.

ephemeris: a tabulation of the positions of a celestial object in an orderly sequence for a number of dates.

ephemeris second: the second defined in 1960 as 1/31556925.9747 of the tropical year for 1900 January 0 12 hours ET.

Ephemeris Time (ET): the timescale used from 1960 to 1984 as the independent variable in gravitational theories of the solar system, with its unit and origin conventionally defined. It was superseded by TT and TDB.

epicycle: in representing the motion of planets with epicycles, a circular orbit, whose center moves uniformly over a circular orbit around the Earth called the deferent.

epoch: a fixed instant of time used as a chronological reference datum for calendars, celestial reference systems, star catalogs, or orbital motions.

equation of center: in elliptic motion, the true anomaly minus the mean anomaly. It is the difference between the actual angular position in the elliptic orbit and the position the body would have if its angular motion were uniform.

equation of the equinoxes: the right ascension of the mean equinox referred to the true equator and equinox; alternatively the difference between apparent sidereal time and mean sidereal time (GAST−GMST).

equation of the origins: distance between the CIO and the equinox along the intermediate equator; it is the CIO right ascension of the equinox; alternatively the difference between the Earth Rotation Angle and Greenwich Apparent Sidereal Time (ERA−GAST).

equation of time: the hour angle of the true Sun minus the hour angle of the fictitious mean sun; alternatively, apparent solar time minus mean solar time.

equator: the great circle on the surface of a body formed by the intersection of the surface with the plane passing through the center of the body perpendicular to the reference axis.

equinox: either of the two points on the celestial sphere at which the ecliptic intersects the celestial equator; also, the time at which the Sun passes through either of these intersection points; i.e., when the apparent longitude of the Sun is 0° or 180°.

era: a system of chronological notation reckoned from a given date.

escapement: a device that controls the continuous motion of the clock's driving mechanism using the periodic motion of its regulator.

fictitious mean sun: an imaginary body introduced to define mean solar time; essentially the name of a mathematical formula that defined mean solar time. This concept is no longer used in high-precision work.

flattening: a parameter that specifies the degree by which the shape of an ellipsoid of revolution differs from a sphere; the ratio $f = (a–b)/a$, where a is the equatorial radius and b is the polar radius.

flicker frequency noise: a type of statistical noise observed in the output frequency of a oscillator that has a $1/f$ spectrum, where f is the sampling frequency.

flicker phase noise: a type of statistical noise observed in the timing signal of a clock that has a $1/f$ spectrum, where f is the sampling frequency.

foliot: in a verge and foliot clock, the crossbar on the verge that controls the time it takes for the verge to rotate.

frame bias: the three offsets of the mean equator and (dynamical) mean equinox of J2000.0, provided by the current model, with respect to the GCRS; the first two offsets are the mean pole offsets at J2000.0 and the third is the offset in right ascension of the mean dynamical equinox of J2000.0.

free core nutation (FCN): free retrograde diurnal mode in the motion of the Earth's rotation axis with respect to the Earth, due to nonalignment of the rotation axis of the core and of the mantle; it is a long-period (432 days) free nutation of the CIP in the GCRS.

free-falling frame: an isolated local frame that is electrically and magnetically shielded, sufficiently small that inhomogeneities in external fields can be ignored, and self-gravitating effects are negligible.

frequency: the number of cycles or complete alternations per unit time of a carrier wave, band, or oscillation.

frequency comb: a tool for measuring different colors, or frequencies, of light in which a mode-locked laser provides an optical spectrum consisting of equidistant lines, which can be used as an optical ruler.

frequency standard: a stable oscillator whose output is used as a precise frequency reference; a primary frequency standard is one whose frequency corresponds to the adopted definition of the second, with its specified accuracy achieved without calibration of the device.

GALILEO: a European satellite navigation system.

Gaussian gravitational constant: used as a constant in the definition of the au prior to 2012 (k = 0.017 202 098 95): the gravitational constant specified in units of length (astronomical unit), mass (solar mass), and time (day).

geocentric: with reference to, or pertaining to, the center of the Earth.

Geocentric Celestial Reference System (GCRS): a system of geocentric space-time coordinates within the framework of general relativity with metric tensor specified by the IAU. The GCRS is defined such that the transformation between BCRS and GCRS spatial coordinates contains no rotation component, so that GCRS is kinematically nonrotating with respect to the BCRS. The equations of motion of, for example, an Earth satellite, with respect to the GCRS will contain relativistic Coriolis forces that come mainly from geodesic precession. The spatial orientation of the GCRS is derived from that of the BCRS, that is, by the orientation of the ICRS.

geocentric coordinates: the latitude and longitude of a point on the Earth's surface relative to the center of the Earth; also, celestial coordinates given with respect to the center of the Earth.

Geocentric Coordinate Time (TCG): coordinate time of the GCRS based on the SI second. It is related to Terrestrial Time (TT) by a conventional linear transformation.

Geocentric Terrestrial Reference System (GTRS): a system of geocentric space-time coordinates within the framework of general relativity, co-rotating with the Earth, and related to the GCRS by a spatial rotation that takes into account the Earth orientation parameters. It was adopted by IUGG 2007 Resolution 2. It replaces the previously defined Conventional Terrestrial Reference System.

geodesic precession and nutation: a Coriolis-like effect from relativistic theory in the transformations of the fixed directions of the GCRS referred to the BCRS. They are the largest components of the relativistic rotation of the GCRS with respect to a dynamically nonrotating geocentric reference system in the framework of general relativity. Geodesic precession is the secular part of the rotation and geodesic nutation is the periodic part. Geodesic precession and nutation are included in the IAU 2000 precession/nutation model.

geodetic coordinates: the latitude and longitude of a point on the Earth's surface determined using the geodetic vertical (normal to the specified spheroid).

geoid: an equipotential surface that coincides with mean sea level in the open ocean. On land, it is the level surface that would be assumed by water in an imaginary network of frictionless channels connected to the ocean.

geometric position: the geocentric position of an object on the celestial sphere referred to the true equator and equinox, but without including the displacement due to planetary aberration.

getter: component of a caesium beam tube frequency standard that absorbs caesium atoms.

glacial isostatic adjustment: the ongoing movement of land once covered by ice-age glaciers.

Global Positioning System (GPS): a US satellite navigation system.

GLONASS: a Russian satellite navigation system.

GPS Time: a timescale based on the clocks of the GPS that is maintained within better than one microsecond of UTC, or TAI, modulo one second. It differs from UTC depending on the number of leap seconds, since GPS Time is not adjusted for leap seconds.

gravitational constant: denoted G, is an empirical physical constant used to specify the gravitational attraction between objects.

great circle: the circle formed by the intersection of a sphere with a plane that passes through the center of the sphere.

great empirical term: a periodic term introduced into lunar theories prior to the 20th century in an attempt to make the theories agree with the observations.

Greenwich Apparent Sidereal Time (GAST): the hour angle of the true equinox from the Terrestrial Intermediate Origin (TIO) meridian. Also Greenwich Sidereal Time.

Greenwich Civil Time (GCT): Term used from 1925 until 1952 to refer to Greenwich Mean Time reckoned from midnight. Replaced by Universal Time, officially, in 1952.

Greenwich Mean Astronomical Time (GMAT): the name recommended for use to designate mean solar time reckoned from noon at Greenwich when a 12-hour discontinuity was introduced in 1925 in GMT.

Greenwich Mean Sidereal Time (GMST): Greenwich hour angle of the mean equinox of date defined by a conventional relationship to the Earth rotation angle, or equivalently to UT1.

Greenwich Mean Time (GMT): currently the civil time of the United Kingdom. Once the predecessor of UTC for civil time and UT1 for celestial navigation. Prior to 1925, hours counted from noon.

Greenwich Sidereal Date (GSD): the number of sidereal days elapsed at Greenwich since the beginning of the Greenwich sidereal day that was in progress at Julian date 0.0.

Greenwich Sidereal Day Number: the integral part of the Greenwich Sidereal Date.

Gregorian calendar: the calendar Pope Gregory XIII introduced in 1582 to replace the Julian calendar; the calendar now used as the civil calendar in most countries. Every year that is exactly divisible by four is a leap year, except for centurial years, which must be exactly divisible by 400 to be leap years.

height: elevation above ground or distance upward from a given level (especially sea level) to a fixed point.

heliacal rising: rising of a celestial object at the same time as the Sun rises.

heliocentric: with reference to, or pertaining to, the center of the Sun.

high altitude winds: winds in the upper atmosphere.

horizon: a plane perpendicular to the direction of the zenith. The great circle formed by the intersection of the celestial sphere with a plane perpendicular to the direction of the zenith is called the astronomical horizon.

horizontal parallax: angle subtended by the Earth's radius at the distance of a celestial body.

horologium: historical term for a clock, either a water clock, or verge and foliot clock.

hour angle: angular distance measured westward along the celestial equator from the meridian to the hour circle that passes through a celestial object.

hour circle: a great circle on the celestial sphere that passes through the celestial poles and is therefore perpendicular to the celestial equator.

Hubble constant: the rate of expansion of the universe, expressed as a velocity gradient, km/s/Mpc.

Hubble's law: the observation that very distant objects have a redshift due to relative velocity away from Earth and that redshift is approximately proportional to their distance from Earth.

Hubble time: the reciprocal of the Hubble Constant, 14 G (billion) years.

hydrogen maser: device that produces coherent electromagnetic waves through amplification due to stimulated emission of the hydrogen atom to serve as a precision frequency reference.

hyperfine levels: energy levels in the structure of individual atoms that originate from the interaction of the magnetic moments of the electron and the nucleus.

IAU 2000A precession nutation model: the IAU-recommended precession/nutation model representing the motion of the CIP for those who need a model at 0.2 mas level. An abridged model, designated IAU 2000B, is available for those who require a model at the 1 mas level. Standard programs are available from IERS and SOFA.

inclination: the angle between two planes or their poles; usually the angle between an orbital plane and a reference plane; one of the standard orbital elements that specifies the orientation of an orbit.

International Atomic Time (TAI): the continuous scale resulting from analyses by the Bureau International des Poids et Mesures of atomic time standards in many countries. The fundamental unit of TAI is the SI second, and the epoch is 1958 January 1.

International Celestial Reference Frame (ICRF): a set of extragalactic objects whose adopted positions and uncertainties realize the International Celestial Reference System axes and give the uncertainties of the axes. It is also the name of the radio catalog whose defining sources are currently the most accurate realization of the ICRS. The orientation of the ICRF catalog is within the errors of the standard stellar and dynamic frames at the time of adoption. Successive revisions of the ICRF are intended to minimize rotation from its original orientation. Other realizations of the ICRS have specific names (e.g., Hipparcos Celestial Reference Frame).

International Celestial Reference System (ICRS): the idealized barycentric coordinate system to which celestial positions are referred. It is kinematically nonrotating with respect to the ensemble of distant extragalactic objects. It has no intrinsic orientation but was aligned close to the mean equator and dynamical equinox of J2000.0 for continuity with previous fundamental reference systems. Its orientation is independent of epoch, ecliptic, or equator and is realized by a list of adopted coordinates of extragalactic sources.

International Terrestrial Reference Frame (ITRF): a realization of ITRS by a set of instantaneous coordinates (and velocities) of reference points distributed on the topographic surface of the Earth (mainly space geodetic stations and related markers). Currently the ITRF provides a model for estimating, to high accuracy, the instantaneous positions of these points, which is the sum of conventional corrections provided by the IERS Convention Center (solid Earth tides, pole tides) and of a "regularized" position. At present, the latter is modeled by a piecewise linear function, the linear part accounting for such effects as tectonic plate motion, glacial isostatic adjustment, and the piecewise aspect representing discontinuities such as seismic displacements. The initial orientation of the ITRF is that of the BIH Terrestrial System at epoch 1984.0.

International Terrestrial Reference System (ITRS): the ITRS is the specific GTRS for which the orientation is operationally maintained in continuity with past international agreements. The co-rotation condition is defined as no residual rotation with regard to the Earth's surface, and the geocenter is understood as the center of mass of the whole Earth system, including oceans and atmosphere. For continuity with previous terrestrial reference systems, the first alignment was close to the mean equator of 1900 and the Greenwich meridian. The ITRS was adopted as the preferred GTRS for scientific and technical applications and is the recommended system to express positions on the Earth.

invariable plane: the plane through the center of mass of the solar system perpendicular to the angular momentum vector of the solar system.

ion: atom or molecule with a net electric charge due to the loss or gain of one or more electrons.

ITRF zero-meridian: the plane passing through the geocenter, ITRF pole, and ITRF x-origin.

ITRS CIP coordinates: direction cosines of the CIP in the ITRS, also called pole coordinates. They are currently expressed in the form of x and y angular coordinates, in arcseconds, the values of which represent the corresponding angles with respect to the polar axis of the ITRS. The sign convention is such that x is positive toward the x-origin of the ITRS and y is in the direction 90° to the west of x.

J2000.0: defined in the framework of general relativity by IAU as being the event (epoch) at the geocenter and at the date 2000 January 1.5 TT = Julian Date 245 1545.0 TT. Note that this event has different dates in different timescales.

Julian calendar: the calendar Julius Caesar introduced in 46 BC to replace the Roman calendar. In the Julian calendar, a common year is defined to comprise 365 days, and every fourth year is a leap year of 366 days. The Julian calendar was superseded by the Gregorian calendar.

Julian date (JD): the interval of time in days and fraction of a day since 4713 BC, January 1, Greenwich noon, Julian proleptic calendar. In precise work, the timescale, e.g., dynamical time or Universal Time, should be specified.

Julian date, modified (MJD): the Julian date minus 2400000.5.

Julian day number: the integral part of the Julian date.

Julian proleptic calendar: the calendar system employing the rules of the Julian calendar, but extended and applied to dates preceding the introduction of the Julian calendar.

Julian year: a period of 365.25 days. This period served as the basis for the Julian calendar.

Laplacian plane: for planets, see invariable plane; for a system of satellites, the fixed plane relative to which the vector sum of the disturbing forces has no orthogonal component.

latitude, celestial: angular distance on the celestial sphere measured north or south of the ecliptic along the great circle passing through the poles of the ecliptic and the celestial object.

latitude, terrestrial: angular distance on the Earth measured north or south of the equator along the meridian of a geographic location.

leap second: a second added or subtracted at announced times to maintain UTC within 0.9 s of UT1. Generally, leap seconds are applied at the end of June or December.

length of day: strictly the number of fixed length seconds in the day determined from the rotation of the Earth, but most often used to refer to the excess length of day or the difference between the length of day and 86,400 SI seconds.

Lense-Thirring effect: also referred to as frame dragging. The rotation of an object alters the space and time, dragging a nearby object out of its position compared to the predictions of Newtonian physics.

librations: (1) variations in the orientation of the Moon's surface with respect to an observer on the Earth. Physical librations are due to variations in the orientation of the Moon's rotational axis in inertial space. The much larger optical librations are due to variations in the rate of the Moon's orbital motion, the obliquity of the Moon's equator to its orbital plane, and the diurnal changes of geometric perspective of an observer on the Earth's surface. (2) variations in pole coordinates corresponding to motions with periods less than two days in space that are not part of the IAU 2000 nutation model.

light-time: the interval of time required for light to travel from a celestial body to the Earth. During this interval, the motion of the body in space may cause an angular displacement of its apparent place from its geometric place.

light-year: the distance that light traverses in a vacuum during one year (9.461×10^{15} m).

limb: the apparent edge of the Sun, Moon, a planet or any other celestial body with a detectable disk.

limb correction: angular correction made to the observed angular distance between the center of mass of the Moon and its limb due to the irregular surface of the Moon.

Liouville equation: basic mathematical formulation used to describe the changes in the angular momentum of a rotating system in response to applied torques.

local sidereal time: the local hour angle of a catalog equinox.

longitude, celestial: angular distance measured eastward along the ecliptic from the dynamical equinox to the great circle passing through the poles of the ecliptic and the celestial object.

longitude, terrestrial: angular distance measured along the Earth's equator from the Greenwich meridian to the meridian of a geographic location.

Lorentz transformation: mathematical formulation to convert between two different observers' measurements of space and time, where one observer is in constant motion with respect to the other.

lunar phases: cyclically recurring apparent forms of the Moon. New moon, first quarter, full moon, and last quarter are defined as the times at which the excess of the apparent celestial longitude of the Moon over that of the Sun is 0°, 90°, 180°, and 270°, respectively.

lunar secular acceleration: see **Secular acceleration of the Moon**.

lunation: the period of time between two consecutive new moons.

lunisolar tidal deceleration: the deceleration of the rotation of the Earth due to tidal interactions. Conservation of angular momentum is maintained by the acceleration of the Moon's orbital motion.

Markowitz wobble: a small-amplitude, irregular, long-period (~30 years) feature in the Earth's polar motion reported by Wm. Markowitz in 1960, thought to be due to geophysical causes.

magnitude, stellar: a measure on a logarithmic scale of the brightness of a celestial object considered as a point source.

maser: an instrument that uses the monochromatic emission from a narrow band in the spectrum of a suitable molecule or atom to control the frequency of a radio-resonant circuit.

mean anomaly: in undisturbed elliptic motion, the product of the mean motion of an orbiting body and the interval of time since the body passed the pericenter. The mean anomaly is the angle from pericenter of a hypothetical body moving with a constant angular speed that is equal to the mean motion.

mean distance: the semi major axis of an elliptic orbit.

mean elements: elements of an adopted reference orbit that approximates the actual, perturbed orbit and may serve as the basis for calculating perturbations.

mean equator and equinox: equator and equinox associated with a celestial pole whose direction is determined only by the precession portion of the precession/nutation transformation.

mean motion: in undisturbed elliptic motion, the constant angular speed required for a body to complete one revolution in an orbit of a specified semi major axis.

mean place: position of an object on the celestial sphere referred to the mean equator and equinox at a standard epoch.

mean solar time: a measure of time based conceptually on the diurnal motion of the fictitious mean sun, under the assumption that the Earth's rate of rotation is constant.

megaparsec: one million parsecs.

meridian: a great circle passing through the reference poles and through the zenith of any location on Earth. For planetary observations, a meridian is half the great circle passing through the planet's poles and through any location on the planet.

metric space: a set where a notion of distance between elements is defined. The choice of the metric defines the geometric properties of the space.

microwave standard: device providing frequency based on energy-level transitions in atoms with wavelengths ranging from 1 mm to 1 m.

moment of inertia: measure of an object's resistance to changes in rotational velocity.

month: the period of one complete synodic or sidereal revolution of the Moon around the Earth; also, a calendrical unit that approximates the period of revolution of the Moon.

moonrise, moonset: the times at which the apparent upper limb of the Moon is on the astronomical horizon.

nadir: the point on the celestial sphere diametrically opposite to the zenith.

Newcomb's Theory of the Sun: The theory of the Sun developed by Simon Newcomb and published in *The Astronomical Papers for the American Ephemeris*, volume VI, in 1898. It was the basis of the ephemeris of the Sun, timescales, and astronomical constants for about 80 years.

Newtonian mechanics: description of motions based on Newton's Universal Law of Gravity.

node: either of the points on the celestial sphere at which the plane of an orbit intersects a reference plane. The position of a node is one of the standard orbital elements used to specify the orientation of an orbit.

non-rigid Earth: model of the Earth that includes the effects of core flattening, core-mantle interactions, oceans, etc.

nonrotating origin: in the context of the GCRS or the ITRS, the point on the intermediate equator such that its instantaneous motion with respect to the system (GCRS or ITRS as appropriate) has no component along the intermediate equator (i.e., its instantaneous motion is perpendicular to the intermediate equator). It is called the CIO and TIO in the GCRS and ITRS, respectively.

numerical integration: the process of calculating the motion of a body by integrating the equations of motion.

nutation: the short-period oscillations in the motion of the pole of rotation of a freely rotating body that is undergoing torque from external gravitational forces.

obliquity: in general, the angle between the equatorial and orbital planes of a body or, equivalently, between the rotational and orbital poles. For the Earth, the obliquity of the ecliptic is the angle between the planes of the equator and the ecliptic.

occultation: the obscuration of one celestial body by another of greater apparent diameter; especially the passage of the Moon in front of a star or planet, or the disappearance of a satellite behind the disk of its primary. If the primary source of illumination of a reflecting body is cut off by the occultation, the phenomenon is also called an eclipse. The occultation of the Sun by the Moon is a solar eclipse.

opposition: a configuration of the Sun, Earth, and a planet in which the apparent geocentric longitude of the planet differs by 180° from the apparent geocentric longitude of the Sun.

optical clock: time standards based on energy-level transitions in atoms that provide precise frequencies in the optical range.

optical molasses: a cloud of atoms slowed in three dimensions by laser cooling.

optical pumping: process in which electromagnetic energy is used to raise (or "pump") electrons from a lower energy level in an atom or molecule to a higher one.

orbit: the path in space followed by a celestial body.

orientation: the set of direction angles made by the axes of one coordinate system with the axes of the other.

oscillation: the repeated movement of an object, at equal time intervals, from one point to another and back again.

osculating elements: a set of parameters that specifies the instantaneous position and velocity of a celestial body in its perturbed orbit. Osculating elements describe the unperturbed (two-body) orbit that the body would follow, if perturbations were to cease instantaneously.

parallax: the difference in apparent direction of an object as seen from two different locations; annual parallax refers to the difference in directions as seen from the barycenter and the geocenter, while diurnal parallax refers to the component of parallax due to the observer's separation from the geocenter.

parsec: the distance at which one astronomical unit subtends an angle of one second of arc; equivalently, the distance to an object having an annual parallax of one second of arc (3.09 $\times 10^{19}$ km).

Paul trap: a quadrupole ion trap that exists in both linear and three-dimensional varieties that uses constant DC and radio frequency oscillating fields to trap ions.

pendulum clock: clock whose timing is based on the swinging of a pendulum.

Penning trap: a quadrupole ion trap using a constant static magnetic field and spatially inhomogeneous static electric field to trap ions.

penumbra: the portion of a shadow in which light from an extended source is partially but not completely cut off by an intervening body; the area of partial shadow surrounding the umbra.

pericenter: the point in an orbit that is nearest to the center of force.

perigee: the point at which a body in orbit around the Earth most closely approaches the Earth. Perigee is sometimes used with reference to the apparent orbit of the Sun around the Earth.

perihelion: the point at which a body in orbit around the Sun most closely approaches the Sun.

period: the interval of time required to complete one revolution in an orbit or one cycle of a periodic phenomenon.

perturbations: deviations between the actual orbit of a celestial body and an assumed reference orbit; also, the forces that cause deviations between the actual and reference orbits. Perturbations, according to the first meaning, are usually calculated as quantities to be added to the coordinates of the reference orbit to obtain the precise coordinates.

petabytes: 2^{50} bytes, 1,024 terabytes, a million gigabytes where a byte is a unit of data that is eight binary digits long.

phase: in astronomical phenomena, the ratio of the illuminated area of the apparent disk of a celestial body to the area of the entire apparent disk taken as a circle. For eclipses, phase designations (total, partial, penumbral, etc.) provide general descriptions of the phenomena. More generally, for use with oddly shaped bodies, phase might be defined as 0.5(1 + cos (phase angle)); the phase of an oscillation or wave is the fraction of a complete cycle corresponding to an offset in the displacement from a specified reference point.

phase angle: the angle measured at the center of an illuminated body between the light source and the observer.

photometry: a measurement of the intensity of light usually specified for a specific frequency range.

piezoelectric effect: the ability of some materials (e.g., quartz crystals) to generate an electric potential in response to applied mechanical stress.

Planck's Constant: (denoted h) a physical constant to describe the sizes of quanta in quantum mechanics. It is the proportionality constant between energy of a photon and the frequency of its associated electromagnetic wave.

polar motion: the motion of the Earth's pole with respect to the ITRS.

post-glacial rebound: the rise of land masses that were depressed by the weight of ice sheets during the last glacial period. Also called continental rebound, glacial isostatic rebound, post-ice age isostatic recovery.

precession/nutation: the ensemble of effects of external torques on the motion in space of the rotation axis of a freely rotating body or alternatively, the forced motion of the pole of rotation due to those external torques. In the case of the Earth, a practical definition is that precession/nutation is the motion of the CIP in the GCRS, including the free core nutation and other corrections to the standard models. **Precession** is the secular part of this motion plus the term of a 26,000-year period and **nutation** is that part of the CIP motion not classed as precession.

precession of the ecliptic: the secular part of the motion of the ecliptic with respect to the fixed ecliptic.

precession of the equator (and CIP): the uniformly progressing motion of the pole of rotation of a freely rotating body, undergoing torque from external gravitational forces. In the case of the Earth, the precession of the equator is caused by solar system objects acting on the Earth's equatorial bulge making the pole of rotation describe an approximately 26,000-year orbit around the ecliptic pole.

precession of the equinox: combination of the precession of the equator and the precession of the ecliptic.

precision: a measure of the tendency of a set of random numbers to cluster about a number determined by the set.

prograde annual wobble: the component of polar motion of the Earth with a period of one year.

prograde semiannual wobble: the component of polar motion of the Earth with a period of one half a year.

proper motion: the projection onto the celestial sphere of the space motion of a star relative to the solar system; thus, the angular transverse component of the space motion of a star with respect to the solar system. Proper motion is usually tabulated in star catalogs as changes in right ascension and declination per year or century.

proper time: time measured along the trajectory of an observer in space-time and invariant in any coordinate change.

pulsar: highly magnetized rotating neutron star that is observed to emit periodically a beam of radio waves with a period ranging between 1.5 ms and 8.5 s.

quality factor (Q): the quality factor for oscillating systems is defined as the ratio of total energy in a system to the energy lost per cycle. Also, the ratio that provides an indication of the sharpness of the peak of a resonance curve.

quantum logic clock: a frequency standard based on optical frequencies produced by energy-level transitions in single ions.

quartz crystal clock: a clock based on the use of quartz crystal to regulate the oscillations in an electrical circuit.

radial velocity: the rate of change of the distance to an object.

Ramsey fringe: the sinusoidal fringe pattern in the excited-state population, as a function of detuning from resonance, caused by the interaction between an atomic wave function and a probe, which is transferred between the ground and excited states, depending upon the phase between the excitation field and the atomic oscillation.

random walk frequency noise: variations in frequency that can be described by a succession of random changes.

redshift: change in the frequency of electromagnetic radiation, toward longer wavelengths. Examples are the Doppler effect, as seen in the observed expansion of the universe, and gravitational redshift, which is a relativistic effect in electromagnetic radiation in gravitational fields. The redshift is described by the relative difference between the observed and emitted wavelengths of the radiation, and specified by the dimensionless quantity z, where z is determined from $1 + z =$ observed wavelength/emitted wavelength.

refraction, astronomical: the bending of a ray of light as it passes through the Earth's atmosphere, depending on atmospheric pressure, temperature, humidity, and wavelength.

residuals: the difference between actual observations and theoretical estimates of a variable.

retrograde motion: for orbital motion in the solar system, motion that is clockwise in the orbit as seen from the north pole of the ecliptic; for an object observed on the celestial sphere, motion that is from east to west, resulting from the relative motion of the object and the Earth.

Riefler clock: precision pendulum clock using an invar pendulum in a low-pressure tank.

Riemannian space time: the four-dimensional coordinates of space and time in Riemannian geometry, which is the study of smooth, curved surfaces, as opposed to Euclidean geometry, which is the study of flat spaces.

right ascension: angular distance given either in arc or time units measured eastward along the celestial equator from the equinox, or CIO, to the hour circle passing through the celestial object, usually given in combination with declination.

Sagnac effect: the interference observed in two beams of light sent in opposite directions around a rotating loop, caused by the fact that the pulse sent in the same direction as the rotation of the loop must travel further than the pulse sent in the opposite direction of the rotation.

seasonal hours: hours of variable length due to the seasonal change in the length of daylight.

second, Système International (SI): the duration of 9 192 631 770 cycles of radiation corresponding to the transition between two hyperfine levels of the ground state of caesium 133.

Second Law of Thermodynamics: statement that the total entropy of a closed system increases with time, or remains constant in ideal cases, when the system is in a steady state or undergoing a reversible process.

secular acceleration of the Moon: the acceleration of the mean motion of the Moon due to gravitational perturbations and tidal retardation of Earth rotation.

secular deceleration of Earth: change in the length of the day by about 1.7 ms per day per century, due to a combination of lunar tides and glacial isostatic adjustment of portions of the Earth's crust.

secular polar motion: a non-periodic motion of the Earth's pole toward the direction of approximately 75° west longitude.

selenocentric: with reference to, or pertaining to, the center of the Moon.

semidiameter: the angle at the observer subtended by the equatorial radius of the Sun, Moon, or a planet.

semimajor axis: half the length of the major axis of an ellipse; a standard element used to describe an elliptical orbit.

Shortt clock: a pendulum clock in which a "slave pendulum" performs the mechanical work such as turning the hands of the clock, opening and closing contacts, and giving impulses to the master pendulum, thus leaving the master pendulum free of most perturbations.

sidereal day: the interval of time between two consecutive transits of the equinox.

sidereal hour angle: angular distance measured westward along the celestial equator from the catalog equinox to the hour circle passing through the celestial object. It is equal to 360° minus the right ascension in degrees.

sidereal time: the measure of time defined by the apparent diurnal motion of the catalog equinox; hence, a measure of the rotation of the Earth with respect to the stars rather than the Sun.

signal-to-noise ratio: the ratio between the amplitude of the signal to that of the noise.

Sisyphus cooling: a mechanism for laser cooling of atoms using light forces. The mechanism involves a polarization gradient that introduces non-conservative light forces, which can reduce the average kinetic energy of atoms.

solar time: time based on the rotation of the Earth with respect to the position of the Sun.

solstice: either of the two points on the ecliptic at which the apparent longitude of the Sun is 90° or 270°; also, the time at which the Sun is at either point.

spacetime: the combination of three dimensions of space and a time dimension.

spacetime metric: a chosen specification defining the properties of time and distance in the general theory of relativity.

stability: ability of a standard to maintain its synchronization, or syntonization, over time. A stable clock would produce the same measures over a range of time intervals.

standard epoch: a date and time used to specify a reference frame. Prior to 1984, coordinates of star catalogs were commonly referred to the mean equator and equinox of the beginning of a Besselian year. Beginning with 1984, the Julian year has been used, as denoted by the prefix J, e.g., J2000.0.

Stark effect: the shifting and splitting of spectral lines of atoms and molecules due to the presence of an external electric field. The Stark effect is the electric analog of the Zeeman effect, and is responsible for pressure broadening (Stark broadening) of spectral lines by charged particles.

sunrise, sunset: the times at which the apparent upper limb of the Sun is on the astronomical horizon; i.e., when the true zenith distance, referred to the center of the Earth, of the central point of the disk is 90° 50′, based on adopted values of 34′ for horizontal refraction and 16′ for the Sun's semidiameter.

synchronization: the process of setting two standards to read the same time.

synodic period: time taken for an object to reappear at the same point in the sky, relative to the Sun, as observed from Earth.

syntonization: the process of setting two standards to the same frequency.

Teph: the independent time argument of the JPL and MIT/CfA solar-system ephemerides. It differs from Barycentric Coordinate Time (TCB) by an offset and a constant rate. The linear drift between Teph and TCB is such that the rates of Teph and TT are as close as possible for the time span covered by the particular ephemeris. Each ephemeris defines its own version of Teph; the Teph of the JPL ephemeris DE405 is for practical purposes the same as TDB defined earlier.

terminator: the boundary between the illuminated and dark areas of the apparent disk of the Moon, a planet, or a planetary satellite.

Terrestrial Dynamical Time (TDT): timescale for apparent geocentric ephemerides defined by a 1979 IAU resolution and replaced by Terrestrial Time (TT) in 1991.

Terrestrial Ephemeris Origin (TEO): the original name for the Terrestrial Intermediate Origin (TIO) given in the IAU 2000 resolutions.

Terrestrial Intermediate Origin (TIO): origin of longitude in the Intermediate Terrestrial Reference System. It is the nonrotating origin in the ITRS that is recommended by the IAU, where it was originally designated Terrestrial Ephemeris Origin. The name Terrestrial Intermediate Origin was adopted by IAU in 2006. The TIO was originally set at the ITRF origin of longitude and throughout 1900–2100 it stays within 0.1 mas of the ITRF zero meridian.

Terrestrial Intermediate Reference System (TIRS): a geocentric reference system defined by the intermediate equator of the CIP and the TIO. It is related to the ITRS by polar motion and s' (TIO locator). It is related to the Celestial Intermediate Reference System by a rotation of ERA around the CIP, which defines the common z-axis of the two systems.

Terrestrial Time (TT): a coordinate time whose mean rate is close to the mean rate of the proper time of an observer located on the rotating geoid. At 1977 January 1.0 TAI exactly, the value of TT was 1977 January 1.0003725 exactly. It is related to the Geocentric Coordinate Time (TCG) by a conventional linear transformation. An accurate realization of TT is TT (TAI) = TAI + 32s.184. In the past, TT was called Terrestrial Dynamical Time (TDT).

tidal acceleration, tidal deceleration: the rate of slowing of the Earth's rotation caused by tidal forces between the Moon and the Earth causing a gradual increase of the distance between the Earth and the Moon.

tidal friction: the frictional force caused by the interaction between the tides and the Earth's surface.

tides, ocean: periodic rise and fall of ocean waters due to the attraction of the Moon and Sun.

tides, solid Earth: periodic rise and fall of areas of the Earth crust due to the attraction of the Moon and Sun.

time dilation: difference of elapsed time between two events as measured by observers moving with respect to each other or being situated in different gravitational fields. In special relativity clocks moving with respect to an inertial system are measured to be running slower. In general relativity clocks at lower potentials in a gravitational field are measured to be running slower.

time variation (TVAR): a statistic used to describe clock noise.

TIO locator (denoted s'): the difference between the ITRS longitude and the instantaneous longitude of the intersection of the ITRS and intermediate equators. The TIO was originally set at the ITRF origin of longitude. As a consequence of polar motion, the TIO moves according to the kinematical property of the nonrotating origin. The TIO is currently located using the quantity s', the rate of which is of the order of 50 μas/cy and is due to the current polar motion.

TIO meridian: moving plane passing through the geocenter, the CIP, and the TIO.

topocentric: with reference to, or pertaining to, a point on the surface of the Earth.

transit: the passage of the apparent center of the disk of a celestial object across a meridian; also, the passage of one celestial body in front of another of greater apparent diameter (e.g., the passage of Mercury or Venus across the Sun or Jupiter's satellites across its disk); however, the passage of the Moon in front of the larger apparent Sun is called an annular eclipse. The passage of a body's shadow across another body is called a shadow transit; however, the passage of the Moon's shadow across the Earth is called a solar eclipse.

trapped ions: ions trapped in an electromagnetic field to provide a timing source.

true anomaly: the angle, measured at the focus nearest the pericenter of an elliptical orbit, between the direction of the pericenter and the radius vector from the focus to the orbiting body; one of the standard orbital elements.

true equator and equinox: the celestial coordinate system determined by the instantaneous positions of the celestial equator defined by the Celestial Intermediate Pole and the ecliptic.

twilight: the interval of time preceding sunrise and following sunset during which the sky is partially illuminated. Civil twilight comprises the interval when the zenith distance, referred to the center of the Earth, of the central point of the Sun's disk is between 90° 50′ and 96°; nautical twilight comprises the interval from 96° to 102°; astronomical twilight comprises the interval from 102° to 108°.

umbra: the portion of a shadow cone in which none of the light from an extended light source (ignoring refraction) can be observed.

uniform time scale: a timescale having the same unit of time consistently.

Universal Time (UT): a measure of time that conforms, within a close approximation, to the mean diurnal motion of the Sun and serves as the basis of all civil timekeeping. The term "UT" is used to designate a member of the family of Universal Time scales (e.g., UTC, UT1).

Universal Time (UT1): angle of the Earth's rotation about the CIP axis defined by its conventional linear relation to the Earth Rotation Angle (ERA). It is related to Greenwich apparent sidereal time through the ERA, and determined by observations. UT1 can be regarded as a time determined by the rotation of the Earth, and can be obtained from the uniform timescale UTC by using the quantity UT1−UTC.

UT1−UTC: difference between the UT1 parameter derived from observations and the uniform time scale UTC, the latter being currently defined as: UTC = TAI + n, where n is an integer number of seconds, such that |UT1 − UTC|< 0.9s.

verge: in a verge and foliot clock, a lever, with projections, which intermittently locks the escape wheel and transmits impulses from the escape wheel to the pendulum.

vernal equinox: the ascending node of the ecliptic on the celestial sphere; also, the time at which the apparent longitude of the Sun is 0°.

vertical: the apparent direction of gravity at the point of observation (normal to the plane of a free level surface).

week: an arbitrary period of days, usually seven days; approximately equal to the number of days counted between the four phases of the Moon.

white frequency noise: a type of statistical noise observed in the output frequency of a oscillator that has no dependency on sampling frequency.

white phase noise: a type of statistical noise observed in the timing signal of a clock that has no dependency on sampling frequency.

year: a period of time based on the revolution of the Earth around the Sun. The **calendar year** is the time between two dates with the same designation in a calendar. The **tropical year** is the period of one complete revolution of the mean longitude of the Sun through 360°. The **anomalistic year** is the mean interval between successive passages of the Earth through perihelion. The **sidereal year** is the mean period of revolution with respect to the background stars.

year, Besselian: the period of one complete revolution in right ascension of the fictitious mean sun, as defined by Newcomb. The beginning of a Besselian year, traditionally used as a standard epoch, is denoted by the suffix ".0". Since 1984, standard epochs have been defined by the Julian year rather than the Besselian year. For distinction, the beginning of the Besselian year is now identified by the prefix B (e.g., B1950.0).

Zeeman effect: the splitting of a spectral line into several components in the presence of a static magnetic field.

zenith: in general, the point directly overhead on the celestial sphere. The astronomical zenith is the extension to infinity of a plumb line. The geocentric zenith is defined by the line from the center of the Earth through the observer. The geodetic zenith is the normal to the geodetic ellipsoid at the observer's location.

zenith distance: angular distance measured along the great circle from the zenith to the celestial object. Zenith distance is 90° minus altitude.

zij: a generic name for an Arabic astronomical book that includes tabular parameters for calculating positions of astronomical bodies.

Index